T0239753

Essentials of Topology with Applications

TEXTBOOKS in MATHEMATICS

Series Editor: Denny Gulick

PUBLISHED TITLES

COMPLEX VARIABLES: A PHYSICAL APPROACH WITH APPLICATIONS AND MATLAB®
Steven G. Krantz

ESSENTIALS OF TOPOLOGY WITH APPLICATIONS
Steven G. Krantz

INTRODUCTION TO ABSTRACT ALGEBRA
Jonathan D. H. Smith

INTRODUCTION TO MATHEMATICAL PROOFS: A TRANSITION
Charles E. Roberts, Jr.

LINEAR ALBEBRA: A FIRST COURSE WITH APPLICATIONS
Larry E. Knop

MATHEMATICAL AND EXPERIMENTAL MODELING OF PHYSICAL AND BIOLOGICAL PROCESSES
H. T. Banks and H. T. Tran

FORTHCOMING TITLES

ENCOUNTERS WITH CHAOS AND FRACTALS
Denny Gulick

Essentials of Topology with Applications

Steven G. Krantz

Washington University
St. Louis, Missouri, U.S.A.

CRC Press
Taylor & Francis Group
Boca Raton   London   New York

CRC Press is an imprint of the
Taylor & Francis Group an **informa** business

A CHAPMAN & HALL BOOK

Chapman & Hall/CRC
Taylor & Francis Group
6000 Broken Sound Parkway NW, Suite 300
Boca Raton, FL 33487-2742

First issued in paperback 2017

© 2010 by Taylor and Francis Group, LLC
Chapman & Hall/CRC is an imprint of Taylor & Francis Group, an Informa business

No claim to original U.S. Government works

ISBN-13: 978-1-4200-8974-5 (hbk)
ISBN-13: 978-1-138-11445-6 (pbk)

This book contains information obtained from authentic and highly regarded sources. Reasonable efforts have been made to publish reliable data and information, but the author and publisher cannot assume responsibility for the validity of all materials or the consequences of their use. The authors and publishers have attempted to trace the copyright holders of all material reproduced in this publication and apologize to copyright holders if permission to publish in this form has not been obtained. If any copyright material has not been acknowledged please write and let us know so we may rectify in any future reprint.

Except as permitted under U.S. Copyright Law, no part of this book may be reprinted, reproduced, transmitted, or utilized in any form by any electronic, mechanical, or other means, now known or hereafter invented, including photocopying, microfilming, and recording, or in any information storage or retrieval system, without written permission from the publishers.

For permission to photocopy or use material electronically from this work, please access www.copyright.com (http://www.copyright.com/) or contact the Copyright Clearance Center, Inc. (CCC), 222 Rosewood Drive, Danvers, MA 01923, 978-750-8400. CCC is a not-for-profit organization that provides licenses and registration for a variety of users. For organizations that have been granted a photocopy license by the CCC, a separate system of payment has been arranged.

Trademark Notice: Product or corporate names may be trademarks or registered trademarks, and are used only for identification and explanation without intent to infringe.

Visit the Taylor & Francis Web site at
http://www.taylorandfrancis.com

and the CRC Press Web site at
http://www.crcpress.com

To the memory of Paul Halmos.

Table of Contents

Appendices   283

Preface

The nineteenth century saw the birth of Bolyai and Lobachevsky's non-Euclidean geometry, followed by Riemann's powerful and broad conception of geometry—the culmination of two thousand years of deep thought about the axioms of Euclid. The result was a theory designed to describe the *rigid geometry of space*. Certainly Einstein's general relativity is modeled on this broad concept of what a geometry should be. But new mathematical and physical theories, including the dynamical systems of Poincaré and the fixed-point theory of Brouwer, required a more flexible framework for our thinking. Thus topology was born.

Topology is a child of twentieth century mathematical thinking. It allows us to consider the shape and structure of an object without being wedded to its size or to the distances between its component parts. Knot theory, homotopy theory, homology theory, and shape theory are all part of basic topology. It is often quipped that a topologist does not know the difference between his coffee cup and his donut—because each has the same abstract "shape" without looking at all alike.

Topology is by its very nature abstract. Traditionally it follows the Euclidean paradigm of formulating axioms and definitions and then proving theorems. The famous R. L. Moore method for teaching mathematics (which is still alive and well today) is based on the teaching of point-set topology, just because the essential structure of the subject lends itself well to a formulaic presentation.

But one thing that we have learned in twentieth century mathematics—and France's Bourbaki as well as Germany's Hilbert both helped in this realization—is that, while the axiomatic method is ideal for *formulating* and recording mathematics, it is not the way that we *learn* mathematics. Mathematics is best learned from the ground up, by way of examples.

The purpose of the present book will be to present topology in a new way: as the natural evolution of ideas that the student has already seen in multivariable calculus, differential equations, and real analysis. Certainly we shall present the traditional concepts of topological space, open and closed set, the separation axioms, and so forth. All the basic cornerstones of the subject will be treated. But they will be presented in a familiar context, bolstered by examples and many applications (both to other parts of mathematics and to physics, economics, engineering, and other disciplines).

Our goal is to bring not only the mathematics student, but also the physics student and the engineeer and the computer scientist, up to speed in this ever-more-important and rapidly growing subject area. After laying the groundwork for the fundamental notions of topology, we shall present applications of the ideas in Morse theory, manifold theory, and homotopy theory. An entree to homology theory will also be provided.

Although Bourbaki prided himself on never providing pictures—because pictures can be inaccurate and misleading—we feel that pictures are an essential teaching tool. We never pretend that the picture is the final rendition of the truth, but we insist that a picture can summarize a key idea and present it in a succinct and memorable form. Since topology is a part of geometry—broadly construed—a book like this should have a surfeit of pictures. And the student should be encouraged to draw his/her own pictures. That is part of the learning process.

The book will also have a generous selection of exercises at all levels. Certainly drill exercises are important so that the student can master the basic concepts. And thought and exploration problems are also necessary so that the student can stretch his/her abilities and look to the development of the ideas.

This book is both logical and practical in its organization. The first two chapters are substantial, and constitute in and of themselves an incisive introduction to the key ideas of topology. Every student will want to study these two chapters carefully. Chapters 3 and 4 treat the more advanced, but central, topics of algebraic topology and manifold theory. The chapters after that are dessert, and may be dipped into as interest dictates. We have endeavored to provide meaningful applications at a number of junctures; these include the traveling salesman problem, digital imaging, mathematical economics, dynamical systems, and several others. The result, we hope, is a rich and varied textbook that will have appeal for many different audiences.

In an effort to make this book self-contained, and accessible to a broad audience, we have provided Appendices on basic background material. This includes logic, set theory, the properties of the real numbers, the Axiom of Choice, and basic algebraic structures. Some readers will have already mastered these fundamental ideas, and will rarely glance at the Appendices. Other readers will find them to be a valuable resource.

This book will be a fresh and accessible approach to a venerable subject, a subject with which every math student and many other mathematical science students should be acquainted. It will present key ideas succinctly and briefly, adequately illustrated by examples and illustrations. We hope that it will provide a foundation and an invitation to further mathematical study.

This is one of many books that I have written with editor Robert Stern. Working with him has always been a special privilege, and I am grateful to him for his guidance and counsel over the years.

— SGK

Chapter 1

Fundamentals

1.1 What Is Topology?

In mathematics and the physical sciences it is important to be able to compare the *shapes* or *forms* of objects. Just what do we mean by the concept of "shape"? What does it mean to say that an object has a "hole" in it (certainly having a hole is part of the concept of shape)? And is the hole in the center of a basketball the same as the hole in the center of a donut? Is it correct to say that a ruler and a sheet of paper both have the same shape (both are, after all, rectangles)? What is a rigorous and mathematical means of establishing that two objects are equivalent—from the point of view of shape or form?

These are the types of questions that we consider in the subject of topology. Topology found its genesis in the mathematics work of Jules Henri Poincaré (1854–1912). It has been a burgeoning part of mathematics ever since. The first actual use of the word "topology" was in the treatise *Vorstudieren zur Topologie* by Johann Listing (1808–1882).

In the past fifty years topology has played an increasingly prominent role in theoretical physics. Engineers and even theoretical computer scientists take advantage of topological thought in developing their ideas. A very recent development is that the modern theory of data mining is founded in topology.

We begin our journey with a thorough grounding in the most fundamental ideas in the subject. Later on we can explore some of the byways and applications that continue to make topology a dynamic and exciting part of modern thought.

1.2   First Definitions

The entire subject of topology is based on this first definition:

Definition 1.2.1 A *topological space* is a set X together with a collection of subsets $\mathcal{U} = \{U_\alpha\}_{\alpha \in A}$ that we call the *open sets*. The open sets are mandated to satisfy these properties:

(a) The entire space X is open.

(b) The empty set $\emptyset$ is open.

(c) If $\{U_\beta\}_{\beta \in B}$ are some of the open sets, then $\cup_{\beta \in B} U_\beta$ is another open set.

(d) If U and V are open sets, then $U \cap V$ is an open set.

The entire subject of topology is based on this simple idea: if you know the open subsets of a space, then you know everything about its form. We sometimes write our topological space as $(X, \mathcal{U})$.

EXAMPLE 1.2.2 Let $X = \mathbb{R}$, the Euclidean line. We say that U is open if, whenever $x \in U$, then there is an $\epsilon > 0$ so that the interval $(x - \epsilon, x + \epsilon)$ lies in U. Certainly the entire real line is open according to this definition, and so is the empty set.

Any interval of the form $J = (a, b)$ with $a < b$ is open. For let $x \in J$. Set $\epsilon = \min\{|x - a|, |x - b|\}$. Then the interval $(x - \epsilon, x + \epsilon)$ lies in J.

It can be shown (exercise) that *any* open set in the topology described here can be written as the union of open intervals as in the last paragraph. It is thus straightforward to check that the open sets described here form a topology on $\mathbb{R}$.

EXAMPLE 1.2.3 Let $X = \mathbb{R}^N$, the standard Euclidean space of N dimensions. Let us say that U is an open set if, whenever $x \in U$ then there is an $\epsilon > 0$ such that $B(x, r) = \{t \in \mathbb{R}^N : |t - x| < \epsilon\}$ lies in U. Certainly the entire set $U = \mathbb{R}^N$ is an open set according to this definition, and so is the empty set. Also the open sets are closed under the union operation and under finite intersection. So $\mathbb{R}^N$, equipped with the notion of openness specified here, is a topological space.

As an instance of an open set, let $U = \{x = (x_1, x_2, \ldots, x_N) \in \mathbb{R}^N : x_1 > 0\}$. Let us verify explicitly that U is open. To do this, let $P =$

$(p_1, p_2, \ldots, p_N) \in U$. Then $p_1 > 0$. Let $\epsilon = p_1/2$. Then the ball $B(P, \epsilon)$ lies in U; for if $x = (x_1, \ldots, x_N)$ lies in this ball, then it is plain that $x_1 > p_1 - \epsilon = \epsilon > 0$. Thus U is open.

EXAMPLE 1.2.4 Let X be the unit interval $[0, 1]$ and let the open sets be just the empty set $\emptyset$ and the entire interval $[0, 1]$. So there are just two open sets. One may check directly that all the axioms for a topological space are satisfied.

EXAMPLE 1.2.5 Let X be the set of integers $\mathbb{Z}$. Call a set open if it is the complement of a finite set. Then it is straightforward to confirm that this is a topological space.

We have now seen four examples of topological spaces. What we have *not* seen is how the concept of form grows out of these ideas. That will become clearer in the next several chapters, as our ideas develop. One thing that we certainly must note is that the idea of a topology is an abstract mathematical construct. It can arise in a variety of contexts, and in many different forms (such as Example 1.2.4). Some of these are not intuitive, and do not correspond to any heuristic concept of form.

Now let us consider a topological space $(X, \mathcal{U})$. A set $E \subseteq X$ is called *closed* if its complement $X \setminus E$ is open. In Example 1.2.2, any interval $[a, b]$ is closed (though these are certainly not all the closed sets!—see Section 1.11 on the Cantor set). In Example 1.2.4, the only closed sets are the entire interval $[0, 1]$ and the empty set. In Example 1.2.5, the closed sets are the finite sets (or the empty set).

One mistake that students commonly make early on is that they assume that, in a topological space, any set is either open or closed. This is like meeting a blonde person and a brunette and assuming therefore that all people are either blonde or brunette. It is an error of logic. Look at Example 1.2.2. Then the set $[0, 1)$ is neither open nor closed. We leave it to the reader to create, in the context of Examples 1.2.4 and 1.2.5, an instance of a set that is neither open nor closed.

It is in fact possible for a set to be *both* open and closed. In Example 1.2.4, the set $[0, 1]$ is both open and closed. So is the empty set. Some texts call a set of this type "clopen," though we shall not use that terminology.

Now let us prove our first result about topology. In an effort at elegance and conciseness, we shall not generally state explicitly that we are working in a topological space X with $\mathcal{U}$ the collection of open sets. This will be understood from context.

Proposition 1.2.6 *The union of two closed sets is closed.*

Proof: Let the two closed sets be E and F. Then $X \setminus E$ and $X \setminus F$ are open. So certainly

$$S \equiv (X \setminus E) \cap (X \setminus F)$$

is open. But then

$$^{c}S \equiv X \setminus S = E \cup F$$

is closed.   □

Proposition 1.2.7 *Let E_β be closed sets. Then $\cap_\beta E_\beta$ is also closed.*

Proof: Exercise for the reader.   □

We want to develop a language for describing and analyzing the parts of a set. Consider the closed disc $\{(x,y) \in \mathbb{R}^2 : x^2 + y^2 \leq 1\}$ depicted in Figure 1.1. We can see intuitively that this set has a boundary—this is just the circle (see Figure 1.2). And it has an interior—this is the open disc $\{(x,y) \in \mathbb{R}^2 : x^2 + y^2 < 1\}$—see Figure 1.3. We would like a precise description of the boundary and interior of any set. For example consider the set S of integers in the topological space $\mathbb{R}$ (with the usual topology, described in Example 1.2.2). What is its interior and what is its boundary? The answer to this last question is not entirely obvious, and requires some study.

First we need a bit of terminology. If X is a topological space and $x \in X$, then a *neighborhood U* of x is simply an open set that has x as an element.

Definition 1.2.8 Let S be any set in the topological space X. A point $p \in X$ is said to be an *interior point* of S if

(a) $p \in S$;

(b) There is a neighborhood U of p such that $p \in U \subset S$.

The collection of all interior points of S is called the *interior* of S and is denoted by $\overset{\circ}{S}$.

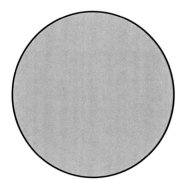

Figure 1.1: The closed disc.

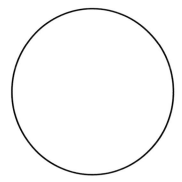

Figure 1.2: The boundary of the disc.

Figure 1.3: The interior of the disc.

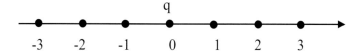

Figure 1.4: The point q is in the boundary of $\mathbb{Z}$.

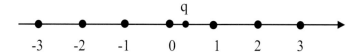

Figure 1.5: The point q is not in the boundary of $\mathbb{Z}$.

Definition 1.2.9 Let S be any set in the topological space X. A point $q \in X$ is said to be a *boundary point* of S if any neighborhood U of q intersects both S and $^cS \equiv X \setminus S$. The collection of all boundary points of S is called the *boundary* of S and is denoted by ∂S.

These new concepts are best elucidated through some examples.

EXAMPLE 1.2.10 Let us return to the example of the integers $\mathbb{Z}$ as a subset of the real line. If q is *any* point of $\mathbb{Z}$ and U is *any* neighborhood of q then it is clear that U will contain points that do not lie in $\mathbb{Z}$. See Figure 1.4. Therefore there is no point that is an interior point of $\mathbb{Z}$. In other words, the interior of $\mathbb{Z}$ is empty.

   If q is any point of $^c\mathbb{Z}$, then let $\delta > 0$ be the distance of q to the nearest integer. See Figure 1.5. Then the interval $U = (q - \epsilon, q + \epsilon)$ is a neighborhood of q that intersects $^c\mathbb{Z}$ but does *not* intersect $\mathbb{Z}$ itself. So q cannot be a boundary point of $\mathbb{Z}$. If instead q is an element of $\mathbb{Z}$ and U is *any* neighborhood of q, then $U \cap \mathbb{Z} \ni q$ and, plainly, U will also contain nearby points that are not in $\mathbb{Z}$ (see Figure 1.4). Thus q is a boundary point of $\mathbb{Z}$. We see then that the boundary of $\mathbb{Z}$ is just $\mathbb{Z}$ itself.

EXAMPLE 1.2.11 Consider X the real line equipped with the usual topology (as in Example 1.2.2). Let S be the set $[0, 1)$. Then the boundary of S is the pair $\partial S = \{0, 1\}$. And the interior is the interval $\overset{\circ}{S} = (0, 1)$.

EXAMPLE 1.2.12 Consider the topological space in Example 1.2.4. Let S be any proper subset of the interval $X = [0, 1]$. If p is *any* point of S then the only possible neighborhood of p is the entire interval $[0, 1]$ (since that is the only open set available). And of course that interval cannot lie in S. We conclude that S has no interior points.

Now if q is *any point* in the entire space X, then the only neighborhood of q is the entire interval $[0, 1]$ (since that is the only open set available). And that interval intersects both S and its complement. We conclude that ∂S is the entire space $[0, 1]$.

Now the interior and the boundary of a set have many interesting properties. We shall record a few of them here. We begin with a lemma that has independent interest.

Lemma 1.2.13 *Let S be a set in a topological space X. If each point $s \in S$ has a neighborhood that lies in S, then S is open.*

Proof: Let $s \in S$ and let U_s be the neighborhood of s that lies in S. We have

$$S = \bigcup_{s \in S} \{s\} \subseteq \bigcup_{s \in S} U_s \subseteq S.$$

We conclude that $\bigcup_{s \in S} U_s = S$. But the set on the left of this last equality is open. Hence S itself is open. $\square$

Proposition 1.2.14 *The interior of any set S is open.*

Proof: First note that if $p \in \overset{\circ}{S}$ and U is a neighborhood of p that lies inside S, then any point x of U is also in the interior of S. For U will be the required neighborhood of x that lies in S.

Now let $S \subset X$ be any set and let p be a point of its interior. We know that there is a neighborhood U_p of p such that U_p lies entirely in S. And in fact we then know (by the last paragraph) that $U_p \subseteq \overset{\circ}{S}$. Therefore

$$\overset{\circ}{S} = \bigcup_{p \in \overset{\circ}{S}} \{p\} \subseteq \bigcup_{p \in \overset{\circ}{S}} U_p \subseteq \overset{\circ}{S}.$$

We conclude that

$$\bigcup_{p \in \overset{\circ}{S}} U_p = \overset{\circ}{S}.$$

But the set on the left-hand side of this last expression, being the union of open sets, is open. Therefore $\overset{\circ}{S}$ itself is open. $\qquad\square$

Proposition 1.2.15 *The boundary of any set S is closed.*

Proof: Let x be a point that is not in the boundary of S. Then there is some neighborhood U of x that does not intersect both S and ^cS. It follows that any point $t \in U$ also has such a neighborhood—namely U itself. So U lies in the complement of ∂S. It follows, by Lemma 1.2.13, that the complement of ∂S is open. So ∂S is closed. $\qquad\square$

Definition 1.2.16 The *closure* of a set S is defined to be the intersection of all closed sets that contain S. We denote the closure of S by $\overline{S}$.

Of course the closure $\overline{S}$ of S is closed.

Proposition 1.2.17 *The set $\overline{S}$ equals the union of S and ∂S.*

Proof: Suppose that x is a point that is *not* in $S \cup \partial S$. Since x is not in ∂S there is a neighborhood U of x that either does not intersect S or does not intersect ^cS. We know that $x \notin S$, so it must be that $U \subseteq {}^cS$. So in fact we see that every point of U is not in $S \cup \partial S$. Thus the complement of $S \cup \partial S$ is open, and $S \cup \partial S$ is closed. We conclude that $S \cup \partial S \supseteq \overline{S}$.

Now if instead $x \notin \overline{S}$, then (since $^c\overline{S}$ is open) there is a neighborhood U of x that lies in $^c\overline{S}$ hence in ^cS. So certainly U is disjoint from $S \cup \partial S$. In particular, $x \notin (S \cup \partial S)$. We conclude that $S \cup \partial S \subseteq \overline{S}$.

The two inclusions taken together give our result. $\qquad\square$

Of course the statement of the last proposition is consistent with our intuition. Examine Figure 1.1—the closed disc. We see plainly that the closure is simply the closed disc, which is the interior plus the boundary circle. In other words the closure is the interior set S union its boundary ∂S.

If X is a topological space, $S \subseteq X$, and $p \in X$, then we say that p is an *accumulation point* of S if every neighborhood of p contains elements of S (other than p itself).

EXERCISE FOR THE READER 1.2.18 If X is a topological space and $S \subseteq X$, then S is closed if and only if S contains all its accumulation points.

EXERCISE FOR THE READER 1.2.19 Let X be the real number line with the usual topology. Let $S \subseteq X$ be the rational numbers $\mathbb{Q}$. What is the interior of S? What is the boundary of S? What is the closure of S?

EXERCISE FOR THE READER 1.2.20 Let X be a topological space and $S \subseteq X$. Show that $\overline{S} = \overset{\circ}{S} \cup \partial S$.

If $(X, \mathcal{U})$ is a topological space and $(Y, \mathcal{V})$ is another topological space with $Y \subseteq X$ and $V \in \mathcal{V}$ implies $V \in \mathcal{U}$, then we say that Y is a *topological subspace* of X. Often we simply say that Y is a subspace of X.

Suppose now that $(X, \mathcal{U})$ is a topological space. We say that a subset $E \subseteq X$ is *dense* if each open set $U \subseteq X$ contains a point of E. The intuition here is that points of E get arbitrarily close to every point of X.

If X is any set then the topology just consisting of X itself and the empty set $\emptyset$ is the smallest topology on X. By contrast, the topology in which each singleton $\{x\}$ for $x \in X$ is open is the largest topology on X. We call the latter topology the *discrete topology*, and we say that the space is discrete.

1.3 Mappings

Our principal tool for comparing and contrasting topological spaces will be mappings. The mappings that carry the most information for us are the continuous mappings (here a *mapping* is a function that takes values in a space rather than in the real numbers or the complex numbers). We shall formulate our notion of continuity in terms of the inverse image of a mapping. Let $f : A \to B$ be a mapping. Let $S \subseteq B$. Then $f^{-1}(S) \equiv \{x \in A : f(x) \in S\}$. We call this set the *inverse image of the set S under f*.

Definition 1.3.1 Let $(X, \mathcal{U})$ and $(Y, \mathcal{V})$ be topological spaces. A function (or mapping) $f : X \to Y$ is said to be *continuous* if, whenever $V \subseteq Y$ is open, then $f^{-1}(V) \subseteq X$ is open.

Remark 1.3.2 This definition requires some discussion. When we first learn about the notion of continuity—perhaps in high school—we see that a function is continuous if its graph can be drawn in a single stroke, without lifting the pencil from the paper. Later on we learn the limitations of that heuristic definition, and we are taught the famous ϵ-δ definition of continuity. One advantage of this last, more technical approach, is that it treats continuity point by point.

But now we are working in an abstract topological space. We do not necessarily have a notion of distance, so we cannot formulate the concept of "if the variable x is less than δ distant from c then $f(x)$ is less than ϵ distant from $f(c)$." We instead rely on our most fundamental structure—the open sets—to express the idea of continuity. Of course we will have to do some work to see that the new notion of continuity is equivalent to the old one.

Remark 1.3.3 For the record, let us record here the rigorous definition of continuity that we learn in calculus and real analysis:

> Let I be an interval in the real line and $f : I \rightarrow \mathbb{R}$ a function. Fix a point $c \in I$. We say that f is continuous at c if, for any $\epsilon > 0$, there is a $\delta > 0$ such that whenever $|x - c| < \delta$ then $|f(x) - f(x)| < \epsilon$.

This definition, thought about properly, makes good intuitive sense.

Proposition 1.3.4 *On the real line (using the standard topology, as in Example 1.2.2), the traditional definition of continuity formulated in terms of ϵs and δs is equivalent to the new definition given in Definition 1.3.1.*

Proof: Suppose that I is an interval in $\mathbb{R}$ and $f : I \rightarrow \mathbb{R}$. Assume that f is continuous according to the classical definition in 1.3.3. Now let V be an open subset of $\mathbb{R}$ and consider the set $f^{-1}(V)$. Let $x \in f^{-1}(V)$. Of course $f(x) \in V$ and (since V is open) there exists an $\epsilon > 0$ such that the interval $(f(x) - \epsilon, f(x) + \epsilon) \subseteq V$. By the classical definition, there exists a $\delta > 0$ such that if $t \in (x - \delta, x + \delta)$ then $f(t) \in (f(x) - \epsilon, f(x) + \epsilon)$. But this says that the interval $(x - \delta, x + \delta)$ lies in $f^{-1}(V)$. In other words, $f^{-1}(V)$ is open. We have shown that the inverse image of an open set is open, and that is the new definition of continuity.

For the converse, assume that $f : I \rightarrow \mathbb{R}$ satisfies the new definition of continuity given in Definition 1.3.1. Fix a point $x \in I$. Let $\epsilon > 0$. The interval $(f(x) - \epsilon, f(x) + \epsilon)$ is an open subset of the range $\mathbb{R}$. Thus, by hypothesis, the inverse image $f^{-1}((f(x) - \epsilon, f(x) + \epsilon))$ is open. In fact it is an open neighborhood of the point x. Thus there exists a $\delta > 0$ such that $(x - \delta, x + \delta) \subseteq f^{-1}((f(x) - \epsilon, f(x) + \epsilon))$. This means that if $|t - x| < \delta$ then $|f(t) - f(x)| < \epsilon$. And that is the classical definition of continuity. $\square$

One thing that is nice about our new definition of continuity is that it is simple and natural to use in contexts where the traditional definition would be awkward. We look at some examples.

EXAMPLE 1.3.5 Let $f : \mathbb{R} \to \mathbb{R}$ be given by $f(x) = x^2$. Discuss the continuity of f.

Of course we know from experience that this f is continuous. After all, all polynomial functions are continuous. But it is instructive to examine the new definition of continuity in this context.

Now let V be an open subset of the range space $\mathbb{R}$. We may take V to be an interval $I = (a, b)$ since any open set is simply a union of such intervals (exercise). Then

- If $0 < a < b$, then $f^{-1}(I) = (\sqrt{a}, \sqrt{b})$, and that is an open set.

- If $a < 0 < b$, then $f^{-1}(I) = (0, \sqrt{b})$, and that is an open set.

- If $a < b < 0$, then $f^{-1}(I) = \emptyset$, and that is an open set.

We have verified directly that f is continuous according to the new definition.

EXAMPLE 1.3.6 Say that a set $V \subseteq \mathbb{Q}$ is open if there is an open $U \subseteq \mathbb{R}$ in the usual topology so that $U \cap \mathbb{Q} = V$. Consider the function $f : \mathbb{Q} \to \mathbb{Q}$ that is defined as follows. If p/q is a rational number expressed in lowest terms (i.e., p and q have no prime factors in common), with q positive, then set $f(p/q) = 1/q$. Determine whether f is continuous at any point.

In point of fact f is discontinuous everywhere. Notice first that the values of f are $1/1$, $1/2$, $1/3$, $1/4$, etc. Now let us take a neighborhood V of $1/2$ in the image—this is a typical open set. We take the neighborhood to be an interval that is small enough that it does not contain any of the other image points ($1/1$, $1/3$, $1/4$, etc.). We see that

$$f^{-1}(V) = \left\{ \ldots, -\frac{5}{2}, -\frac{3}{2}, -\frac{1}{2}, \frac{1}{2}, \frac{3}{2}, \frac{5}{2}, \ldots \right\}.$$

In particular, this is *not* an open set. So f is not continuous.

EXERCISE FOR THE READER 1.3.7 Use the traditional form of the definition of continuity to verify that the function f in the last example is not continuous.

EXERCISE FOR THE READER 1.3.8 Let $(X, \mathcal{U})$ and $(Y, \mathcal{V})$ be topological spaces. Show that a mapping $f : X \to Y$ is continuous if and only if the inverse image of any closed set is closed.

   In practice in this book, and in topology in general, when we say "Let $f : X \to Y$ be a mapping," we mean that f is a *continuous* mapping. [There is little use in this subject for looking at a mapping that is not continuous.] We shall follow that custom in what follows. The word "map" is used interchangeably with "mapping."

1.4   The Separation Axioms

The richness of the subject of topology begins to become evident when we begin to examine and to classify the different types of topological spaces. We do so by way of the so-called *separation axioms*.
   We begin with a sample separation axiom that is particularly intuitively appealing—just to give a flavor of this circle of ideas.

Definition 1.4.1 We say that a topological space X is a *Hausdorff space* if, for any two distinct points P, Q in X, there are open sets U and V such that

- $P \in U$ and $Q \in V$;

- $U \cap V = \emptyset$.

EXAMPLE 1.4.2 Let X be the real line with the usual topology (as in Example 1.2.2). This is a Hausdorff space. For if P and Q are distinct points in $\mathbb{R}$ and if $\epsilon = |P - Q|$, then the intervals $U = (P - \epsilon/3, P + \epsilon/3)$ and $V = (Q - \epsilon/3, Q + \epsilon/3)$ are neighborhoods of P and Q, respectively, that are disjoint.

EXAMPLE 1.4.3 Let X be the integers with the topology that U is open if it is the complement of a finite set. Then this X is *not* a Hausdorff space. For if P, Q are distinct points, and if U, V are neighborhoods of P and Q, respectively, then $U \cap V$ will always be an infinite set—certainly not empty.

EXERCISE FOR THE READER 1.4.4 Give an example of a topology on $\mathbb{R}$ that is not Hausdorff.

In fact the separation axioms are so important that they are numbered. Traditionally we call a Hausdorff space a $\boldsymbol{T_2}$ space. Let us in fact now lay out all the separation axioms. In this description, we shall use the terminology "neighborhood of a set S" to simply mean an open set that contains S. After we enunciate the axioms we shall present several examples to show that they are distinct.

$\boldsymbol{T_0}$ **Space:** The space X is $\boldsymbol{T_0}$ if, whenever $P, Q \in X$ are distinct points, then either there is a neighborhood U of P such that $Q \notin U$ or else there is a neighborhood V of Q such that $P \notin V$.

$\boldsymbol{T_1}$ **Space:** The space X is $\boldsymbol{T_1}$ if, whenever $P, Q \in X$ are distinct points, then there are neighborhoods U of P and V of Q such that $Q \notin U$ and $P \notin V$.

$\boldsymbol{T_2}$ **Space:** The space X is $\boldsymbol{T_2}$ (also called *Hausdorff*) if, whenever $P, Q \in X$ are distinct points, then there are neighborhoods U of P and V of Q such that $U \cap V = \emptyset$.

$\boldsymbol{T_3}$ **Space:** The space X is $\boldsymbol{T_3}$ (also called *regular* if points are closed, i.e., the space is $\boldsymbol{T_1}$) if, whenever $P \in X$ and $F \subseteq X$ is a closed subset not containing P, then there are a neighborhood U of P and V of F so that $U \cap V = \emptyset$.

$\boldsymbol{T_4}$ **Space:** The space X is $\boldsymbol{T_4}$ (also called *normal* if points are closed, i.e., the space is $\boldsymbol{T_1}$) if, whenever E and F are disjoint closed sets in X, there are neighborhoods U of E and V of F such that $U \cap V = \emptyset$.

Now let us examine some examples which show that these different notions of separation are really different. The entire subject of point-set topology is built on examples like these. You should master them and make them part of your toolkit. It should be noted that the separation axioms (at least for spaces that are assumed to be $\boldsymbol{T_1}$) increase in strength as the index increases. So a $\boldsymbol{T_3}$ space is certainly $\boldsymbol{T_2}$. And so forth.

EXAMPLE 1.4.5 Consider X the real line with the open sets being the half-lines of the form (a, ∞). Then this space is $\boldsymbol{T_0}$ but not $\boldsymbol{T_1}$. To see this, note that if P and Q are distinct points of X and if $P < Q$ then $U = \{x : x > P\}$ is an open set in X. Also $Q \in U$ but $P \notin U$. So certainly X is $\boldsymbol{T_0}$. But it is easy to see that there is no open neighborhood of P that will separate it from Q. So X is not $\boldsymbol{T_1}$.

EXAMPLE 1.4.6 Let X be the integers equipped with the topology that U is open if its complement is finite. We have already seen that this space is not Hausdorff. It *is*, however, of type $\boldsymbol{T_1}$. Because if $P, Q \in X$ are distinct then let U be the complement of $\{Q\}$ and let V be the complement of $\{P\}$. Since the space is $\boldsymbol{T_1}$, it is certainly also $\boldsymbol{T_0}$.

EXAMPLE 1.4.7 Let X be the real line and equip it with the following topology. If $x \in X$ is a point *other than* 0, then let the neighborhoods of x be the usual intervals $(x - \alpha, x + \alpha)$. If $x = 0$ then let a neighborhood of x have the form

$$U_\alpha = \{t \in \mathbb{R} : -\alpha < t < \alpha, t \neq 1, 1/2, 1/3, \dots \}.$$

It is easy to see that this space is $\boldsymbol{T_2}$, because any two distinct points can be separated by intervals in the usual fashion.

But the space is *not* $\boldsymbol{T_3}$. For observe that $E = \{1, 1/2, 1/3, \dots \}$ is a closed set in this topology. And it cannot be separated from the point 0 with open sets.

EXAMPLE 1.4.8 The *Moore plane* $\mathcal{P}$ (named after R. L. Moore (1882–1974)) is the usual closed upper halfplane $\{(x, y) : x \in \mathbb{R}, y \in \mathbb{R}, y \geq 0\}$ with the topology generated by these open sets: **(a)** If $(x, y) \in \mathcal{P}$ and $y > 0$, then a "generating" element is any disc of the form $\{(s, t) \in \mathcal{P} : (s - x)^2 + (t - y)^2 < r^2\}$ for $r < y$; **(b)** If $(x, 0) \in \mathcal{P}$ then a "generating" element is the singleton $\{(x, 0)\}$ union any disc of the form $\{(s, t) : (s - x)^2 + (t - r)^2 < r^2\}$ for $r > 0$. As usual (the concept of sub-basis is treated in detail in Section 2.1), we generate the topology by taking finite intersections and arbitrary unions of the indicated generating sets. See Figure 1.6. This space is $\boldsymbol{T_3}$ but not $\boldsymbol{T_4}$.

To see that the space is $\boldsymbol{T_3}$, first note that *any* subset of the real axis is closed. But it can easily be separated by open sets from a disjoint point in the real axis using the special open sets of type **(b)**. Points in the open upper halfplane are easily separated from closed sets in the usual fashion.

For non-normality, examine $E = \{(x, 0) \in \mathcal{P} : x \text{ is rational}\}$ and $F = \{(x, 0) \in \mathcal{P} : x \text{ is irrational}\}$. Then each of these sets is closed (exercise), they are clearly disjoint, but they cannot be separated by open sets.

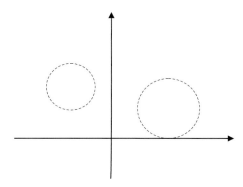

Figure 1.6: Open sets in the Moore plane.

EXAMPLE 1.4.9 Consider the interval $X = [0,1] \subseteq \mathbb{R}$ with the usual topology (see Example 1.2.2). Thus a set $U \subseteq X$ is open if and only if there is an open set V in the real line such that $V \cap X = U$. This X is a $\boldsymbol{T_4}$ space. To see this let E and F be disjoint closed sets in $[0,1]$. We claim that there is a number $\delta > 0$ such that if $e \in E$ and $f \in F$ then $|e - f| > \delta$. If not, then there are $e_j \in E$ and $f_j \in F$ with $|e_j - f_j| \to 0$. But then the common limit point x of the two sequences $\{e_j\}$ and $\{f_j\}$ would have to lie in the boundary both of E and of F. Since E and F are disjoint, that is impossible. So δ exists. Now let $U = \{x \in X : \mathrm{dist}(x, E) < \delta/3\}$ and $V = \{x \in X : \mathrm{dist}(x, F) < \delta/3\}$. Here $\mathrm{dist}(x, S)$ denotes the distance of x to the set S, defined to be $\inf_{s \in S} |x - s|$. Of course $E \subseteq U$ and $F \subseteq V$. Now it is easy to see that U, V are both open, and they are certainly disjoint (by the triangle inequality).

We refer the reader to the discussion of compactness in Section 1.5—especially Proposition 1.5.8—for further consideration of some of the ideas in the last example.

Certainly one of the most important applications of the idea of separation is the following basic result of Urysohn (commonly known as *Urysohn's lemma*—Pavel Urysohn (1898–1924)):

THEOREM 1.4.10 *Let $(X, \mathcal{U})$ be a normal space and let E and F be disjoint, closed sets in X. Then there is a continuous function $f : X \to [0,1]$ such that $f(E) = \{0\}$ and $f(F) = \{1\}$ (that is, the image of each point in E is 0 and the image of each point if F is 1).*

Proof: By normality, there are disjoint open sets U and V such that $E \subseteq U$ and $F \subseteq V$. For technical (and also traditional) reasons we shall denote this set U by $U_{1/2}$. Now we see that E and $X \setminus U_{1/2}$ are closed and disjoint. Also $\overline{U}_{1/2}$ and F are closed and disjoint. Therefore open sets $U_{1/4}$, $U_{3/4}$ exist such that

$$E \subseteq U_{1/4} \quad, \quad \overline{U}_{1/4} \subseteq U_{1/2} \quad, \quad \overline{U}_{1/2} \subseteq U_{3/4} \quad, \quad \overline{U}_{3/4} \cap F = \emptyset.$$

Suppose now inductively that sets $U_{j/2^n}$, $j = 1, 2, \ldots, 2^n - 1$, have been defined in such a fashion that

$$E \subseteq U_{1/2^n} \quad, \quad \cdots \quad, \quad \overline{U}_{(j-1)/2^n} \subseteq U_{j/2^n} \quad, \quad \cdots \quad, \quad \overline{U}_{(2^n-1)/2^n} \cap F = \emptyset.$$

Then we may continue and select sets $U_{j/2^{n+1}}$, $j = 1, \ldots, 2^{n+1} - 1$ with analogous properties.

The result of our construction is that we have, for each dyadic rational number r of the form $j/2^n$, for some $n > 0$ and $j = 1, 2, \ldots, 2^n - 1$, an open set U_r satisfying

- $E \subseteq U_r$ and $\overline{U}_r \cap F = \emptyset$ for each dyadic r;

- $\overline{U}_r \subseteq U_s$ whenever $r < s$ are dyadic as above.

Now define a function $f : X \rightarrow [0, 1]$ by

$$f(x) = \begin{cases} 1 & \text{if} \quad x \text{ belongs to no } U_r \, ; \\ \inf\{r : x \in U_r\} & \text{if} \quad x \text{ belongs to some } U_r \, . \end{cases}$$

Plainly $f(E) = 0$ and $f(F) = 1$. It remains to show that f is continuous. But observe that

Continuity at points x with $f(x) = 1$: If $x \notin \overline{U}_r$, then $f(x) \geq r$.

Continuity at points x with $f(x) = 0$: If $x \in U_r$, then $f(x) \leq r$.

Continuity at all other points: If $x \in U_s \setminus \overline{U}_r$, where $r < s$ are dyadic, then $r \leq f(x) \leq s$.

The existence of the continuous function f is now established. $\square$

EXERCISE FOR THE READER 1.4.11 Prove that the converse of the theorem is true as well: If such a function f exists for any two disjoint, closed sets E and F, then the space is normal.

If X is a topological space, then a collection $\mathcal{U} = \{U_\alpha\}_{\alpha \in A}$ is called an *open cover* for X if $X \subseteq \cup_\alpha U_\alpha$. We say that $\mathcal{U}$ *covers* X. We call $\{U_{\alpha_j}\}_{j=1}^\infty$ a *countable subcover* if it is a countable subcollection of $\mathcal{U}$ that still covers X. More will be said about open covers in the next section and in later parts of the book.

Definition 1.4.12 Let us say that a topological space X is *Lindelöf* if every open cover of X has a countable subcover.

Proposition 1.4.13 *A regular, Lindelöf space is normal.*

Proof: Let X be as in the hypothesis, and let E, F be disjoint, closed sets in X. For each point $e \in E$, let U_e be an open set containing e such that $\overline{U}_e \cap F = \emptyset$. [This is of course by the regularity hypothesis.] Similarly, for each $f \in F$ we find an open set V_f such that $f \in V_f$ and $\overline{V}_f \cap E = \emptyset$. Now of course E and F are Lindelöf subspaces of X. It follows that there is a countable subcover $U_{e_1}, U_{e_2}, \ldots$ of E and a countable subcover $V_{f_1}, V_{f_2}, \ldots$ of F. Now we inductively construct open sets S_j and T_j as follows:

$$S_1 = U_{e_1} \qquad\qquad T_1 = V_{f_1} \setminus \overline{S}_1$$
$$S_2 = U_{e_2} \setminus \overline{T}_1 \qquad\qquad T_2 = V_{f_2} \setminus \overline{(S_1 \cup S_2)}$$
$$S_3 = U_{e_3} \setminus \overline{(T_1 \cup T_2)} \qquad\qquad T_3 = V_{f_3} \setminus \overline{(S_1 \cup S_2 \cup S_3)}.$$

It is now easily seen that $S \equiv \cup_j S_j$ and $T = \cup_j T_j$ are disjoint open sets containing E and F, respectively. So X is normal. $\square$

Definition 1.4.14 Let $(X, \mathcal{U})$ be a topological space. We say that X is *completely regular* if, whenever E is a closed set in X and p is a point of X that does not lie in E, then there is a continuous function $f : X \to I$ (where $I = [0, 1]$ is the unit interval) such that $f(p) = 0$ and $f(E) = 1$. Any such function f will be said to *separate* E and p.

We record now a technical result which we shall not prove, but refer the reader to [WIL, p. 57] for details.

Proposition 1.4.15 (Tychanoff) *If $(X, \mathcal{U})$ is a T_1 space and $\{f_\alpha\}_{\alpha \in A}$ is a collection of functions on X (mapping to spaces X_α) which separates points from closed sets, then the evaluation mapping $e : X \to \prod_\alpha X_\alpha$ is an embedding.*[1]

For later reference, we now provide a definition and an elegant result about embedding.

Definition 1.4.16 A completely regular T_1 space will be termed a *Tychanoff space*.

Proposition 1.4.17 (Tychanoff) *Every Tychanoff space X can be embedded as a subspace of a (possibly infinite-dimensional) cube.*

Proof: Let C denote the family of all continuous real functions from X to I. The complete regularity tells us that C can distinguish points from closed sets in X. The T_1 property tells us that every singleton set is closed, hence C can also distinguish pairs of points in X. It follows that the mapping

$$
\begin{aligned}
f : X &\to I^C \\
x &\mapsto [f \mapsto f(x)]
\end{aligned}
$$

is an embedding. [Here I^C denotes the collection of all functions from C to I—see Appendix 1.]   □

1.5   Compactness

Compact sets are a fundamental idea in basic topology. A compact set is a (possibly infinite) set that behaves like a finite set. How is this possible? The concept of compact set evolved over a period of fifty or more years. It is subtle, but it is very important. You should spend some time to master the idea. We begin by reiterating the basic terminology of open covers.

---

[1]Here an embedding is a continuous, one-to-one mapping with a continuous inverse. We discuss embeddings in more detail in Section 1.6.

Definition 1.5.1 Let X be a topological space and $S \subseteq X$. An *open covering* for S is a collection $\mathcal{W} = \{W_\alpha\}_{\alpha \in A}$ of open sets such that $S \subseteq \cup_\alpha W_\alpha$.

Definition 1.5.2 Let S be a set and $\mathcal{W} = \{W_\alpha\}$ an open covering of S. A *finite subcovering* of $\mathcal{W}$ is a finite collection W_{α_1}, W_{α_2}, ..., W_{α_k} of the W_α such that $S \subseteq \cup_{j=1}^{k} W_{\alpha_j}$.

Definition 1.5.3 Let X be a topological space and $K \subseteq X$. We say that K is *compact* provided that any open covering $\mathcal{W} = \{W_\alpha\}_{\alpha \in A}$ of K has a finite subcovering.

The definition of "compact" is subtle, and requires some discussion and some examples.

EXAMPLE 1.5.4 Let X be the real numbers with the usual topology. Let $K = \{0\}$. So K is a set with a single point. Then K is compact.

This is almost obvious. If $\mathcal{W} = \{W_\alpha\}_{\alpha \in A}$ is an open covering of K, then one of the W_α, say W_{α_1}, contains 0. Then the finite subcovering $\{W_{\alpha_1}\}$ will do the job.

EXAMPLE 1.5.5 Let $X = \mathbb{R}$ be the real numbers with the usual topology. Let K be the entire space X. Then K is *not* compact. For let $\mathcal{W}$ be the open cover consisting of the intervals $W_k = (k - 2/3, k + 2/3)$ for $k = \cdots - 3, -2, -1, 0, 1, 2, 3, \ldots$. Certain $\mathcal{W}$ is a covering of the set K. But each integer k lies just in W_k and in no other element of the cover. So there is no subcovering that will still cover K. In particular, there is no finite subcovering. Thus K is not compact.

EXAMPLE 1.5.6 Let X be the real numbers with the usual topology. Let

$$K = \left\{ 1, \frac{1}{2}, \frac{1}{3}, \frac{1}{4}, \ldots \right\} \bigcup \{0\}.$$

Then K is compact. To see this, let $\mathcal{W} = \{W_\alpha\}$ be an open covering of K. Then some W_{α_0} contains 0. But then W_{α_0} will contain all $1/j$ for $j > J_0$, some J_0 sufficiently large. See Figure 1.7. Now select W_{α_1} that contains $1/1$, W_{α_2} that contains $1/2$, on up to $W_{\alpha_{J_0}}$ that contains $1/J_0$. We may conclude that the open sets

$$W_{\alpha_0}, W_{\alpha_1}, W_{\alpha_2}, \ldots, W_{J_0}$$

form a finite subcover of the original covering $\mathcal{W}$. Thus K is compact.

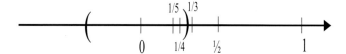

Figure 1.7: Compactness of the set K.

Figure 1.8: Compactness of the unit interval.

EXAMPLE 1.5.7 Let X be the real numbers with the usual topology. Let $K = [0,1]$. Then K is compact. This assertion is not so easy to prove, though you will recognize elements of Example 1.5.6 in the argument that we are about to present.

Now fix an open cover $\mathcal{W} = \{W_\alpha\}$ of $K = [0,1]$. Let

$$S = \{b \in [0,1] : \text{the interval } [0,b] \text{ has a finite subcover}\}.$$

Then S is not empty since $0 \in S$; that is to say, the singleton $\{0\}$ has a finite subcover (again see Example 1.5.4). Also S is bounded above—by 1! Let s_0 be the supremum of S (see Appendix 3 for basic ideas about the real number system, including completeness and the idea of supremum). Of course $s_0 \in [0,1]$. Seeking a contradiction, we suppose that $s_0 < 1$.

Let W_{α_0} be a member of the original open covering that contains s_0. Look at Figure 1.8. Since s_0 is the supremum of S, there are elements of S (to the left of s_0) that are arbitrarily close to s_0. Choose one such that lies inside W_{α_0}. Call it s. The interval $[0,s]$ has a finite subcovering W_{α_1}, ..., W_{α_k} by the definition of S. But then $[0, s_0]$ has the finite subcovering $W_{\alpha_0}, W_{\alpha_1}, \ldots, W_{\alpha_k}$. In point of fact there are points s' to the right of s_0 that lie in W_{α_0}, and we may note that the interval $[0, s']$ has a finite subcover. But this contradicts the choice of s_0 as the supremum of S. This contradiction implies that $s_0 = 1$ and the entire interval $[0,1]$ has a finite subcover.

The last example illustrates the important point that it is not always easy to see that even a simple set like the unit interval is compact. We need some theorems to help us in the process of identifying compact sets.

Having looked at some concrete examples, we now begin to assemble some ideas that will help us to understand compact sets.

Proposition 1.5.8 *Let K be a compact set in a Hausdorff (i.e., T_2) space and let x be a point that is not in K. Then there are disjoint open sets U and V such that $U \supseteq K$ and $V \ni x$.*

Proof: This proof is a nice illustration of how compactness works. For each point k in K, there is a neighborhood U_k of k and a neighborhood V_k of x such that $U_k \cap V_x = \emptyset$. The sets $\{U_k\}$ of course form an open cover of K. So there is a finite subcover $\{U_{k_1}, U_{k_2}, \ldots, U_{k_m}\}$ of K. But then $\mathcal{V} \equiv V_{k_1} \cap V_{k_2} \cap \cdots \cap V_{k_m}$ is an open neighborhood of x and $\mathcal{U} \equiv U_{k_1} \cup U_{k_2} \cup \cdots \cup U_{k_m}$ is an open neighborhood of K, and $\mathcal{V}$ and $\mathcal{U}$ separate x and K. $\square$

Proposition 1.5.9 *A compact set in a Hausdorff space is closed.*

Proof: Let K be a compact set and x a point that is not in K. By the preceding proposition, there is a neighborhood U of x that is disjoint from K. That shows that the complement of K is open. So K is closed. $\square$

Remark 1.5.10 The hypothesis of "Hausdorff" is definitely needed in this last proposition. Consider, for instance, the real line equipped with the topology consisting of sets of the form $(a, \infty) = \{x \in \mathbb{R} : a < x\}$ (together of course with the empty set and the whole space). Then the space is certainly not Hausdorff. And the set $[0, 1]$ is compact (exercise) but it is not closed.

EXERCISE FOR THE READER 1.5.11 Show that the union of two compact sets is compact.

EXERCISE FOR THE READER 1.5.12 Show that, in a Hausdorff space, the intersection of two compact sets is compact.

Proposition 1.5.13 *Let $f : X \to Y$ be a mapping of topological spaces. If $K \subseteq X$ is compact, then $f(K) \equiv \{f(k) : k \in K\}$ is compact.*

Proof: Let $\mathcal{W} = \{W_\alpha\}_{\alpha \in A}$ be an open covering of $f(K)$. Then, since f is continuous, $\{f^{-1}(W_\alpha)\}_{\alpha \in A}$ is an open covering of K. Therefore there is a finite subcovering $f^{-1}(W_{\alpha_1}), f^{-1}(W_{\alpha_2}), \ldots, f^{-1}(W_{\alpha_m})$. It follows then that $W_{\alpha_1}, W_{\alpha_2}, \ldots, W_{\alpha_m}$ is a finite subcovering of K. Thus K is compact. $\square$

THEOREM 1.5.14 (Heine-Borel) *A set $E \subseteq \mathbb{R}$ is compact if and only if it is closed and bounded.*

Proof: If the set is closed and bounded, then compactness follows precisely as in the proof of Example 1.5.7. We leave the details to the reader.

Now suppose that $E \subseteq \mathbb{R}$ is compact. Since $\mathbb{R}$ is Hausdorff, we can be sure by Proposition 1.5.9 that E is closed. It remains to show that E is bounded. If E is not bounded, then let $E \ni e_j \to +\infty$ (the case $e_j \to -\infty$ is handled in exactly the same fashion). We may assume, passing to a subsequence if necessary, that $|e_j - e_{j+1}| \geq j$. Let $\mathcal{W} = \{W_\alpha\}$ be a covering of E such that no W_α has diamater greater than 1. It follows that any subcovering will have to have distinct elements that cover e_1, e_2, etc. So there must be infinitely many elements in any subcovering. There is no finite subcovering.
$\square$

The Heine-Borel theorem is an extremely useful result, as it makes it a straightforward matter to check whether any set in $\mathbb{R}$ is compact. The result also holds in $\mathbb{R}^N$ for any N, and we shall say more about that matter later.

Corollary 1.5.15 (The Extreme Value Property) *Let $[a, b] \subseteq \mathbb{R}$ be a closed, bounded interval and $f : [a, b] \to \mathbb{R}$ a continuous function. Then there is a point $m \in [a, b]$ at which f takes its minimum value and there is a point $M \in [a, b]$ at which f takes its maximum value.*

Proof: The Heine-Borel theorem tells us that $[a, b]$ is compact. By Proposition 1.5.13, the image of $[a, b]$ under f is compact. Thus, by the Heine-Borel theorem, $f([a, b])$ is closed and bounded. Hence it is easy to see (exercise) that $f([a, b])$ has a least element α and a greatest element β. It follows from the definition of $f([a, b])$ that there is an $m \in [a, b]$ such that $f(m) = \alpha$ and an $M \in [a, b]$ so that $f(M) = \beta$. That proves the result. $\square$

Proposition 1.5.16 *Let X be a topological space. A closed subset of a compact set in X is compact.*

Proof: Let K be a compact set and $E \subseteq K$ a closed subset. Let $\mathcal{W} = \{W_\alpha\}_{\alpha \in A}$ be an open covering of E. Let $W' = X \setminus E$. Then of course W'

is open and $\mathcal{W} \cup \{W'\}$ is an open covering of K. Hence there is a finite subcovering of K given by

$$W', W_{\alpha_1}, W_{\alpha_2}, \ldots, W_{\alpha_m}.$$

But then

$$W_{\alpha_1}, W_{\alpha_2}, \ldots, W_{\alpha_m}$$

is a finite subcover of E itself. Thus E is compact. $\qquad\square$

Proposition 1.5.17 *A one-to-one, continuous map from a compact space X onto a Hausdorff space Y must be bicontinuous.*

Proof: Let f be such a map. Let U be an open subset of X. Then $E = X \backslash U$ is closed. Hence, by 1.5.16, it is compact. It follows from 1.5.13 then that $f(E)$ is compact in Y. So $f(E)$ is closed. But then $f(U) = Y \backslash f(E)$ is open. So f is an open mapping. But that just says that f^{-1} is continuous. $\qquad\square$

We next turn to one of the more profound and useful results about compactness. It is necessary to begin with a definition.

Definition 1.5.18 Let X be a topological space. Let $\mathcal{F} = \{F_\alpha\}_{\alpha \in A}$ be a family of sets in X. We say that $\mathcal{F}$ has the *finite intersection property* if, whenever $F_{\alpha_1}, F_{\alpha_2}, \ldots, F_{\alpha_m}$ is a finite collection of elements of $\mathcal{F}$ then $\cap_{j=1}^m F_{\alpha_j}$ is nonempty.

THEOREM 1.5.19 *A topological space $(X, \mathcal{U})$ is compact if and only if any family $\mathcal{F} = \{F_\alpha\}_{\alpha \in A}$ of closed sets in X with the finite intersection property actually satisfies $\cap_{\alpha \in A} F_\alpha \neq \emptyset$.*

Proof: First suppose that X is compact. Let $\mathcal{F} = \{F_\alpha\}_{\alpha \in A}$ be a family of closed sets in X and suppose that $\cap_{\alpha \in A} F_\alpha = \emptyset$. Now look at $\{X \backslash F_\alpha\}$. This must then be an open cover of X. Since X is compact, there is a finite subcover $X \backslash F_{\alpha_1}, X \backslash F_{\alpha_2}, \ldots, X \backslash F_{\alpha_m}$. But this says that $F_{\alpha_1} \cap F_{\alpha_2} \cap \cdots \cap F_{\alpha_m} = \emptyset$. So $\mathcal{F}$ does not have the finite intersection property. That proves one direction of the theorem.

Now suppose that whenever $\mathcal{F} = \{F_\alpha\}_{\alpha \in A}$ is a family of closed sets with the finite intersection property then $\cap_{\alpha \in A} F_\alpha \neq \emptyset$. Our job then is to show that X is compact. Let $\mathcal{U} = \{U_\alpha\}_{\alpha \in A}$ be an open cover of X. Now consider the family $\mathcal{F} \equiv \{X \setminus U_\alpha : \alpha \in A\}$. Then $\mathcal{F}$ is a family of closed sets and, by de Morgan's law, $\cap_\alpha X \setminus U_\alpha = \emptyset$. So there must be finitely many $X \setminus U_{\alpha_1}$, $X \setminus U_{\alpha_2}, \ldots, X \setminus U_{\alpha_m}$ with empty intersection. But, again by de Morgan's law, this says that $U_{\alpha_1}, U_{\alpha_2}, \ldots, U_{\alpha_m}$ is a finite subcover of the family $\mathcal{U}$. Thus X is compact. $\square$

We close this section with a notion that will come up later in the book.

Definition 1.5.20 Let $(X, \mathcal{U})$ be a topological space. We say that X is *locally compact* if each point of X has a neighborhood base consisting of compact sets. [Here a neighborhood base of a point x is a collection $\mathcal{U}$ of neighborhoods of x such that any neighborhood of x contains an element of $\mathcal{U}$.]

EXAMPLE 1.5.21 Let X be the real numbers equipped with the usual topology. Let $x \in X$. Then the sets $[x - \epsilon, x + \epsilon]$ for $\epsilon > 0$ form a neighborhood base for the point x, and each of these sets is compact. So X is locally compact.

1.6   Homeomorphisms

The most fundamental tool in the subject of point-set topology is the homeomorphism. This is the device by means of which we measure the equivalence of topological spaces. This section will introduce you to the idea, and provide several examples.

Definition 1.6.1 Let X and Y be topological spaces. A mapping $f : X \to Y$ is said to be a *homeomorphism* if

- The mapping f is one-to-one and onto;

- The mapping f is continuous;

- The mapping f^{-1} is continuous.

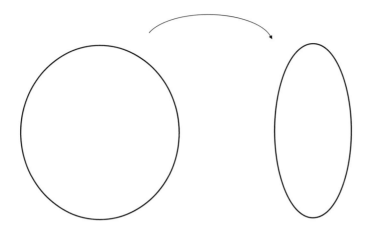

Figure 1.9: Homeomorphism of the circle and the ellipse.

As always in this text, mappings are assumed to be continuous. We are a bit redundant in the last definition just to be explicit and clear.

In the case that the mapping f satisfies all the properties of a homeomorphism except onto-ness then we call it an *embedding*. Another way to say this is that an embedding is a homeomorphism onto its image. The mapping

$$\Phi : \mathbb{R} \rightarrow \mathbb{R}^2$$
$$x \mapsto (x, 0)$$

is an embedding of $\mathbb{R}$ into $\mathbb{R}^2$.

It is plain that a homeomorphism f preserves open sets, closed sets, and compact sets. And so does f^{-1}. Thus all the essential features of a topology are transferred naturally under a homeomorphism. If $f : X \rightarrow Y$ is a homeomorphism, then we say that the spaces X and Y are *homeomorphic*.

EXAMPLE 1.6.2 The set $S = \{(x, y) \in \mathbb{R}^2 : x^2 + y^2 = 1\}$ and the set $T = \{(x, y) \in \mathbb{R}^2 : 4x^2 + y^2 = 1\}$ are homeomorphic. In fact the mapping $f : S \rightarrow T$ given by $(x, y) \mapsto (x/2, y)$ is the needed homeomorphism. See Figure 1.9.

Example 1.6.2 is a very simple example, but it captures the spirit of what a homeomorphism is. Look at Figure 1.9. We see that the circle S and the ellipse T definitely have different shapes. But there is some essential sameness

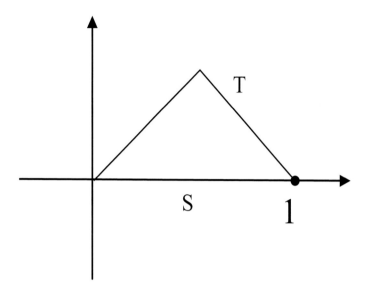

Figure 1.10: Homeomorphism of a segment with a bent, piecewise linear curve.

to them. You can get one from the other by a continuous deformation—a *bending and stretching without tearing.* That is what homeomorphism is all about.

EXAMPLE 1.6.3 Let

$$S = \{(x, y) \in \mathbb{R}^2 : 0 \leq x \leq 1, y = 0\}$$

and

$$
\begin{aligned}
T \;=\; & \{(x, y) \in \mathbb{R}^2 : 0 \leq x \leq 1/2, y = 1/2 - x\} \\
& \cup \{(x, y) \in \mathbb{R}^2 : 1/2 \leq x \leq 1, y = x - 1/2\}\,.
\end{aligned}
$$

Then S and T are homeomorphic. See Figure 1.10.

In fact the homeomorphism is

$$
\begin{aligned}
f : S \;&\rightarrow\; T \\
(x, 0) \;&\mapsto\; \begin{cases} (x, 1/2 - x) & \text{if} \quad 0 \leq x \leq 1/2 \\ (x, x - 1/2) & \text{if} \quad 1/2 < x \leq 1\,. \end{cases}
\end{aligned}
$$

We leave it to the reader to verify the details that this is indeed a homeomorphism. Once again we see that a homeomorphism represents a bending and stretching (without tearing).

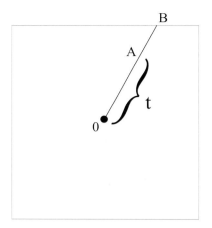

Figure 1.11: The fractional distance $t = |0A|/|0B|$.

EXAMPLE 1.6.4 Let $S = \{(x, y) \in \mathbb{R}^2 : |x| \leq 1, |y| \leq 1\}$ and $T = \{(x, y) \in \mathbb{R}^2 : x^2 + y^2 \leq 1\}$. Then S and T are homeomorphic.

To see this, assign to each point $(x, y) \in S$ a number $t = t(x, y), 0 \leq t \leq 1$ that represents the fraction of the distance that the point (x, y) is from the origin to the boundary. See Figure 1.11. Now define

$$
\begin{aligned}
f : S &\rightarrow T \\
(x, y) &\mapsto \frac{t(x, y)}{\sqrt{x^2 + y^2}} \cdot (x, y).
\end{aligned}
$$

It is easy to check that f maps S to T in a one-to-one, onto bicontinuous fashion. See Figure 1.12.

The next result is at first a bit surprising.

EXAMPLE 1.6.5 The spaces $X = (0, 1) \subseteq \mathbb{R}$ and $Y = (0, \infty)$ are homeomorphic. For let $f : X \rightarrow Y$ be given by $f(x) = x/[1 - x]$. Then it is easy to verify that f satisfies all the properties of a homeomorphism of these spaces.

This last example illustrates the idea that a homeomorphism does not preserve size or distance. Instead, it preserves the essential topological nature (or the shape) of a space.

EXAMPLE 1.6.6 The spaces $[0, 1]$ and $(0, 1)$ (considered as subspaces of $\mathbb{R}$ with the usual inherited topology given by intersection) are *not* homeomorphic. For the first of these is compact and the second not.

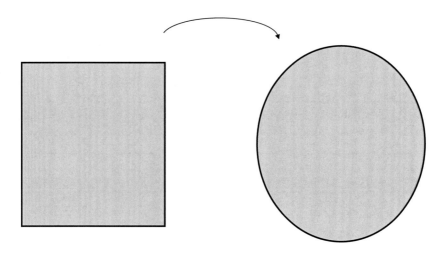

Figure 1.12: A homeomorphism of the square to the disc.

EXAMPLE 1.6.7 The unit circle $S = \{(x, y) \in \mathbb{R}^2 : x^2 + y^2 = 1\}$ and the closed annulus $A = \{(x, y) \in \mathbb{R}^2 : 1 \le x^2 + y^2 \le 4\}$ are *not* homeomorphic. For the first has boundary with one connected component, or one piece— namely $\partial S = S$ (see the next section for the concept of "connected compo- nent"). But the second set has boundary with two connected components, or two pieces—namely $\partial A = \{(x, y) : x^2 + y^2 = 1\} \cup \{(x, y) : x^2 + y^2 = 4\}$.

1.7 Connectedness

Certainly we have an intuitive idea of what it means for a space to be con- nected. A space is connected if it is "all of one piece"; the space is discon- nected if it is in "several pieces." Now it is time for us to make these ideas precise.

Definition 1.7.1 Let X be a topological space and $E \subseteq X$. We say that E is *disconnected* if we can write $E = F \cup G$ and if there are open sets U and V so that $F \subseteq U$, $G \subseteq V$, and $U \cap V = \emptyset$. If there are no such U and V, then we say that E is *connected*.

EXAMPLE 1.7.2 Let X be the real line with the standard topology and $I = [0, 1]$. Then I is connected. For suppose to the contrary that $I = F \cup G$, $F \subseteq U$ open, $G \subseteq V$ open, and $U \cap V = \emptyset$. We derive a contradiction as follows.

Certainly the point 1 will lie in one of the two open sets; say that $1 \in U$. Let $c = \sup_{v \in V} v$. Since there is a neighborhood of 1 that lies entirely in U, we can be sure that c is not in U. So $c \in V$. Thus c has a neighborhood that lies entirely in V. But, since c is the supremum of V, c has points to the immediate left that lie in V. Also c has points to the immediate right that lie in U. That contradicts the neighborhood of c that is entirely in V. In conclusion, U and V do not exist and the interval I is connected.

Proposition 1.7.3 *Let $f : X \rightarrow Y$ be a mapping. Let $E \subseteq X$ be connected. Then $f(E)$ is connected.*

Proof: Suppose to the contrary that $f(E)$ is disconnected. Write $f(E) = A \cup B$ with disjoint open sets U and V so that $A \subseteq U$ and $B \subseteq V$. Then $f^{-1}(U)$ and $f^{-1}(V)$ are disjoint open sets in X that separate E, meaning that E is disconnected. This of course is a contradiction. So $f(E)$ must be connected. $\square$

EXAMPLE 1.7.4 Consider the topology on the real line generated by intervals of the form $[a, b)$ or $[a, +\infty)$ (here we mean "generated" in the sense of taking finite intersection and arbitrary union—this is the language of a sub-basis, which we explore further in Section 2.1). This is called the *Sorgenfrey line*, named after Robert Sorgenfrey (1915-1996). The Sorgenfrey line is one of the most important examples in topology. We show here that the Sorgenfrey line is disconnected.

First note that if (c, d) is any open interval, then

$$(c, d) = \bigcup_{\epsilon > 0} [c + \epsilon, d) \,.$$

Thus (c, d) is the union of Sorgenfrey open sets. So any standard open interval is open in the Sorgenfrey topology. We see, then, that the Sorgenfrey topology contains all the usual open sets and some new ones as well.

We conclude then that $(-\infty, 0)$ and $[0, \infty)$ are both open. And certainly $\mathbb{R} = (-\infty, 0) \cup [0, \infty)$. So $\mathbb{R}$ is disconnected in the Sorgenfrey topology.

EXAMPLE 1.7.5 The *topologist's sine curve* (another famous example) is the set

$$S = \{(0, y) : y \in \mathbb{R}\} \bigcup \left\{ \left(x, \sin \frac{1}{x}\right) : x > 0 \right\} \,.$$

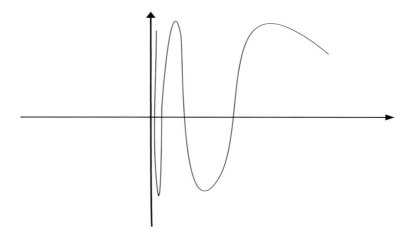

Figure 1.13: The topologist's sine curve.

See Figure 1.13.

It is connected. For certainly the left-hand portion of S, which is the y-axis, is connected. And any open set that contains that portion will contain a neighborhood of the origin and hence intersect the right-hand portion (which gets arbitrarily close to the origin).

Corollary 1.7.6 (The Intermediate Value Property) *Let $[a, b]$ be a closed bounded interval in $\mathbb{R}$. Let f be a continuous, real-valued function on $[a, b]$. Let γ be a real number that lies between $f(a)$ and $f(b)$. Then there is a number c between a and b such that $f(c) = \gamma$.*

Proof: The image of $[a, b]$ under f is a compact, connected set J. So it is an interval. Obviously $f(a)$ and $f(b)$ lie in J. So all points in between $f(a)$ and $f(b)$ lie in J. In particular, $\gamma \in J$. So γ in the image of f, and the point c exists. $\qquad \square$

Proposition 1.7.7 *If A and B are connected sets with a common point p then $A \cup B$ is connected.*

Proof: Suppose not. Say that the disjoint open sets U and V disconnect $A \cup B$. Then p must lie in one of these two open sets. Say that it lies in U. Since A cannot be disconnected, it follows that $A \subseteq U$. A similar argument shows that $B \subseteq U$. Thus $A \cup B \subseteq U$ and it is not the case that U and

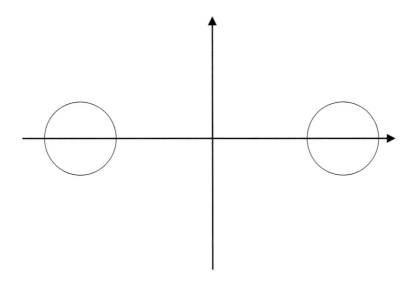

Figure 1.14: A set with two connected components.

V disconnect $A \cup B$. This is a contradiction. We conclude that $A \cup B$ is connected.   □

We say that two points x and y in a topological space X are *connected in X* if there is a connected set $S \subseteq X$ which contains both x and y. It is easy to verify that this is an equivalence relation. The resulting equivalence classes are called *connected components*.

Of course a connected topological space X has just one connected component. As a counterpoint, the set

$$X = \{(x, y) \in \mathbb{R}^2 : |(x, y) \pm (4, 0)| = 1\}$$

has two connected components. See Figure 1.14.

EXAMPLE 1.7.8 The space $X = [0, 1]$ and the space $Y = S^1 = \{(x, y) \in \mathbb{R}^2 : x^2 + y^2 = 1\}$ are *not* homeomorphic. For if x is an interior point of X, then $X \setminus \{x\}$ is disconnected. But if s is any point of Y, then $Y \setminus \{s\}$ is connected.

EXAMPLE 1.7.9 The spaces $X = \mathbb{Q} \cap (0, \infty)$ (where $\mathbb{Q}$ is the rationals) and $Y = \mathbb{Q} \cap (0, 1)$ (both equipped with the usual topology inherited from $\mathbb{R}$) are both disconnected. But they are still homeomorphic by way of the mapping

$$f : X \rightarrow Y$$
$$x \mapsto \frac{1}{x+1} .$$

EXERCISE FOR THE READER 1.7.10 In the next section we shall encounter the first of many "connectedness arguments." In such an argument, we prove that an entire connected set S has a certain property $\mathcal{P}$ by reasoning as follows: Let $T = \{s \in S : s \text{ has Property } \mathcal{P}\}$. We prove that T is nonempty, open, and closed. This being the case, the sets T and $^c T$ disconnect S—which is a contradiction unless T is all of S. Fill in the details of this argument so that you will be comfortable using it in the sections that follow.

1.8   Path-Connectedness

There is a stronger concept of connectedness that is particularly useful in Euclidean space. We treat it in this section.

Definition 1.8.1 Let $(X, \mathcal{U})$ be a topological space. We say that X is *path-connected* if, given any two points $P, Q \in X$, there is a continuous path $\gamma : [0,1] \rightarrow X$ such that $\gamma(0) = P$ and $\gamma(1) = Q$.

EXAMPLE 1.8.2 Let X be the unit disc in the Euclidean plane. Then X is path connected. For if $P, Q \in X$ then the path

$$\gamma(t) = (1-t)P + tQ$$

will connect the two points.

EXAMPLE 1.8.3 Consider the topologist's sine curve as in Example 1.7.5:

$$S = \{(0, y) : y \in \mathbb{R}\} \bigcup \left\{ \left(x, \sin \frac{1}{x} \right) : x > 0 \right\} .$$

We know that this set is connected. But it is *not* path-connected. For certainly the point $(0, 0)$ lies in S. And so does the point $(2/\pi, 1)$. Suppose that γ is a continuous path-connecting the two points. We may take it that $\gamma(0) = (0, 0)$. But then there are points t arbitrarily closed to 0 (of the form $2/[(2k+1)\pi])$ at which the function $\sin \frac{1}{x}$ takes the values ± 1. So γ cannot be continuous.

Proposition 1.8.4 Let $(X, \mathcal{U})$ be a topological space. If X is path-connected, then X is connected.

Proof: Suppose to the contrary that X is disconnected. So there are disjoint open sets U, V that disconnect X. Let P be a point of $U \cap X$ and Q be a point of $V \cap X$ and $\gamma : [0, 1] \to X$ a path that connects them. Then $\gamma^{-1}(U)$ and $\gamma^{-1}(V)$ disconnect the unit interval $[0, 1]$, and that is impossible. $\quad\square$

Proposition 1.8.5 Every connected open set U in Euclidean space is path-connected.

Proof: Fix a point $P \in U$. Define

$$S = \{u \in U : \text{the point } u \text{ can be connected to } P \text{ by a path}\} .$$

Then S is nonempty because P itself is in the set. Also S is open because if $s \in S$ then $s \in U$ so there is a small ball $\mathbf{b}$ around s that lies in U. We know that there is a path γ_1 that connects P to s. And if $x \in \mathbf{b}$, then there is certainly a linear path γ_2 that connects s to x. The concatenation of γ_1 and γ_2 connects P to x. Hence the ball $\mathbf{b}$ lies in S and S is open. Finally, if $t \notin S$ then t cannot be connected to the point P by a path. But of course $t \in U$, so there is a small ball $\mathbf{b}'$ about t that lies in U. And points of $\mathbf{b}'$ can be connected to t by a linear path. Thus they cannot be connected to P. So the entire ball is in the complement of S. We conclude that the complement of S is open hence S is closed.

Since S is nonempty, open, and closed, we conclude by the connectedness of Euclidean space that S is all of space. That gives the desired conclusion. $\square$

EXERCISE FOR THE READER 1.8.6 Prove that the continuous image of a path-connected space is path-connected.

A topological space is said to be locally path-connected if it has a neighborhood basis consisting of path-connected sets. See Section 2.1 for the concept of neighborhood basis.

1.9   Continua

The heuristic model for a continuum is (the image of) a curve in the plane. A continuum is a set that should have no cuts or breaks. A formal definition is as follows.

Definition 1.9.1 Let X be a compact, connected Hausdorff space. Then X is a *continuum*.

EXAMPLE 1.9.2 Certainly the unit interval $I = [0, 1]$, the unit circle in the plane, and the torus are all continua.

Our goal in this discussion is to develop useful topological characterizations of continua.

Definition 1.9.3 Let X be a connected T_1 space. A *cut point* of X is a point $p \in X$ such that $X \setminus \{p\}$ is disconnected. If p is *not* a cut point of X, then we say that p is a *noncut point*. A *cutting* of X is a triple (p, U, V) where p is a cut point for X and the open sets U, V separate $X \setminus \{p\}$.

Lemma 1.9.4 *If K is a metric space that is a continuum, and if K has exactly two noncut points, then K is homeomorphic to the unit interval.*

Proof: Although this result is intuitively appealing, it is remarkably tricky to prove (relying as it does on the construction of the real numbers and other subtle ideas). We refer the reader to [WIL, pp. 206–207] for the details.   □

THEOREM 1.9.5 *Let K be a metric space and assume that K is a continuum. Further suppose that, for any two distinct points $p, q \in K$, $K \setminus \{p, q\}$ is disconnected. Then K is homeomorphic to the unit circle.*

The theorem teaches us that the property of having cuts induces an order on the topological space in question. It is a remarkable characterization of the circle.

Proof of Theorem 1.9.5: First let us show that K has no cut points. Suppose to the contrary that (p, U_0, V_0) is a cutting. Then, since $U_0 \cup \{p\}$ and $V_0 \cup \{p\}$ are both continua, each contains at least some noncut points. Say that y is a noncut point of $U_0 \cup \{p\}$ and z is a noncut point of $V_0 \cup \{p\}$. Observe that the connected sets $(U_0 \cup \{p\}) \setminus \{y\}$ and $(V_0 \cup \{p\}) \setminus \{z\}$ intersect; and their union $K \setminus \{y, z\}$ is therefore connected. This contradicts our hypothesis. So K has no cut points.

Now, following the statement of the theorem, let p and q be arbitrary distinct points of K. Then $K \setminus \{p, q\} = U \cup V$, where U and V are nonempty, disjoint, open subsets of K. Let $U^* = U \cup \{p, q\}$ and $V^* = V \cup \{p, q\}$. We claim that U^*, V^* are arcs with p, q as endpoints. Moreover, $U^* \cap V^* = \{p, q\}$. This will clearly show that $K = U^* \cup V^*$ is homeomorphic to a circle.

Certainly U^* and V^* are both connected. To see this, suppose that $U^* = S \cup T$, with S, T open, nonempty, and disjoint in U^*. If S contains both p and q, , then T is open in U and hence in K. This is impossible, since T is closed in U^* hence closed in K (and K is connected). As a result, we may suppose that $p \in S$ and $q \in T$. But now the same reasoning shows that $S \setminus \{p\}$ is open and closed in the connected set $K \setminus \{p\}$. That is impossible. As a result, U^* and V^* are connected.

Next we assert that p and q are both noncut points of $U*$ (and also of V^*). For if S and T disconnect $U^* \setminus \{p\}$, and if $q \in S$, then (again by the arguments that we presented above) T is both open and closed in $K \setminus \{p\}$. That is impossible.

Finally, we wish to show that U^* and V^* each has precisely two noncut points (namely, p and q). We proceed as follows:

- Say that each of U^* and V^* has a third noncut point. Say that p' is a noncut point of U^* and q' is a noncut point of V^* and moreover these two new points are distinct from p and q. Then the sets $U^* \setminus \{p'\}$ and $V^* \setminus \{q'\}$ are connected, they intersect, and their union is $K \setminus \{p', q'\}$ (which is a disconnected set). This is a contradiction, so the existence of these third noncut points is impossible.

- Suppose that just one of our two sets, say U^*, has a third noncut point p''. Then, if q'' is any point in V, we have a cutting (q'', L, M) of V^*, where L and M are connected and $p \in L$, $q \in M$. [Clearly not both p and q can belong to the same one of these sets.] Now $U^* \setminus \{p'\}$, L, and M form a chain of connected sets whose union is $K \setminus \{y, z\}$, a contradiction.

So we see that each of U^* and V^* is a metric continuum with precisely two noncut points, p and q, and $U^* \cap V^* = \{p, q\}$. It now follows from the lemma that $K = U^* \cup V^*$ is homeomorphic to the circle.   □

1.10   Totally Disconnected Spaces

Of course a connected topological space has just one connected component. A totally disconnected space is just the opposite. We explore that new idea here.

Definition 1.10.1 A topological space X is *totally disconnected* if the connected components of X are single points.

EXAMPLE 1.10.2 Let X be the rational numbers $\mathbb{Q}$. Then X is totally disconnected. For certainly any subset of X that contains at least two points is disconnected.

Similary, if X is the irrational numbers $\mathbb{R} \setminus \mathbb{Q}$, then X is totally disconnected.

EXAMPLE 1.10.3 Refer ahead to the construction of the Cantor set $\mathbf{C}$ in the next section.

Let Q be the set of endpoints of the intervals that are deleted from $I = [0, 1]$ in order to construct $\mathbf{C}$. Set $P = \mathbf{C} \setminus Q$. Fix the point $p = (1/2, 1/2)$ in the plane. Of course the Cantor set lives in the x-axis. If $x \in \mathbf{C}$ then let L_x be the line segment joining p to x. Define

$$L_x^* = \{(x_1, x_2) \in L_x : x_2 \text{ is rational}\} \quad \text{for} \quad x \in Q ,$$
$$L_x^* = \{(x_1, x_2) \in L_x : x_2 \text{ is irrational}\} \quad \text{for} \quad x \in P .$$

Now set
$$\mathbf{K} = \bigcup_{x \in \mathbf{C}} L_x^* .$$

It is plain that $\mathbf{K}$ is connected, while $\mathbf{K} \setminus \{p\}$ is totally disconnected.

Proposition 1.10.4 *The product of totally disconnected spaces is totally disconnected. Also every subspace of a totally disconnected space is totally disconnected.*

Proof: Let X_1 and X_2 be totally disconnected and let π_j be the projection from $X = X_1 \times X_2$ to X_j. If S is a nonempty, connected subset of X, then $\pi_j(S)$ is a connected subset of X_j. So $\pi_j(S)$ is a point, $j = 1, 2$. It follows then that S is a point.

The second statement of the proposition is nearly obvious, and we leave it as an exercise. □

In dimension theory, we define topological dimension inductively. Suppose that $(X, \mathcal{U})$ is a separable metric space (that is, X has a countable dense subset). We begin by positing that the empty set has dimension -1. Next, X has dimension $\leq N$ if each point of X has a neighborhood basis of open sets U such that ∂U has dimension $\leq (N-1)$.

EXAMPLE 1.10.5 The set

$$X = \{1, 1/2, 1/3, \dots\} \cup \{0\}$$

in $\mathbb{R}$ has dimension 0. The proof is obvious. Any of the points $1/j$ has a neighborhood basis $(1/j - 1/4^j, 1/j + 1/4^j) \cap X$ with boundaries that are empty—hence they are of dimension -1. Also the point 0 has a neighborhood basis $(0 - \pi/10^j, 0 + \pi/10^j)$ with boundaries that are empty—hence they are of dimension -1. We conclude that X has dimension ≤ 0. But the set obviously is *not* of dimension -1. So it has dimension $= 0$.

EXAMPLE 1.10.6 The set

$$X = \{(x, y) \in \mathbb{R}^2 : x^2 + y^2 < 1\}$$

has dimension 2. We calculate as follows: If $(x, y) \in \mathbb{R}^2$ then let $\delta = \sqrt{1 - x^2 - y^2}$. Then $U_j = \{(s, t) : |(x, y) - (s, t)| < 1/j\}$ for $j > 2/\delta$ is a neighborhood system with ∂U_j having dimension 1 (that is a separate, but plausible, calculation). Hence the dimension of X is ≤ 2. But it is easy to check that the dimension of X is *not* ≤ 1. So the dimension of X is precisely 2.

Lemma 1.10.7 *A nonempty subset S of a 0-dimensional space X is 0-dimensional.*

Proof: The set S clearly has dimension *at most* 0. It is not empty, so does not have dimension -1. So the dimension must be 0. □

Proposition 1.10.8 *A nonempty space X has dimension 0 if and only if every point $p \in X$ and every closed set $C \subseteq X$ such that $p \notin C$ can be separated by open sets.*

Proof: First suppose that X has dimension 0 according to our original definition. Certainly $X \setminus C$ is a neighborhood of p. So there is a set V with $p \in V \subseteq X \setminus C$ and V is both open and closed. But $V \cap C = \emptyset$, so p and C are separated.

The converse is proved similarly.   □

Lemma 1.10.9 *A connected, 0-dimensional space consists of just one point.*

Proof: Suppose instead that the space contains two points. Then Proposition 1.10.3 tells us that these points are separated. So the space is in fact disconnected, and that is a contradiction.   □

Proposition 1.10.10 *A 0-dimensional space is totally disconnected.*

Proof: This follows from 1.10.8 and 1.10.9.   □

1.11   The Cantor Set

The Cantor set is one of the most remarkable sets in all of mathematics. It has permeated many areas of analysis, topology, and geometry, and frequently arises as an example or counterexample in various theories. We describe it here, and derive some of its properties. The set only begins to suggest the richness of the structure of the real number system.

We begin with the unit interval $S_0 = [0, 1]$. We extract from S_0 its open middle third; thus $S_1 = S_0 \setminus (1/3, 2/3)$ (Figure 1.15). Observe that S_1 consists of two closed intervals of equal length $1/3$.

Now we construct S_2 from S_1 by extracting from each of its two intervals the middle third: $S_2 = [0, 1/9] \cup [2/9, 3/9] \cup [6/9, 7/9] \cup [8/9, 1]$. Figure 1.16 shows S_2.

Figure 1.15: First step in the construction of the Cantor set.

Figure 1.16: Second step in the construction of the Cantor set.

Continuing in this fashion, we construct S_{j+1} from S_j by extracting the middle third from each of its component subintervals. We define the Cantor set C to be

$$C = \bigcap_{j=1}^{\infty} S_j \,.$$

Notice that each of the sets S_j is closed and bounded, hence compact. By Proposition 1.5.19, C is therefore not empty. The set C is closed and bounded, hence compact.

Proposition 1.11.1 *The Cantor set C has zero length, in the sense that the complementary set $[0,1] \setminus C$ has length 1.*

Proof: In the construction of S_1, we removed from the unit interval one interval of length 3^{-1}. In constructing S_2, we further removed two intervals of length 3^{-2}. In constructing S_j, we removed 2^{j-1} intervals of length 3^{-j}. Thus the total length of the intervals removed from the unit interval is

$$\sum_{j=1}^{\infty} 2^{j-1} \cdot 3^{-j} \,.$$

This last equals

$$\frac{1}{3} \sum_{j=0}^{\infty} \left(\frac{2}{3}\right)^j \,.$$

The geometric series sums easily and we find that the total length of the intervals removed is

$$\frac{1}{3}\left(\frac{1}{1-2/3}\right) = 1 \,.$$

Thus the Cantor set has length zero because its complement in the unit interval has length one.   □

Proposition 1.11.2 *The Cantor set is uncountable.*

Proof: We assign to each element of the Cantor set a "label" consisting of a sequence of 0s and 1s that identifies its location in the set.

Fix an element x in the Cantor set. Then certainly x is in S_1. If x is in the left half of S_1, then the first digit in the "label" of x is 0; otherwise it is 1. Likewise $x \in S_2$. By the first part of this argument, it is either in the left half S_{21} of S_2 (when the first digit in the label is 0) or the right half S_{22} of S_2 (when the first digit of the label is 1). Whichever of these is correct, that half will consist of two intervals of length 3^{-2}. If x is in the leftmost of these two intervals, then the second digit of the "label" of x is 0. Otherwise the second digit is 1. Continuing in this fashion, we may assign to x an infinite sequence of 0s and 1s.

Conversely, if $a, b, c, \ldots$ is a sequence of 0s and 1s, then we may locate a unique corresponding element y of the Cantor set. If the first digit is a zero then y is in the left half of S_1; otherwise y is in the right half of S_1. Likewise the second digit locates y within S_2, and so forth.

Thus we have a one-to-one correspondence between the Cantor set and the collection of all infinite sequences of zeroes and ones. [Notice that we are in effect thinking of the point assigned to a sequence $c_1 c_2 c_3 \ldots$ of 0s and 1s as the limit of the points assigned to $c_1, c_1 c_2, c_1 c_2 c_3, \ldots$ Thus we are using the fact that C is closed.] However, as we learned in Appendix 1, the set of all infinite sequences of zeroes and ones is uncountable. Thus the Cantor set is uncountable.   □

The Cantor set is quite thin (it has zero length) but it is large in the sense that it has uncountably many elements. Also it is compact. The next result reveals a surprising, and not generally well known, property of this "thin" set:

THEOREM 1.11.3 *Let C be the Cantor set and define*

$$S = \{x + y : x \in C, y \in C\}.$$

Then $S = [0, 2]$.

Proof: We sketch the proof.

Since $C \subseteq [0, 1]$ it is clear that $S \subseteq [0, 2]$. For the reverse inclusion, fix an element $t \in [0, 2]$. Our job is to find two elements c and d in C such that $c + d = t$.

First observe that $\{x + y : x \in S_1, y \in S_1\} = [0, 2]$. Therefore there exist $x_1 \in S_1$ and $y_1 \in S_1$ such that $x_1 + y_1 = t$.

Similarly, $\{x + y : x \in S_2, y \in S_2\} = [0, 2]$. Therefore there exist $x_2 \in S_2$ and $y_2 \in S_2$ such that $x_2 + y_2 = t$.

Continuing in this fashion, we may find for each j numbers x_j and y_j such that $x_j, y_j \in S_j$ and $x_j + y_j = t$. Of course $\{x_j\} \subseteq C$ and $\{y_j\} \subseteq C$ hence there are subsequences $\{x_{j_k}\}$ and $\{y_{j_k}\}$ which converge to real numbers c and d, respectively. Since C is compact, we can be sure that $c \in C$ and $d \in C$. But the operation of addition respects limits, thus we may pass to the limit as $k \to \infty$ in the equation

$$x_{j_k} + y_{j_k} = t$$

to obtain

$$c + d = t\,.$$

Therefore $[0, 2] \subseteq \{x + y : x \in C\}$. This completes the proof. $\qquad\square$

In the exercises at the end of the chapter we shall explore constructions of other Cantor sets, some of which have zero length and some of which have positive length. The Cantor set that we have discussed in detail in the present section is sometimes distinguished with the name "the Cantor ternary set" because of the role of the number 3 in the construction). We shall also consider in the exercises other ways to construct the Cantor ternary set.

Observe that, whereas any open set is the union of open intervals, the existence of the Cantor set shows us that there is no such structure theorem for closed sets. We often think of a closed interval $[a, b]$ as a "typical" closed set. But in fact closed intervals are atypically simple when considered as examples of closed sets.

1.12 Metric Spaces

Certainly one of the most important examples of topological spaces is metric spaces. Metric spaces are rather special. They have more structure than most topological spaces. But they are the typical spaces for mathematical analysis so they are important. They provide a rich panoply of examples.

Definition 1.12.1 Let X be a topological space equipped with a function

$$d : X \times X \to \mathbb{R}$$

and satisfying these conditions:

   (i) $d(x, y) = d(y, x)$;

  (ii) $d(x, y) \geq 0$;

 (iii) $d(x, y) = 0$ if and only if $x = y$;

(iv) $d(x, y) \leq d(x, z) + d(z, y)$.

We call such a space a *metric space*, and we call d the *metric*. Property **(iv)** is called the *triangle inequality*.

When a space satisfies **(i)**, **(ii)**, and **(iv)** (but not necessarily **(iii)**), then we call it a *pseudometric space*, and we call d a *pseudometric*.

EXAMPLE 1.12.2 Let $X = \mathbb{R}^N$ and let

$$d(x, y) = |x - y| = \sqrt{(x_1 - y_1)^2 + \cdots + (x_N - y_N)^2} \, .$$

Then (X, d) is a metric space. The verification of the triangle inequality is standard, and we leave the details for the reader to verify.

EXAMPLE 1.12.3 Let $X = \mathbb{R}^N$ and let

$$d(x, y) = \left\{ \begin{array}{ll} 1 & \text{if} \quad x \neq y \\ 0 & \text{if} \quad x = y \, . \end{array} \right)$$

Then it is straightforward to check that this is a metric space.

EXAMPLE 1.12.4 Let $X = \mathbb{R}^N$ and let

$$d(x, y) = \text{distance of } x \text{ to } 0 + \text{ distance of } 0 \text{ to } y \, .$$

This is affectionately known as the "New York subway metric" because the way one calculates distance from A to B in New York City is that one calculates the distance from A to Grand Central Station and then the distance from Grand Central Station to B.

Verify for yourself that d is a metric.

EXERCISE FOR THE READER 1.12.5 On $\mathbb{R}^N$, define

$$d(x, y) = \max\{|x_1 - y_1|, |x_2 - y_2|, \ldots, |x_N - y_N|\}.$$

Verify that this d is a metric.

EXAMPLE 1.12.6 Let X be the space of all continuous functions on the interval $I = [0, 1]$. Let

$$d(f, g) = \max_{x \in I} |f(x) - g(x)|.$$

Then d is a metric on X.

First note that $d \geq 0$. Second, $d(f, g) = 0$ if and only if $f \equiv g$. Lastly, the triangle inequality for d is inherited from the triangle inequality for the real numbers.

EXAMPLE 1.12.7 Let X be the space of all continuous functions on the interval $I = [0, 1]$. Let

$$d(f, g) = |f(1/2) - g(1/2)|.$$

Then it is easy to check that this d satisfies all the axioms of a metric except **(iii)**. For if we let $f(x) = x^2/4$ and $g(x) = x^4$, then $d(f, g) = 0$ yet $f \neq g$. So this d is a pseudometric.

Definition 1.12.8 Let (X, d) be a metric space. If $x \in X$ and $r > 0$, then we let

$$B(x, r) = \{t \in X : d(x, t) < r\}.$$

We call $B(x, r)$ the *open ball* with center x and radius r. Likewise

$$\overline{B}(x, r) = \{t \in X : d(x, t) \leq r\}$$

is the *closed ball* with center x and radius r.

Remark 1.12.9 The reader may verify as an exercise that, in an arbitary metric space, the closed ball $\overline{B}(x, r)$ is not necessarily equal to the closure of the open ball $B(x, r)$. However, it will contain that closure.

Definition 1.12.10 Let (X, d) be a metric space. A set $U \subseteq X$ is said to be *open* if, for each $u \in U$, there is an $\epsilon > 0$ such that $B(u, \epsilon) \subseteq U$.

It is easy to check that the open sets U specified in the last definition form a topology on X in the usual sense.

Definition 1.12.11 Let (X, d) be a metric space. A *sequence* $\{a_j\}$ in X is a function $\alpha : \mathbb{N} \to X$. We let $\alpha(1) = a_1$, $\alpha(2) = a_2$, etc. We say that the sequence *converges* if there is an element $\ell \in X$ such that, for every $\epsilon > 0$ there is an $N > 0$ such that $j \geq N$ implies that $d(a_j, \ell) < \epsilon$.

If $\{a_j\}$ is a sequence in the metric space X, then a *subsequence* is a function $A : \mathbb{N} \to \{a_j\}$ such that $\ell < k$ implies $A(\ell)$ has lower index than $A(k)$. In other words, a subsequence is an ordered list of some of the elements of the original sequence $\{a_j\}$. As an example, if the original sequence is

$$1, 2, 4, 8, 16, \ldots$$

then a subsequence is

$$1, 4, 16, \ldots .$$

Any given sequence has, in general, infinitely many distinct subsequences.

THEOREM 1.12.12 *Let (X, d) be a metric space. A set $K \subseteq X$ is compact if and only if every sequence $\{a_j\} \subseteq X$ has a convergent subsequence (we call this last condition sequential compactness).*

Proof: Suppose that $K \subseteq X$ is compact according to our usual definition and let $\{a_j\}$ be a sequence in K. Seeking a contradiction, we assume that $\{a_j\}$ does *not* have a convergent subsequence. That being the case, each element $k \in K$ has a neighborhood U_k that contains only finitely many elements of the sequence. But then there is a finite subcover U_{k_1}, U_{k_2}, $\ldots$, U_{k_m}. This in turn implies that there are only finitely many elements in the sequence (as each U_{k_j} has only finitely many sequential elements). That is a contradiction.

The converse result is more work. Assume that K is sequentially compact and we want to show that it is compact (according to the original definition). First we claim that, for $\epsilon > 0$, there is a finite subset $S \subseteq K$ such that $K \subseteq \cup_{s \in S} B(s, \epsilon)$. If this were not the case, then we may choose $x_1 \in K$ and then choose $x_2 \in K$ with $d(x_1, x_2) \geq \epsilon$ and then choose $x_3 \in K$ with both $d(x_3, x_2) \geq \epsilon$ and $d(x_3, x_1) \geq \epsilon$, and so forth. But then the sequence $\{x_j\}$ has no convergent subsequence, contradicting sequential compactness. We call S a *finite ϵ-net for K*.

Now let $\mathcal{W} = \{W_\alpha\}_{\alpha \in A}$ be an open cover of K. We claim that there is a $\delta > 0$ such that if $x \in K$ then the ball $B(x, \delta)$ lies completely inside some W_α. If this were not the case then, for each positive integer j, there is a point $x_j \in K$ such that $B(x_j, 1/j)$ does not lie in any W_α. Consider the sequence $\{x_j\}$ and let $\{x_{j_k}\}$ be a convergent subsequence. Of course the limit point ℓ of this subsequence belongs to some W_α. And there is certainly some $\delta > 0$ such that $B(\ell, \delta) \subseteq W_\alpha$. But then it is surely the case that, for k large enough, $B(x_{j_k}, 1/j_k) \subseteq B(\ell, \delta) \subseteq W_\alpha$. That is a contradiction. We call the number $\delta > 0$ a *Lebesgue number* (Henri L. Lebesgue (1875–1941)) for the covering $\mathcal{W}$. We say more about Lebesgue numbers in Section 1.15. We shall give an application of the concept of Lebesgue number at the end of Section 2.11.

Now, to complete the proof, let K be sequentially compact and let $\mathcal{W} = \{W_\alpha\}_{\alpha \in A}$ be an open cover of K. Let $\delta > 0$ be a Lebesgue number for $\mathcal{W}$. Further let $\{x_1, x_2, \ldots, x_m\}$ be a finite δ-net for K. Thus for each $j = 1, \ldots, m$ there is a W_{α_j} such that $B(x_j, \delta) \subseteq W_{\alpha_j}$. Finally we see that

$$K \subseteq \bigcup_{j=1}^{m} B(x_j, \delta) \subseteq \bigcup_{j=1}^{m} W_{\alpha_j}.$$

Since the $B(x_j, \delta)$ cover K, we can thus be sure that the $\{W_{\alpha_j}\}_{j=1}^{m}$ cover K. Thus the original open cover $\mathcal{W}$ has a finite subcover. $\square$

Proposition 1.12.13 *Let (X, d) be a metric space and $f : X \to \mathbb{R}$ a function. Then f is continuous if and only if, for each $x \in X$ and each $\epsilon > 0$, there is a $\delta > 0$ such that if $d(x, t) < \delta$ then $|f(x) - f(t)| < \epsilon$.*

Proof: This proof is just the same as that of Proposition 1.3.4. $\square$

EXAMPLE 1.12.14 Let X be the real numbers equipped with the usual topology. This is of course a metric space, with the familiar metric

$$d(x, y) = |x - y|.$$

Let $I = [0, 1] = \{x \in \mathbb{R} : 0 \leq x \leq 1\}$. Then I is compact according to the last theorem. For if $\{a_j\}$ is any sequence in I then we may use the method of bisection to find a convergent subsequence. Namely, divide I into the two subintervals $[0, 1/2] \cup [1/2, 1]$. Then one of those subintervals must contain infinitely many elements of the sequence. Say it is the first subinterval. Now subdivide that interval into two: $[0, 1/2] = [0, 1/4] \cup [1/4, 1/2]$. Then one of these two sub-sub-intervals must contain infinitely many elements of the sequence.

Continue in this fashion to obtain a decreasing nest of intervals $I_1 \supseteq I_2 \supseteq I_3 \supseteq \cdots$, each of which contains infinitely many elements of the sequence. Certainly the intersection of these intervals will be a single point that is the limit of a subsequence of the original sequence.

Proposition 1.12.15 *In a metric space (X, d), the metric function d is continuous.*

Proof: In fact

$$d(x, y) \leq d(x, z) + d(z, y)$$

hence

$$d(x, y) - d(x, z) \leq d(z, y) = d(y, z).$$

By symmetry,

$$d(x, z) - d(x, y) \leq d(z, y).$$

It follows that

$$|d(x, y) - d(x, z)| \leq d(y, z).$$

In like manner,

$$|d(x, y) - d(z, y)| \leq d(x, z).$$

This uniform continuity condition (called a *Lipschitz condition*) is much stronger than ordinary continuity. $\square$

1.13   Metrizability

It is natural to ask when a given topological space can be equipped with a metric (such that the metric topology is equivalent to the original topology). Of course metric spaces are Hausdorff. So if the given topological space is not Hausdorff then it cannot be metrized. What we need is a necessary and sufficient condition for metrizability.

In the ensuing discussion we shall make use of the *Hilbert cube*, which is $I^{\aleph_0}$ (where $\aleph_0$ is the first infinite cardinal). That is, the Hilbert cube consists of all sequences with elements in the interval $I = [0, 1]$. In practice it is often more convenient to consider the homeomorph of $I^{\aleph_0}$ consisting of sequences $\{a_j\}$ with $0 \leq a_j \leq 1/j$. This latter is clearly a metric space with metric

$$d(\{a_j\}, \{b_j\}) = \left[\sum_j (a_j - b_j)^2 \right]^{1/2}.$$

One of the most fundamental results about metrizability of a topological space is this theorem of Urysohn:

THEOREM 1.13.1 *Let $(X, \mathcal{U})$ be a T_1 topological space. Then the following are equivalent:*

(a) *X is regular and second countable.*

(b) *X is separable[2] and metrizable.*

(c) *X can be embedded as a subspace of the Hilbert cube $I^{\aleph_0}$.*

Proof: We divide the proof into three natural parts.

(a) $\Rightarrow$ **(c):** Let $\mathcal{B}$ be a countable basis for X and define $\mathcal{C} = \{(U, V) : U, V \in \mathcal{B}, \overline{U} \subseteq V\}$. Of course $\mathcal{C}$ is a countable set. We know from Proposition 1.4.2 that X is in fact normal, so there is (by Urysohn's lemma) a continuous function $f_{UV} : X \to [0, 1]$ such that $f(\overline{U}) = 0$ and $f(X \setminus V) = 1$. Let $\mathcal{F} = \{f_{UV} : (U, V) \in \mathcal{C}\}$. Then $\mathcal{F}$ is countable, and the pairs in $\mathcal{C}$ will separate points from (disjoint) closed sets in X. Now Proposition 1.4.3 tells us that if I_f is a copy of $I = [0, 1]$ for each $f \in \mathcal{F}$, then the evaluation mapping $e : X \to \prod_{f \in \mathcal{F}} I_f$ defined by

$$[e(x)]_f = f(x)$$

is an embedding. Since $\mathcal{F}$ is countable, we see that $\prod_{f \in \mathcal{F}} I_f = I^{\aleph_0}$. Thus we have proved **(c)**.

---

[2]A separable space is one with a countable dense set—see Section 2.4.

(c) $\Rightarrow$ **(b):** Of course $I^{\aleph_0}$ is separable and metric, hence so is every subspace of $I^{\aleph_0}$.

(b) $\Rightarrow$ **(a):** This is obvious. $\square$

1.14   Baire's Theorem

One of the great triumphs of basic analysis is the Baire category theorem. This is a truly powerful result which is elegant in its simplicity of formulation. We begin by introducing some terminology.

Definition 1.14.1 Let $\{x_j\}$ be a sequence in the metric space (X, d). We say that $\{x_j\}$ is *Cauchy* provided that, for each $\epsilon > 0$, there is an $N > 0$ such that whenever $j, k \geq N$ then $d(x_j, x_k) < \epsilon$.

Definition 1.14.2 Let (X, d) be a metric space. We say that X is *complete* if, whenever $\{x_j\}$ is a Cauchy sequence in X, then there is a limit $x_0 \in X$ so that $x_j \to x_0$.

Definition 1.14.3 Let (X, d) be a metric space. We say that $S \subseteq X$ is *dense* in X if the closure of S is X itself. In other words, S is dense if each metric ball $B(x, \epsilon)$ contains an element of S.

Definition 1.14.4 Let (X, d) be a metric space. We say that $S \subseteq X$ is *nowhere dense* if the closure $\overline{S}$ of S contains no metric ball.

THEOREM 1.14.5 (Baire) *Let (X, d) be a complete metric space. If each of the sets S_j, $j = 1, 2, \dots$ is dense and open in X, then $\cap_j S_j$ is dense in X.*

Corollary 1.14.6 *Let (X, d) be a complete metric space. If each of the sets T_j, $j = 1, 2, \dots$ is nowhere dense in X, then $\cup_j T_j$ is nowhere dense in X.*

In fact there is some classical terminology connected with Baire's theorem that is worth belaboring (because it is so commonly used, and it aids in one's understanding).

Definition 1.14.7 Let (X, d) be a metric space. We say that a set $S \subseteq X$ is of *first category* if S can be written as the countable union of nowhere dense sets. All other sets are called *second category*.

Definition 1.14.8 A metric space (X, d) is called a *Baire space* if the intersection of each countable family of dense open sets is still dense.

Proof of Theorem 1.14.5: Let W be any nonempty open set in X. We need to show that $\cap_j S_j \cap W \neq \emptyset$.

As usual, we let $B(x, r)$ denote the open ball with center x and radius r and $\overline{B}(x, r)$ the corresponding closed ball. Since S_1 is dense, we can be sure that
$$W \cap S_1 \neq \emptyset. \tag{1.14.5.1}$$
So there exist $x_1 \in X$ and $1 > r_1 > 0$ such that

$$\overline{B}(x_1, r_1) \subseteq W \cap S_1. \tag{1.14.5.2}$$

Inductively, if $j \geq 2$ and x_{j-1}, r_{j-1} have been selected, then $S_j \cap B(x_{j-1}, r_{j-1}) \neq \emptyset$. So we can find x_j, $1/j > r_j > 0$ such that

$$\overline{B}(x_j, r_j) \subseteq S_j \cap B(x_{j-1}, r_{j-1}). \tag{1.14.5.3}$$

Thus we have a sequence $\{x_j\} \subseteq X$. If now $\ell, m > j$, then we can be sure that x_ℓ and x_m both lie in $B(x_j, r_j)$. Thus $d(x_\ell, x_m) < 2r_j < 2/j$. Thus $\{x_j\}$ is a Cauchy sequence. Since X is complete, there is a limit point x such that $x_j \to x$.

Since x_ℓ lies in $\overline{B}(x_j, r_j)$ if $\ell > j$, we may conclude that x itself lies in $\overline{B}(x_j, r_j)$. As a result, by (1.14.5.3), x lies in S_j. But then (1.14.5.2) says that $x \in W$. That is what we wished to show. □

The next result is a classical theorem of Weierstrass. In fact he proved that such a function exists by constructing the function rather explicitly—using the theory of Fourier series. The proof that we present here is more abstract, and relies on Baire's theorem.

Proposition 1.14.9 *There is a continuous, real-valued function on the interval $I = [0, 1]$ that is nowhere differentiable on I.*

In fact what is remarkable about this proof is **(i)** it is nonconstructive, **(ii)** it shows that nowhere differentiable functions are generic.

Proof of Proposition 1.14.9: Let $C(I)$ be the space of all real, continuous functions on the interval I. Equip this space with the uniform metric:

$$d(f, g) = \max_{x \in I} |f(x) - g(x)| \,.$$

Also define $\mathcal{F}$ to be the set of functions in $C(I)$ which have a derivative at some point of I.

We make two claims:

(i) The space $C(I)$ is complete.

(ii) The set $\mathcal{F}$ is of first category in $C(I)$.

Proof of (i): Let $\{f_j\}$ be a Cauchy sequence in $C(I)$. We know from real analysis that there is a limit function f for this sequence that is continuous. We conclude therefore that $C(I)$ is complete.

Proof of (ii): For each $j = 1, 2, \ldots$ let us define

$$\mathcal{F}_j = \left\{ f \in C(I) : \text{for some } x \in [0, 1 - 1/j], \right.$$
$$\left. \text{whenever } h \in (0, 1/j], \ \left| \frac{f(x+h) - f(x)}{h} \right| \le j \right\}.$$

Observe that if a function $f \in C(I)$ has a derivative at some point of I, then for some j large enough, $f \in \mathcal{F}_j$. Thus $\mathcal{F} = \cup_j \mathcal{F}_j$. We can prove **(ii)** by showing that each $\mathcal{F}_j$ is closed and has no interior.

We first show that $\mathcal{F}_j$ has no interior. Let $f \in \mathcal{F}_j$ and $\epsilon > 0$. We will find a $g \in C(I)$ such that $d(f, g) < \epsilon$ and $g \notin \mathcal{F}_j$. The characterizing property of g is then that, for all $x \in [0, 1 - 1/j]$, there is some $h \in (0, 1/j]$ such that

$$\left| \frac{g(x+h) - g(x)}{h} \right| > j \,.$$

To construct g, first find a polynomial P such that $d(f, P) < \epsilon/2$ (the Weierstrass approximation theorem gives such a P). Let M be the maximum

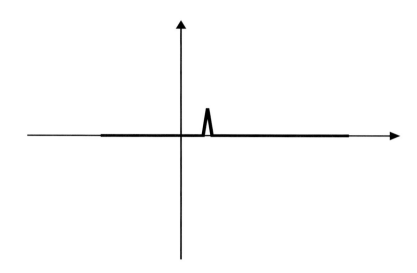

Figure 1.17: The function g.

of $|P'|$ on I. Now let $Q \in C(I)$ be a piecewise linear function with the properties **(a)** $|Q(x)| \leq \epsilon/2$ for all x and **(b)** each line segment in the graph of Q has slope $M + j + 1$. See Figure 1.17.

Now define $g(x) = P(x) + Q(x)$. It is apparent from the triangle inequality that $d(f, g) < \epsilon$. Furthermore,

$$
\begin{aligned}
\left| \frac{g(x+h) - g(x)}{h} \right| &= \left| \frac{P(x+h) + Q(x+h) - P(x) - Q(x)}{h} \right| \\
&\geq \left| \frac{Q(x+h) - Q(x)}{h} \right| - \left| \frac{P(x+h) - P(x)}{h} \right|.
\end{aligned}
$$

We know that, for $x \in [0, 1 - 1/j]$, there is an $h \in (0, 1/n]$ such that the right-hand side of this last display is $\geq (M + j + 1) - M = j + 1$. Therefore $g \notin \mathcal{F}_j$ and $\mathcal{F}_j$ has no interior.

Now we look at the closedness of $\mathcal{F}_j$. If $h_0 \in (0, 1/j]$ is a fixed element then the map $E : C(I) \times [0, 1 - 1/j]$ given by

$$
E(f, x) = \left| \frac{f(x + h_0) - f(x_0)}{h_0} \right|
$$

is clearly continuous. Hence $E^{-1}([0, j])$ is closed in $C(I) \times [0, 1 - 1/j]$. Let

$$
D_{h_0} = \{ f \in C(I) : (f, x) \in E^{-1}([0, j]) \text{ for some } x \in [0, 1 - 1/j] \}.
$$

Then D_{h_0} is closed in $C(I)$. Now let $f_\ell \in D_{h_0}$ for $\ell = 1, 2, \ldots$ with $f_\ell \to f$. Select points $x_\ell \in [0, 1 - 1/j]$ so that $(f_\ell, x_\ell) \in E^{-1}([0, j])$. Since I is compact, the sequence $\{x_\ell\}$ has a cluster point $x \in I$. Clearly $(f, x) \in E^{-1}([0, j])$, hence $f \in D_{h_0}$. We further see that

$$D_{h_0} = \left\{ f \in C(I) : \text{for some } x \in [0, 1 - 1/j], \ \left| \frac{f(x + h_0) - f(x)}{h_0} \right| \leq j \right\}.$$

It follows that $\mathcal{F}_j = \cap_j \{ D_{h_0} : h_0 \in (0, 1/j] \}$. So $\mathcal{F}_j$ is closed. $\quad\square$

1.15 Lebesgue's Lemma and Lebesgue Numbers

We discovered Lebesgue numbers in the proof of Theorem 1.12.12. Here we discuss the matter in more explicit detail.

Let X be a compact metric space and let $\mathcal{U} = \{U_\alpha\}_{\alpha \in A}$ be an open covering of X. Of course we may extract a finite subcover $U_{\alpha_1}, U_{\alpha_2}, \ldots, U_{\alpha_k}$. Intuitively, we see that there is a certain amount of overlap among the U_{α_j}. Our sense is that a sufficiently small ball $B(x, \epsilon)$ must of necessity lie entirely inside one of the U_{α_j}, no matter what the location of the center x. That is the content of the next lemma of Lebesgue (Henri L. Lebesgue (1875–1941)).

Lemma 1.15.1 *Let X be a compact metric space and $U_{\alpha_1}, U_{\alpha_2}, \ldots, U_{\alpha_k}$ a finite open cover. There is a number $\epsilon > 0$—called the Lebesgue number of the covering—so that any ball $B(x, \epsilon)$ must lie entirely inside some U_{α_j}.*

Proof: There are many different ways to prove this result. We provide a proof by contradiction.

Suppose that the assertion is not true. Then there are points $x_\ell \in X$ and numbers $\epsilon_\ell \searrow 0$ so that, for each ℓ, the ball $B(x_\ell, \epsilon_\ell)$ does not lie in any U_{α_j}, $j = 1, \ldots, k$. Since X is compact, we may choose a subsequence x_{ℓ_m} which converges to some point x_0. Of course x_0 must lie in some U_{α_p}.

Now let $\delta = d(x_0, {}^c U_{\alpha_p}) \equiv \inf_{t \in {}^c U_{\alpha_p}} d(x_0, t)$. Certainly $\delta > 0$. If m is large enough then both

- $d(x_{\ell_m}, x_0) < \delta/2$;

- $\epsilon_{\ell_m} < \delta/2$.

Then it follows from the triangle inequality that $B(x_{\ell_m}, \epsilon_{\ell_m}) \subseteq U_{\alpha_p}$. That is a contradiction. □

Exercises

1. Let X be the interval $[1, 5]$ and define

$$\mathcal{U} = \{[1, 4], [3, 5], [3, 4], [1, 5], \emptyset\} .$$

   Explain why $\mathcal{U}$ is a topology on X.

2. Let X be the real numbers. Let a subset of X be open if it consists of irrational numbers only, or if it is all of X, or if it is the empty set. Explain why these open sets form a topology on X.

3. Declare a collection of polynomials in the variable x with real coefficients to be open if the set of all their coefficients forms an open set in $\mathbb{R}$ according to Example 1.2.2. As an instance, the set of polynomials $a + bx + cx^2$ with $a, b, c \in (0, 1)$ is an open set of polynomials. Verify that these open sets form a topology on the collection of all real-coefficient polynomials of a single variable.

4. Declare a set in $\mathbb{Z}$ to be open if it is finite. Verify that these open sets do *not* form a topology. What happens if the word "finite" is replaced by "infinite"?

5. Let $X = \{a, b\}$ be a set with just two points. Describe all possible topologies on X.

6. Show that if a finite topological space is T_1, then it must be discrete.

7. Let (X, d) be a metric space and $E \subseteq X$ a closed set. Show that the function

$$\rho(x) = \inf\{d(x, e) : e \in E\}$$

   is continuous. Indeed it is Lipschitz continuous.

8. Let $(X, \mathcal{U})$ be a topological space. A set $S \subseteq X$ is said to be *dense* if, for every point $x \in X$ and every neighborhood U of x, there is a point $s \in S$ that lies in U. Show that the polynomials are a dense subset of the continuous functions on $[0, 1]$ (equipped with the sup-norm topology). Show that the rational numbers $\mathbb{Q}$ are dense in the reals $\mathbb{R}$. Show that the integers $\mathbb{Z}$ are *not* dense in the reals $\mathbb{R}$.

9. Prove that a set E in a topological space X is open if and only if E contains none of its boundary points.

10. Prove that a set E in a topological space X is open if and only if E equals its interior.

11. Let (X, d) be a metric space. Let $K \subseteq X$ be compact and $E \subseteq X$ be closed and disjoint from K. Show that there is a positive distance between K and E. That is to say, there is a number $\epsilon > 0$ such that if $k \in K$ and $e \in E$ then $d(k, e) > \epsilon$.

12. Show that the result of the last exercise is false if the two sets are assumed only to be closed.

13. Declare a set in $\mathbb{R}$ to be open if its complement is an interval $[a, b]$ or the empty set. Does this form a topology?

14. Let X be the space of sequences $\{a_j\}$ of real numbers such that $\sum_j |a_j|^2$ is finite. Define a metric on this space by

$$d(\{a_j\}, \{b_j\}) = \left[\sum_j (a_j - b_j)^2 \right]^{1/2}.$$

Show that this is a metric. Explain why the closed unit ball

$$B = \{\{a_j\} : \sum_j a_j^2 \leq 1\}$$

is not compact. It is nonetheless the case that B is closed and bounded.

15. Give an example of a compact set in a non-Hausdorff space that is not closed.

16. Let X be the real numbers and declare that every singleton set $\{x\}$ is open. Generate a topology with these sets. Now describe all the open sets. Which sets are closed? Which sets are compact?

17. Let $f : X \to Y$ be a continous mapping of topological spaces and let $K \subseteq Y$ be compact. Is it necessarily the case that $f^{-1}(K)$ is compact? Give a proof or a counterexample.

18. Let $X = \{A, B, C, D, E\}$. Describe two distinct topologies on X that are not homeomorphic.

19. Think of the space in Exercise 14 as a vector space. Show that it does not have a basis consisting of finitely many elements.

20. Let B be the closed unit ball defined in Exercise 12. Calculate its interior.

21. Let the topological space X be $\mathbb{R}^2$ equipped with the usual Euclidean topology. Let S be the points that have both integer coordinates. Calculate the closure of S. Calculate the interior of S. Calculate the boundary of S.

22. Let the topological space X be the real line $\mathbb{R}$. An open set is any set whose complement is finite. Let $S = [0, 1]$. Calculate the closure of S. Calculate the interior of S. Calculate the boundary of S.

23. Put a topology on $\mathbb{R}^2$ with the property that the line $\{(x, 0) : x \in \mathbb{R}\}$ is dense in $\mathbb{R}^2$.

24. Define a new Cantor set with the property that the first set removed has length 3^{-10}, the next two sets removed have lengths 3^{-10^2}, the next four sets removed have lengths 3^{-10^4}, and so forth. The resulting set will still be nonempty, compact, and have uncountably many elements. Prove these statements. What will be the length of this new set?

25. Define a subset $E \subseteq \mathbb{R}$ to be *perfect* if it is closed and every point of E is an accumulation point of E. Use the Baire category theorem to prove that E must be uncountable.

26. Look at all the examples in Section 1.2 and determine which of the topologies is metrizable and which is not. In each case explain your answer by citing a theorem.

27. Provide the details of the argument that the sum of the Cantor ternary set with itself is the entire interval $[0, 2]$.

28. The open discs $D((1, 0), 7/6)$, $D((1/2, 1/2), 9/8)$, $D((0, 1), 6/5)$, $D((-1/2, 1/2), 3/2)$, $D((-1, 0), 5/4)$, $D((-1/2, -1/2), 6/5)$, $D((0, -1), 7/6)$, $D((1/2, -1/2), 9/8)$ cover the compact set $\overline{D}(0, 1)$. What is the Lebesgue number of this covering?

29. Prove that, in $\mathbb{R}^N$, any connected open set is path-connected.

30. A countable set in $\mathbb{R}$ (equipped with the usual topology) can never be a continuum. Why not?

31. Prove that the product of two continua is still a continuum.

32. Discuss closure of the property of being a continuum under union and intersection.

33. Suppose that $X \subseteq \mathbb{R}$ and $Y \subseteq \mathbb{R}$ are of the first category in the usual topology of $\mathbb{R}$. What can you say about $X \times Y \subseteq \mathbb{R}^2$?

Chapter 2

Advanced Properties of Topological Spaces

2.1   Basis and Sub-Basis

We have encountered some of the ideas of this section in context in earlier parts of the book. Now we make them more formal.

Definition 2.1.1 Let $(X, \mathcal{U})$ be a topological space. We call a collection of sets $\mathcal{S} = \{S_\alpha\}_{\alpha \in A}$ a *basis* for the topology $\mathcal{U}$ if the collection of all unions of elements of $\mathcal{S}$ equals $\mathcal{U}$. In other words, if $U \in \mathcal{U}$ then there exist $S_\alpha \in \mathcal{S}$ such that $\cup_\alpha S_\alpha = U$.

EXAMPLE 2.1.2 Let X be the real line equipped with the usual topology. Then the collection of open intervals (a, b) forms a basis for this topology. In other words, every open set in $\mathbb{R}$ is a union of open intervals.

   To see this, let U be an open subset of $\mathbb{R}$. Say that two elements a, b of U are related if the interval having endpoints a, b lies in U. Then this is an equivalence relation on U. And the resulting equivalence classes are open intervals whose union is U.

EXAMPLE 2.1.3 Let $\mathbb{R}^2$ be the plane equipped with the usual topology. Then the collection of all open, square boxes

$$S_{(x,y),\epsilon} \equiv \{(s, t) \in \mathbb{R}^2 : |x - s| < \epsilon, |y - t| < \epsilon\}$$

is a basis for the topology.

Certainly, by the definition of open set in $\mathbb{R}^2$, every open set is the union of open discs. But every open disc has a square of the form $S_{(x,y),\epsilon}$ lying inside it.

EXERCISE FOR THE READER 2.1.4 The instance of a basis in the last example is certainly not the only basis for the usual topology in $\mathbb{R}^2$. Give another example of a basis.

The trouble with the concept of basis is that a given collection of sets may or may not prove to be a basis for some topology. It may not be rich enough. We would like an idea that allows us to take *any* collection of sets and use it to generate a topology. That is the notion of "sub-basis."

Definition 2.1.5 Let X be a space. A collection of subsets $\mathcal{S} = \{S_\alpha\}_{\alpha \in A}$ is a *sub-basis* for a topology $\mathcal{U}$ on X if it holds that $\mathcal{U}$ consists of all those sets obtained from $\mathcal{S}$ through finite intersection or arbitrary union or both.

It is a simple matter to check that the process of taking finite intersection or arbitrary union, beginning with *any* collection of sets, will always generate a topology. That is why the concept of sub-basis is such a useful and flexible tool. We may rephrase the definition of sub-basis as saying that $\mathcal{S}$ forms a sub-basis if the family of all finite intersections of elements of $\mathcal{S}$ forms a basis.

EXAMPLE 2.1.6 Let X be the real line with the usual topology. Then the collection of halflines $(-\infty, a)$ and (b, ∞) forms a sub-basis for the topology. This is so because intersections of these sets generate all the open intervals, and those in turn (by Example 2.1.2) form a basis.

EXAMPLE 2.1.7 Let X be the integers, and consider each of those sets which is the complement of a single point. For instance, $\{x \in \mathbb{Z} : x \neq 0\}$ is such a set. These sets form a sub-basis for the topology consisting of all sets with finite complement (see Example 1.2.5).

EXERCISE FOR THE READER 2.1.8 Use the finite sets as the sub-basis for a topology on $\mathbb{Z}$. What topology do you obtain?

EXERCISE FOR THE READER 2.1.9 Use the horizontal lines as the sub-basis for a topology on $\mathbb{R}^2$. What topology do you obtain?

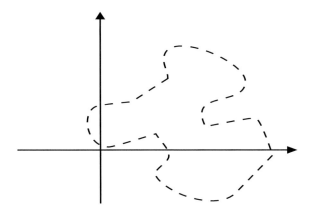

Figure 2.1: An open set in $\mathbb{R}^2$.

2.2 Product Spaces

Let $(X, \mathcal{U})$ and $(Y, \mathcal{V})$ be topological spaces. Then we may consider $X \times Y$ as a topological space. We use the sets $U \times V$, with $U \in \mathcal{U}$ and $V \in \mathcal{V}$, as a sub-basis for the topology on the product space.

EXAMPLE 2.2.1 It would be wrong to think that the open sets in $\mathbb{R}^2$ are just the product sets

$$(a, b) \times (c, d) = \{(x, y) : a < x < b, c < y < d\}.$$

But these sets form a sub-basis for the topology. In fact open sets in $\mathbb{R}^2$ can have quite arbitrary shapes—see Figure 2.1.

For the next fundamental theorem we will want to consider not simply finite products of spaces, but in fact infinite products—even uncountably infinite products. This will require some new definitions.

Definition 2.2.2 Let $\{S_\alpha\}_{\alpha \in A}$ be sets. We define

$$\prod_{\alpha \in A} S_\alpha$$

to be the set of functions $\Phi : A \to \cup_\alpha S_\alpha$ such that $\Phi(\alpha) \in S_\alpha$ for each α.

To understand the last definition, if we are given $\{S_1, S_2, \ldots, S_m\}$ then the product $\prod_{j=1}^m S_j$ is defined to be the set of functions Φ from $\{1, 2, \ldots, m\}$ to $\cup_j S_j$ such that $\Phi(j) \in S_j$. We typically denote an element of this product by $(\Phi(1), \Phi(2), \ldots, \Phi(m)) = (s_1, s_2, \ldots, s_m)$ with $s_j \in S_j$ for each j.

Definition 2.2.3 Let $\{X_\alpha\}_{\alpha \in A}$ be topological spaces. We define the projection

$$\pi_\beta : \prod_\alpha X_\alpha \quad \rightarrow \quad X_\beta$$

$$(x_\alpha) \quad \mapsto \quad x_\beta \,.$$

Then a sub-basis element for the product topology is $\pi_\beta^{-1}(U)$ for any open subset U in X_β.

Notice that a sub-basis element for the product topology as in the last definition is trivial (that is, it is the whole space) in all slots of the product except one. In the one exceptional slot it is an open set in that component space. Because of closure under finite intersection, this new definition is consistent with the simpler definition that we have already considered for finite products.

THEOREM 2.2.4 (Tychanoff) *The product of any number of compact spaces is compact.*

Proof: In fact this result can be shown to be equivalent to the Axiom of Choice. So using the Axiom of Choice (or some equivalent thereof) in our proof will be unavoidable.

Now let $\{X_\alpha\}_{\alpha \in A}$ be a family of compact topological spaces. Let

$$X = \prod_{\alpha \in A} X_\alpha \,,$$

and equip X with the product topology. We shall in fact use Theorem 1.5.19 and show that if $\mathcal{F} = \{F_\alpha\}_{\alpha \in A}$ is a family of closed sets in X with the finite intersection property then $\cap_\alpha F_\alpha$ is non-empty.

Now fix an $\mathcal{F}$ as in the last paragraph. Let $\mathcal{C}$ denote the class of all families of sets in X that **(i)** contain $\mathcal{F}$ and **(ii)** have the finite intersection property. An application of Zorn's lemma, or the Hausdorff maximality principle, shows that $\mathcal{C}$ will have a maximal element.[1] Call that maximal family $\mathcal{F}'$.

The maximality of $\mathcal{F}'$ tells us that the intersection of the members of any finite subfamily of $\mathcal{F}'$ will be a set that is also an element of $\mathcal{F}'$ (or else $\mathcal{F}'$ would not be maximal).

---

[1]Obviously it is here that we use the dreaded Axiom of Choice.

Now let $\beta \in A$ (where A is the index set for the product) and let

$$\pi_\beta : X \to X_\beta$$

as usual. Define

$$\mathcal{F}'_\beta = \{\pi_\beta(F) : F \in \mathcal{F}'\}.$$

So $\mathcal{F}'_\beta$ is a family of subsets of X_β. Certainly $\mathcal{F}'_\beta$ has the finite intersection property. Since we know that X_β is compact, we may use 1.5.7 to conclude that

$$H_\beta \equiv \bigcap \left\{ \overline{\pi_\beta(F)} : F \in \mathcal{F}' \right\}$$

is nonempty. Choose a point $x_\beta \in H_\beta$ for each $\beta \in A$.

Now let x denote the point in the product space X whose β^{th} coordinate is x_β. What we need to do is to prove that $x \in \overline{F}$ for every $F \in \mathcal{F}'$. That will enable us to use 1.5.7 to deduce that X is compact.

So let $\widehat{U}_\beta$ be a sub-basic open set in X which contains the point x. Implicit here is the fact that U_β is the projection of $\widehat{U}_\beta$ into X_β, and $U_\beta \ni x_\beta$. Moreover, the projections of $\widehat{U}_\beta$ into any of the other component spaces X_α will be all of X_α. Since $x_\beta \in \overline{\pi_\beta(F)}$ for each $F \in \mathcal{F}'$, it follows that U_β meets $\pi_\beta(F)$ for each $F \in \mathcal{F}'$. Thus

$$\widehat{U}_\beta = \pi_\beta^{-1}(U_\beta).$$

meets every member F of $\mathcal{F}'$. Therefore the maximality of $\mathcal{F}'$ tells us that $\widehat{U}_\beta$ belongs to $\mathcal{F}'$; thus the intersection of any finite number of sub-basic open sets containing x is a member of $\mathcal{F}'$. This implies that every neighborhood of x in X meets each member F of $\mathcal{F}'$. As a result, $x \in \overline{F}$ for each $F \in \mathcal{F}'$. We have verified the hypothesis of 1.5.7, so X is compact. $\square$

The Tychanoff theorem has many profound consequences, including the Stone-Čech compactification (Section 2.5) and the Banach-Alaoglu theorem (see [RUD]). You will encounter these ideas in a more advanced course.

EXAMPLE 2.2.5 The Hilbert cube is the product $I^{\aleph_0}$ of countably many copies of the closed (compact) interval $I = [0,1]$. [Here $\aleph_0$ is the first infinite cardinal.] Of course the Hilbert cube is compact.

2.3 Relative Topology

We begin this discussion with a definition.

Definition 2.3.1 Let $(X, \mathcal{U})$ be a topological space and let $Y \subseteq X$ be a subset. We define the *relative topology* on Y to simply be the collection of those sets $Y \cap U$ for $U \in \mathcal{U}$. It is straightforward to check that $\{Y \cap U\}_{U \in \mathcal{U}}$ is indeed a topology.

EXAMPLE 2.3.2 Let $X = \mathbb{R}^2$ with the usual topology and consider the subset $Y = \{(x, 0) : x \in \mathbb{R}\}$. The relative topology on Y will just be the topology generated by the intervals $\{(x, 0) : a < x < b\}$.

EXAMPLE 2.3.3 Let $X = \mathbb{R}$ with the usual topology and consider the subset $Y = [0, 1] = \{x \in \mathbb{R} : 0 \leq x \leq 1\}$. Then a basis for the relative topology on Y consists of four types of sets:

- Open intervals (a, b) with $0 < a < b < 1$;

- Half-open intervals of the form $[0, a)$ with $0 < a < 1$;

- Half-open intervals of the form $(b, 1]$ with $0 < b < 1$;

- The entire interval $[0, 1]$.

The relative topology is just part of the language in this subject, and you will encounter it throughout the rest of the book.

2.4 First Countable, Second Countable, and So Forth

There are many different ways to measure the "size" of a set or space. One of these is based on Georg Cantor's (1845–1918) ideas of cardinality (see Appendix 2).

Definition 2.4.1 We say that a topological space $(X, \mathcal{U})$ is *separable* if it contains a countable subset $S \subseteq X$ that is dense in X. This means that every open set in X contains an element of S.

EXAMPLE 2.4.2 Let $X = \mathbb{R}$ with the usual topology. Let $S = \mathbb{Q}$. Then S is countable, and S is dense in X. Therefore the real numbers form a separable topological space.

EXAMPLE 2.4.3 Let X be the continuous functions on the interval $[0, 1]$ with the uniform topology. Let S be the set of polynomials. Then S is dense in X—by the Weierstrass approximation theorem. But S is not countable. So we can make no conclusion about separability.

On the other hand, let T be those polynomials with rational coefficients. Then T is dense in X (exercise), and certainly T is countable. So the space X is indeed separable.

Definition 2.4.4 Let $(X, \mathcal{U})$ be a topological space. We say that a point $x \in X$ has a *countable neighborhood base* if there is a countable collection $\{U_j^x\}$ of open sets such that every neighborhood W of x contains some U_j^x.

EXAMPLE 2.4.5 Let $X = \mathbb{R}$ with the usual topology. Then, for each $x \in \mathbb{R}$, the sets $(x - 1/j, x + 1/j)$, $j = 1, 2, \ldots$, form a countable neighborhood base.

Definition 2.4.6 Let $(X, \mathcal{U})$ be a topological space. We say that X is *first countable* if each point $x \in X$ has a countable neighborhood base.

EXAMPLE 2.4.7 Certainly the intervals $[a, b)$ form a neighborhood base at a in the topology of the Sorgenfrey line (Example 1.7.4). The intervals $[a, b')$, with b' rational, form a countable neighborhood base at a. Thus the Sorgenfrey line is first countable.

Definition 2.4.8 A topological space $(X, \mathcal{U})$ is said to be *second countable* if the topology $\mathcal{U}$ has a countable basis.

The reader should note that first countability is a local property (at each point) while second countability is a global property.

EXAMPLE 2.4.9 Let $X = \mathbb{R}^N$ with the usual topology. If U is any open set and $u \in U$, then there is a ball $B(x, r)$ with rational center (all coordinates rational) and rational radius such that

$$u \in B(x, r) \subseteq U.$$

Thus the balls with rational center and rational radius form a basis for the topology. Therefore this space X is second countable.

EXAMPLE 2.4.10 Let X be the real line equipped with the topology of all sets whose complements are finite. Let $S \subseteq X$ be any infinite set. Then S is dense in X. To see this, let x be any point in X and let U be a neighborhood of x. Then the complement of U is finite, so $U \cap S \neq \emptyset$. Since S intersects every neighborhood of every point, it is dense in X. In particular $\mathbb{Z}$ is an infinite set, so it is dense. So we see that X is separable.

In spite of this, X is *not* second countable. To see this, suppose to the contrary that there is a countable basis for the topology of X. Fix a point x_0 of X. If U is a neighborhood of x_0 and if x is any other point of X, then $U \setminus \{x\}$ is also a neighborhood of x_0. Thus there will be an open set V from the basis such that

$$x_0 \subseteq V \subseteq U \setminus \{x\}.$$

We conclude that the intersection of all basis elements that contain x_0 is simply $\{x_0\}$. If we call the basis elements U_j for $j = 1, 2, \ldots$ then we may let S_j be the complement of U_j. Of course each S_j is finite. So if U_{j_k} are the basis elements that contain x_0, then we have

$$\{x_0\} = \bigcap_k U_{j_k}$$

hence

$$\mathbb{R} \setminus \{x_0\} = \bigcup_k S_{j_k}.$$

But this says that $\mathbb{R} \setminus \{x_0\}$ is countable. That is absurd.

EXERCISE FOR THE READER 2.4.11 Prove that the set of all isolated points of a second countable space is empty or countable. Here a point x is said to be *isolated* if it has a neighborhood with no other points of the space in it.

EXERCISE FOR THE READER 2.4.12 Let X be a set, and define a metric on X by

$$d(x, y) = \begin{cases} 0 & \text{if} \quad x = y \\ 1 & \text{if} \quad x \neq y. \end{cases}$$

For which sets X is this space separable? And when not? [**Hint:** Consider separately the cases of X finite, countable, and uncountable.]

Proposition 2.4.13 *Any second countable space is separable.*

Proof: Let $(X, \mathcal{U})$ be the topological space. Let $\{U_j\}$ be a countable basis for the topology on X. Select a point $p_j \in U_j$ for each j. We claim that the countable set $\{p_j\}$ is dense.

For let $x \in X$ be arbitrary and U any neighborhood of x. Then some $U_j \subseteq U$, and $p_j \in U_j \subseteq U$. That does the job. $\square$

EXAMPLE 2.4.14 Let X be the real number line equipped with the discrete topology (i.e, every singleton is an open set). For any point $x \in X$, the singleton set $\{x\}$ is a neighborhood basis for the point x. So the space is first countable.

Now every set in this space is open. Therefore every set is closed. It follows that the only dense subset of X is X itself—which is certainly un-countable. Therefore X is *not* separable. By Proposition 2.4.13, it follows that X is *not* second countable.

Proposition 2.4.15 *Let X be a separable metric space. Then X is second countable.*

Proof: Let $\{p_j\}$ be a countable dense set in X. The countable base for the topology will be all metric balls with center p_j for some j and rational radius. To see this, let U be any open set and let $x \in U$. Our job is to produce one of the indicated balls that lies in U and contains x.

Now let $\delta > 0$ be such that $B(x, \delta) \subseteq U$. Certainly there is a p_j that lies in $B(x, \delta/3)$. Choose a rational number r that is positive and lies between $\delta/3$ and $\delta/2$. Then the ball $B(p_j, r)$ certainly contains x. And, by the triangle inequality, $B(p_j, r)$ lies in U. $\square$

2.5   Compactifications

As we have seen, compact spaces are often much easier to manipulate, and to prove theorems about, than arbitrary topological spaces. A useful construct is to be able to take a noncompact space and to modify it in a natural fashion to make it compact. We illustrate this idea with our first example.

EXAMPLE 2.5.1 (THE STEREOGRAPHIC PROJECTION) Stereographic pro-jection puts $\widehat{\mathbb{R}^2} \equiv \mathbb{R}^2 \cup \{\infty\}$ into one-to-one correspondence with the two-dimensional sphere S in $\mathbb{R}^3$, $S = \{(x, y, z) \in \mathbb{R}^3 : x^2 + y^2 + z^2 = 1\}$ in such a way that topology is preserved in both directions of the correspondence.

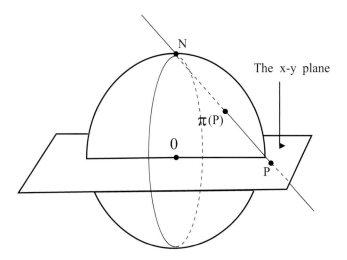

Figure 2.2: Stereographic projection.

In detail, begin by imagining the unit sphere bisected by the Cartesian plane with the center of the sphere $(0,0,0)$ coinciding with the origin in the plane—see Figure 2.2. We define the stereographic projection as follows: if $P = (x, y) \in \mathbb{R}^2$, then connect P to the north pole N of the sphere with a line segment. The point $\pi(P)$ of intersection of this segment with the sphere is called the *stereographic projection* of P. Under stereographic projection, the point at infinity in the plane corresponds to the north pole N of the sphere. For this reason, $\mathbb{R}^2 \cup \{\infty\}$ is often thought of as being a sphere, and is then called, for historical reasons, the *Riemann sphere*.

We see that what we have done here is that we have taken the ordinary Cartesian plane and appended to it a "point at ∞." In effect, the plane (thought of as a large sheet of paper) is gathered up into a sack and the ends or edges coalesced into a point (the north pole, or point at infinity). The construction of the stereographic projection is semi-intuitive (it can be made rigorous, in fact one can write down an explicit formula for the mapping), but compelling. Nonetheless, it seems to be very special to this particular situation. If we wanted to perform a compactification on a fairly arbitrary topological space, this would give us little hint as to how to do it.

The compactification treated in Example 2.5.1 is known as a "one-point compactification" or "Alexandroff compactification." There are in fact several different theories of compactification. We shall treat a few of them here.

First we begin with an abstract definition of what a compactification is.

Definition 2.5.2 Let $(X,\mathcal{U})$ be a non-compact topological space. A *compactification* of X is an embedding $h : X \to Y$ of X into a compact space Y so that $h(X)$ is dense in Y.

Definition 2.5.3 Let $(X,\mathcal{U})$ be a noncompact, locally compact, Hausdorff space. Let p be a point that does not lie in X. Set $X^* = X \cup \{p\}$. Let the topology $\mathcal{U}^*$ on X^* be $\mathcal{U}$ together with sets which are given by $\{p\}$ union the complement of a compact set in X. It is easy to see that $\mathcal{U}^*$ is still a topology. We call $(X^*,\mathcal{U}^*)$ a *one-point compactification* of X.

Again, Example 2.5.1 gives an example of a compactification as just described. Clearly the plane is a dense subspace of the compactification (which is the sphere), because the plane and the sphere only differ by one point and the topology is locally Euclidean.

Proposition 2.5.4 *The space $X^* = X \cup \{p\}$ in the preceding definition is compact.*

Proof: Let $\mathcal{U} = \{U_\alpha\}_{\alpha \in A}$ be an open covering of X^*, with the open sets taken from the topology $\mathcal{U}^*$. Then there is at least one set that covers p—select one of these and call it U_{α_0}. All the sets U_α that lie entirely in X cover ${}^c U_{\alpha_0}$ (which is compact), so there is a finite subcovering $U_{\alpha_1}, \ldots, U_{\alpha_k}$. Then $U_{\alpha_0}, U_{\alpha_1}, \ldots, U_{\alpha_k}$ is a finite subcovering of $\mathcal{U}$ that covers X^*. $\square$

EXAMPLE 2.5.5 This is a reworking of Example 2.5.1. Let X be the Cartesian plane with the usual topology. Let p be some point that does not lie in X. Set $X^* = X \cup \{p\}$. Define a topology on X^* by taking all the usual open sets in X together with all sets of the form ${}^c K \cup \{p\}$, where K is a compact subset of X. Then this topologizes X^* and X^* is automatically compact by the proposition.

Another important compactification procedure is the "Stone-Čech compactification." We shall describe that now.

Let X be a Tychanoff space (Section 1.4). As at the end of Section 1.4, let C denote the set of all continuous real functions from X to the unit interval I. By the Tychanoff embedding theorem (Theorem 1.4.15), the natural mapping

$$f : X \to I^C$$

is an embedding. Of course Tychanoff's compactness theorem tells us that I^C is compact. Let $\beta(X)$ denote the closure of the image $f(X)$ in I^C; certainly $\beta(X)$ is compact. Let

$$h : X \to \beta(X)$$

be the map given by f. Then $h : X \to \beta(X)$ is a compactification of X called the *Stone-Čech compactification.*

EXERCISE FOR THE READER 2.5.6 Calculate the Stone-Čech compactification of the real line.

EXERCISE FOR THE READER 2.5.7 Show that the one-point compactification of the natural numbers $\mathbb{N}$ is homeomorphic to the space $\{0\} \cup \{1/j : j = 1, 2, \ldots\}$.

2.6   Quotient Topologies

Definition 2.6.1 Let $(X, \mathcal{U})$ be a topological space and let Y be any set. Suppose that $f : X \to Y$ is a surjective mapping. Then the collection of subsets

$$\tau_f = \{G \subseteq Y : f^{-1}(G) \text{ is open in } X\}$$

is a topology on Y called the *quotient topology* induced on Y by f. When Y is endowed with such a quotient topology, then it is called a *quotient space* of X, and the inducing map f is called a *quotient map.*

EXAMPLE 2.6.2 Let X be the standard Cartesian plane with the usual topology and let $f : X \to \mathbb{R}^3$ be the map

$$(x, y) \mapsto (\cos x, \sin x, y).$$

Notice that the image of f is a right circular cylinder $\mathcal{C}$ in space. For the purposes of this example, we think of f as a surjective mapping from $\mathbb{R}^2$ to $\mathcal{C}$. The quotient topology is generated by any set of the form $A \times J$, where A is any open arc of the circle and J is any open interval in $\mathbb{R}$.

Definition 2.6.3 Let $(X, \mathcal{U})$ be any topological space. A *decomposition* or *partition* of X is a pairwise disjoint family $\mathcal{Q}$ of subsets of X whose union is all of X. The function $p : X \to \mathcal{Q}$ which assigns to each element of X the unique element of $\mathcal{Q}$ to which it belongs is called the *natural projection* of X onto $\mathcal{Q}$. We endow $\mathcal{Q}$ with the quotient topology under this mapping.

Remark 2.6.4 One of the most important examples of a partition occurs when X is equipped with an equivalence relation $\sim$. The equivalence classes induced by $\sim$ of course form a decomposition. We let $X/\sim$ denote the collection of equivalence classes, and call it the *quotient space*.

Remark 2.6.5 In our original definition of quotient topology in Definition 2.6.1, the sets $f^{-1}(y)$ for $y \in Y$ form a decomposition.

EXAMPLE 2.6.6 Let

$$X = \mathbb{R}^{N+1} \setminus \{0\}.$$

Say that two points p and q in X are related if there is a nonzero real number λ such that $\lambda p = q$. The resulting quotient space $X/\sim$ is called the *N-dimensional real projective space* $\mathbb{P}^N$. We see that the "points" in $\mathbb{P}^N$ are in fact the lines in $\mathbb{R}^{N+1}$ passing through the origin (less the origin itself).

Let $\pi : X \to \mathbb{P}^N$ be the natural projection to the quotient. Let S^N be the unit N-dimensional sphere in $\mathbb{R}^{N+1}$. Thus S^N is a subspace of X. So we may consider

$$\pi : S^N \to \mathbb{P}^N. \tag{$*$}$$

Since every point $x \in X$ can be normalized to $x/\|x\|$, which lies on the same line through the origin, we see that π (in this new form) is still surjective. In fact the map is 2-to-1. Two points a and b of S^N are mapped to the same point under π if and only if $a = -b$. Of course the mapping in line $(*)$ makes the quotient topology on $\mathbb{P}^N$ easy to understand.

EXAMPLE 2.6.7 Let $(X, \mathcal{U})$ be a topological space, and let $E \subseteq X$ be a nonempty subset. The singletons $\{x\}$ for $x \notin E$ together with the set E form a decomposition of the space X. The resulting quotient space is called *the space obtained from X by collapsing the subset E to a point*.

As an instance, let X be the closed unit ball in $\mathbb{R}^N$:

$$X = \{(x_1, x_2, \ldots, x_N) \in \mathbb{R}^N : \sum_j |x_j|^2 \leq 1\}.$$

Let S^{N-1} be the unit sphere in $\mathbb{R}^N$. Then S^{N-1} is the boundary of X. If we take the space obtained from X by collapsing its boundary to a point, we obtain a space that is homeomorphic to S^N (the unit sphere in $\mathbb{R}^{N+1}$).

2.7   Uniformities

In a metric space, the ideas of uniform convergence and uniform continuity are very natural and make good sense. We review them just for a moment on the metric space (X, d):

- The function $f : X \to \mathbb{R}$ is *uniformly continuous* if, given $\epsilon > 0$, there is a $\delta > 0$ such that if $d(x, y) < \delta$ then $|f(x) - f(y)| < \epsilon$.

- The sequence of functions $f_j : X \to \mathbb{R}$ *converges uniformly* to a function $f : X \to \mathbb{R}$ if, given $\epsilon > 0$, there is an $N > 0$ such that if $j \geq N$ then $|f_j(x) - f(x)| < \epsilon$ for all $x \in X$.

The key idea here is that the distance or metric may be applied uniformly to points regardless of their location in space. A general topological space will not have such a structure; but we can sometimes impose a "uniformity" that makes the space look qualitatively like a metric space. That is the topic considered in this section.

Definition 2.7.1 Let S be any set. The *diagonal* of S, denoted $\triangle = \triangle(S)$, is the set $\{(s, s) : s \in S\}$.

Definition 2.7.2 If S and T are sets in $X \times X$ then we let

$$S \circ T = \{(s, t) : (s, u) \in S \text{ and } (u, t) \in T \text{ for some } u \in X\}.$$

This is a generalization of the notion of composition of functions.

In what follows, if $E \subseteq X \times X$, then we let $E^{-1} = \{(y, x) : (x, y) \in E\}$.

In a metric space we may observe that two points x and y are close to each other precisely when (x, y) is close to the diagonal. The idea motivates the following definition:

Definition 2.7.3 Let $(X, \mathcal{U})$ be a topological space. A *diagonal uniformity* on X is a collection $\mathcal{D} = \mathcal{D}(X)$ of subsets of $X \times X$ such that

(a) $D \in \mathcal{D} \Rightarrow \triangle \subseteq D$;

(b) $D_1, D_2 \in \mathcal{D} \Rightarrow D_1 \cap D_2 \in \mathcal{D}$;

(c) $D \in \mathcal{D} \Rightarrow E \circ E \subseteq D$ for some $E \in \mathcal{D}$;

(d) $D \in \mathcal{D} \Rightarrow E^{-1} \subseteq D$ for some $E \in \mathcal{D}$;

(e) $[D \in \mathcal{D},\, D \subseteq E] \Rightarrow E \in \mathcal{D}$.

When X has such a structure then we call X a *uniform space*. The uniformity $\mathcal{D}$ is called *separating* if $\cap_{D \in \mathcal{D}} D = \triangle$.

A *basis* for the uniformity $\mathcal{D}$ is any subcollection $\mathcal{E}$ of $\mathcal{D}$ from which $\mathcal{D}$ can be recovered by applying condition **(e)** above.

EXAMPLE 2.7.4 Let X be the real numbers with the usual topology. The standard uniformity on $\mathbb{R}$ is that having as a basis the sets

$$S_\epsilon = \{(x, y) \in \mathbb{R} \times \mathbb{R} : |x - y| < \epsilon\}.$$

Of course, since $\mathbb{R}$ has a metric, it is very easy to write down a uniformity. A similar uniformity exists on any metric space.

EXAMPLE 2.7.5 Let S be any set. Define $\mathcal{D}$ to be the collection of all subsets of $S \times S$ which contain $\triangle$. This is a uniformity that we call the *discrete uniformity*.

EXAMPLE 2.7.6 Let S be any set. Define $\mathcal{D}$ to be the single set $S \times S$. This uniformity on S is called the *trivial uniformity*.

We see that the uniformity in Example 2.7.5 is essentially the *largest possible* uniformity, while that in Example 2.7.6 is the *smallest*.

Definition 2.7.7 Let X be a set and $\mathcal{D}$ a uniformity on X. For $D \in \mathcal{D}$ and $x \in X$, we set
$$D[x] = \{y \in X : (x, y) \in D\}.$$
See Figure 2.3. Now if $A \subseteq X$ is any subset, we let

$$D[A] = \bigcup_{x \in A} D[x] = \{y \in X : (x, y) \in D \text{ for some } x \in A\}.$$

THEOREM 2.7.8 *Let X be a set and $\mathcal{D}$ a uniformity on X. For each $x \in X$, the collection $\mathcal{U}_x \equiv \{D[x] : D \in \mathcal{D}\}$ forms a neighborhood basis at x. Thus the uniformity induces a topology on X. This topology is Hausdorff if and only if $\mathcal{D}$ is separating.*

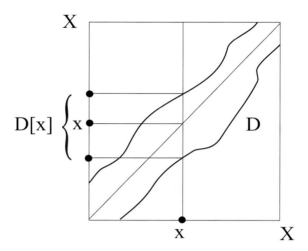

Figure 2.3: The topology induced by a uniformity.

Proof: Certainly $x \in D[x]$ for each x. Second observe that $D_1[x] \cap D_2[x] = (D_1 \cap D_2)[x]$, hence the intersection of neighborhoods is a neighborhood. Lastly, if $D[x] \in \mathcal{U}_x$, then we know by **(c)** of the definition of uniformity that there is a set $E \in \mathcal{D}$ such that $E \circ E \subseteq D$. Thus, for any $y \in E[x]$, we see that $E[y] \subseteq D[x]$. We have confirmed all the properties of a neighborhood basis.

Now suppose that $\mathcal{D}$ is separating. If x and y are distinct points of X, then there is some $D \in \mathcal{D}$ such that $(x, y) \notin D$. Then there is a symmetric $E \in \mathcal{D}$ such that $E \circ E \subseteq D$. Now let $z \in E[x] \cap E[y]$. Then $(x, z) \in E$ and $(y, z) \in E$ hence $(x, y) \in E$. Therefore $(x, y) \in E \circ E \subseteq D$. This is impossible by assumption, so we may conclude that $E[x]$ and $E[y]$ are disjoint neighborhoods of x and y, respectively. Thus the induced topology is Hausdorff.

For the converse to the last assertion, assume that the topology is Hausdorff. If $(x, y) \notin \triangle$ with $x \neq y$, then there are sets $S, T \in \mathcal{D}$ such that $S[x] \cap T[y] = \emptyset$. But then $S \cap T$ is an element of $\mathcal{D}$ that does not contain (x, y). $\qquad \square$

Definition 2.7.9 Let X be a space equipped with a uniformity $\mathcal{D}$. Let $f : X \to \mathbb{R}$ be a function. We say that f is *uniformly continuous* if, whenever $\epsilon > 0$, there is an element $D \in \mathcal{D}$ such that if $x \in X$ and $y \in D[x]$ then $|f(x) - f(y)| < \epsilon$.

EXERCISE FOR THE READER 2.7.10 Verify that, if (X, d) is a metric space and $f : X \to \mathbb{R}$ is uniformly continuous according to the classical metric definition, then f satisfies this new definition of uniform continuity with respect to the obvious uniformity on X.

We close this section by noting that a family $\mathcal{P}$ of pseudometrics (see Section 1.12) for a set X is called a *gage* if there is a uniformity $\mathcal{U}$ on X such that $\mathcal{P}$ is the family of all pseudometrics which are uniformly continuous on $X \times X$ relative to the product uniformity derived from $\mathcal{U}$. The idea of gage will come up later in the book.

2.8 Morse Theory

Building on earlier ideas of Arthur Cayley (1821–1895) and James Clerk Maxwell (1831–1879), Marston Morse (1892–1977) created the elegant and powerful idea of *Morse theory* (otherwise known as the *calculus of variations in the large*). The idea of Morse theory is that the topology/geometry of a manifold can be understood by examining the smooth functions (and their singularities) on that manifold.

In the present section we shall describe the key ideas of basic Morse theory, but we shall not prove the results in any rigorous fashion. The reader will come away with a good intuitive understanding of what the subject is about, and can consult a more definitive work like [MIL] for further details.

As a first example, consider a mountainous terrain in 3-dimensional Euclidean space (Figure 2.4). We consider the function (called the *Morse function*) that is *height h* of a point. Of particular interest are the singular points of the height function.

Now a point P on the surface will be a singular point of the height function if the gradient $\nabla h(P)$ equals 0. This in turn means that the tangent plane to the surface will be horizontal. In Morse theory we want to consider singular points which are *nondegenerate*, meaning that the determinant of the matrix of second derivatives

$$\left(\frac{\partial^2 h}{\partial x_i \partial x_j} \right)^n_{i,j=1}$$

is non-zero. And we want to classify the singular point according to how many (linearly) independent tangent directions there are at the critical point

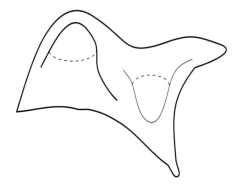

Figure 2.4: A mountainous terrain.

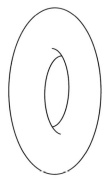

Figure 2.5: The torus.

in which the function h is decreasing. The number of such directions is called the *index* of the critical point.

A favorite example to help understand the aforementioned classification is the torus. Look at Figure 2.5. Consider the level sets $h^{-1}(c)$ of the height function. Say that the lowest point of the torus is at height 0.

If $c < 0$ then $h^{-1}(c)$—the set of points on the torus that have height c—is the empty set. Not very illuminating.

If $c = 0$, then $h^{-1}(c)$ is a single point—see Figure 2.6. Notice that that point is a singular point, for the tangent space is horizontal. And also notice that there are no directions at this point at which the height function is decreasing—in fact the height function is *increasing* in all directions. So we say that this critical point has index 0.

Now the basic rule of Morse theory describes how the surface changes at a critical point, and the answer will depend on the index. The rule is that,

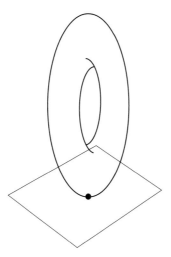

Figure 2.6: The first critical point on the torus.

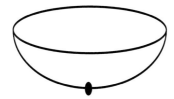

Figure 2.7: Building the torus by examining the critical points—the first critical point.

if the index is γ, then the surface changes by attaching a cell of dimension γ. For us, at this first critical point, the index is 0 so we attach a cell of dimension 0—which is a point. Now a point is homotopy equivalent to a disc (see Section 3.1), so let us instead imagine attaching a disc. What we are doing is *building* the torus from scratch by examining the height function. We began with the empty set (when $c < 0$). Then we increased c until it hit the value 0. At that special stage, we were at a critical point of index 0, and that signals us to paste in a 0-cell—which is a point, or (by homotopy equivalence) a disc. See Figure 2.7.

Now let c increase some more. For a small increase in c, we observe that the sets $f^{-1}(c)$ are circles—see Figure 2.8. This continues until c reaches the level of the bottom of the hole in the middle of the torus. See Figure 2.9. The point at the bottom of the hole in the middle of the torus is a critical point. Notice that $f^{-1}(c)$ is now a Figure 8. Of particular interest is that

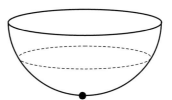

Figure 2.8: Level sets of h for c between the first two critical points.

Figure 2.9: The second critical point.

this is a saddle point. In one direction the function h is increasing and in the other direction the function h is decreasing. So this critical point has index 1. According to the fundamental rule of Morse theory, this means that, as c passes through the critical level at the bottom of the hole, we add a 1-cell to the surface. The result is homotopically equivalent to the surface shown in Figure 2.10.

The next thing to notice—and this is typical in a Morse-theory analysis— is that the nature of the level sets changes when we pass through the second critical point. *Before* the critical point the level sets were all circles. Now the level sets are pairs of circles—see Figure 2.11.

Now the nature of the level sets will not change until we hit the next critical point. And it is clear from the picture that the next critical point is at the *top* of the hole in the middle of the torus. See Figure 2.12. Once again, the level set at that critical point is a Figure 8. The critical point is in fact

Figure 2.10: The level set at the second critical point.

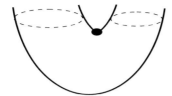

Figure 2.11: Level sets after the second critical point.

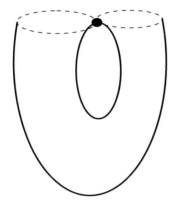

Figure 2.12: The third critical point.

a saddle point, so it has index 1. That means that, as we pass through the critical point, we add a 1-cell to the surface. The result is shown in Figure 2.13. Now the level sets are once again circles.

The final critical point of the height function is at the very top of the torus. This is a critical point of index 2, as both tangential directions yield decrease of the height function. See Figure 2.14. This means that we finish our construction by adding a 2-cell. The result is that we have built—step by step, using the analysis of critical points—the toric surface.

One of the great theorems of nineteenth century geometry (due to August Möbius (1790–1868) and Camille Jordan (1838–1922)) is that any closed, connected surface in $\mathbb{R}^3$ (i.e., any two-manifold embedded in 3-dimensional space) is homeomorphic to a sphere with finitely many handles attached. As an example, a torus is such a surface, and it is homeomorphic to a sphere with just one handle attached. Today there are many proofs of this theorem, but one of the most charming is a proof (following the steps we have just outlined for the torus) using Morse theory.

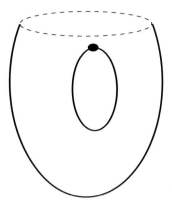

Figure 2.13: Structure of the surface after the third critical point.

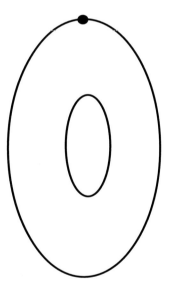

Figure 2.14: The fourth critical point, completing the construction of the torus.

2.9 Proper Mappings

Let X, Y be topological spaces, and let $f : X \to Y$ be a mapping. We say that f is *proper* if $f^{-1}(E)$ is compact whenever E is compact. What does this mean? First let us consider an example of a mapping that is not proper:

EXAMPLE 2.9.1 Let $X = (-1, 1)$ and let $Y = (-1, 1)$. Let $f : X \to Y$ be given by $f(x) = 1 - x^2$. Of course f is continuous. But it is *not* proper because $f^{-1}([-1/2, 1/2]) = (-1, -1/\sqrt{2}] \cup [1/\sqrt{2}, 1)$. So we see that the inverse image of a compact set is not necessarily compact.

The next proposition helps to clarify the concept of properness:

Proposition 2.9.2 *Let X, Y be Euclidean spaces. Let $E \subseteq X$ and $F \subseteq Y$ be bounded, open sets. Suppose that $f : E \to F$ is proper. If $\{e_j\} \subseteq E$ satisfies $e_j \to \partial E$ (that is, the points e_j accumulate at one or more elements of ∂E), then $f(e_j) \to \partial F$ (that is, the points $f(e_j)$ accumulate at one or more elements of ∂F).*

Proof: If not then there are $e_j \in E$ with $e_j \to \partial E$ but $f(e_j) \not\to \partial F$. It follows then that there is a compact set $K \subseteq F$ so that $f(e_j) \in K$ for all j. But then $f^{-1}(K)$ is compact, and $\{e_j\} \subseteq f^{-1}(K)$. This means, of course, that the sequence $\{e_j\}$ cannot converge to ∂E. Contradiction. □

EXERCISE FOR THE READER 2.9.3 Prove the converse of the last proposition.

Proposition 2.9.4 *Let X, Y be topological spaces and let $f : X \to Y$ be a homeomorphism. Then f is proper.*

Proof: Of course $f^{-1} : Y \to X$ is a mapping. If $K \subseteq Y$ is compact then of course $f^{-1}(K)$ must also be compact. So f is proper. □

Thus we see that properness is a generalization of the property of being a homeomorphism. And, at least in the familiar context of Euclidean space, a proper mapping is one that takes the boundary to the boundary.[2]

---

[2]The *Brouwer invariance of domain* theorem (see [HUW]) generalizes this idea even further.

Proposition 2.9.5 *Every mapping f from a compact space X to a Hausdorff space Y is both proper and closed.*

Proof: The closedness is clear, for if $E \subseteq X$ is closed then it is compact. Hence $f(E)$ is compact; since Y is Hausdorff, $f(E)$ is closed.

For the properness, let $F \subseteq Y$ be compact. Then F is certainly closed. Since f is continuous, we may be sure that $f^{-1}(F)$ is closed. But a closed subset of a compact space is compact, hence $f^{-1}(F)$ is compact. Thus f is proper.   □

Proposition 2.9.6 *Let X be a topological space. Then X is compact if and only if the map of X to a single-point space Z is proper.*

Proof: Let $Z = \{z\}$. Suppose that $f : X \to Z$ is proper. Certainly Z is compact, so $f^{-1}(Z)$ is compact. But $f^{-1}(Z) = X$. That proves one direction.

Now suppose that X is compact. Let $f : X \to Z$. Let F be a compact subset of Z. Then either $F = Z$ or F is the empty set. In the first instance, $f^{-1}(F) = X$ and is compact. In the second instance $f^{-1}(F) = \emptyset$ and is compact. So f is proper.   □

2.10   Paracompactness

Recall the concept of open cover that we used to good effect in our study of compactness (Section 1.5). Let us say that an open cover $\mathcal{U} = \{U_\alpha\}_{\alpha \in A}$ of a space X is *locally finite* if each point $x \in X$ has a neighborhood V so that V has nontrivial intersection with only finitely many of the U_α.

A *refinement* of an open cover $\mathcal{U} = \{U_\alpha\}_{\alpha \in A}$ is collection $\mathcal{V} = \{V_\beta\}_{\beta \in B}$ of open sets such that **(i)** $\mathcal{V}$ still covers X and **(ii)** each V_β is a subset of some U_α.

Definition 2.10.1 *Let X be a topological space. We say that X is paracompact if every open cover of X admits a locally finite refinement.*

Clearly paracompactness is a generalization of compactness. Every compact space is paracompact. It turns out—and this is perhaps the fundamental theorem in the subject—that every metric space is paracompact. The reference [RUD] provides a particularly brief and modern proof of that result. It is also a fact (theorem of Jean Dieudonné (1906–1992)) that every paracompact space is normal.

EXAMPLE 2.10.2 One of the most famous examples in topology is the *long line*. This is uncountably many copies of the half-open unit interval pasted end-to-end. We perform the construction as follows:

Let $I = [0, 1)$. Let us consider the product $\mathcal{P} = \mathbb{R} \times I$. If $A = (s, x)$ and $B = (t, y)$ are elements of $\mathcal{P}$, then we say that $A < B$ if either $s < t$ or $s = t$ and $x < y$. This makes $\mathcal{P}$ into a totally ordered space which we call $\widehat{\mathbb{R}}$, the long line. We use sets of the form $\{X \in \widehat{\mathbb{R}} : X < A\}$ and $\{X \in \widehat{\mathbb{R}} : X > B\}$ to form a sub-basis for the topology on $\widehat{\mathbb{R}}$.

We see that the long line is locally just like the real line. But it is *very* long. It is easy to see that $\widehat{\mathbb{R}}$ is not paracompact. Consider the open covering consisting of the sets $U_X = \{T \in \widehat{\mathbb{R}} : X < T\}$ for $X \in \widehat{\mathbb{R}}$. The sets $\{U_X\}$ form an open covering of $\widehat{\mathbb{R}}$. If P is any point of $\widehat{\mathbb{R}}$ and V is any neighborhood of P then of course V will intersect uncountably many of the U_X. And any refinement of this covering will fail to be locally finite as well.

In practice the most important property of paracompactness relates to the concept of partition of unity.

Definition 2.10.3 Let X be a topological space and let $\mathcal{U} = \{U_\alpha\}_{\alpha \in A}$ be a locally finite cover of X by open sets. We call a collection φ_α of continuous functions on X a *partition of unity subordinate to $\mathcal{U}$* if

(i) Each φ_α satisfies $0 \leq \varphi_\alpha(x) \leq 1$ for all x.

(ii) For each α, the set $S_\alpha = \{x \in X : \varphi_\alpha(x) \neq 0\}$ lies entirely in U_α.

(iii) We have the identity

$$\sum_\alpha \varphi_\alpha(x) \equiv 1 .$$

If ψ is a continuous function on a space X, then we define the *support* of ψ to be the complement of the union of all open sets on which ψ vanishes. Essentially, the support of the function ψ is the set where ψ is nonzero. In the definition of partition of unity, each φ_α has support lying in U_α.

Partitions of unity are extremely useful because we can make a local construction on each U_α and then patch them together with the partition of unity. We shall give examples later.

THEOREM 2.10.4 *Let* X *be a paracompact topological space. Let* $\mathcal{U} = \{U_\alpha\}_{\alpha \in A}$ *be an open cover of* X. *Then there is a partition of unity* $\{\varphi_\alpha\}$ *subordinate to* $\mathcal{U}$.

Proof: For simplicity we shall only treat the case when X is a metric space with metric d. Let $\mathcal{V} = \{V_\beta\}_{\beta \in B}$ be a refinement of $\mathcal{U}$ which is locally finite and which still covers X. For each $\beta \in B$, define

$$\psi_\beta(x) = \begin{cases} d(x, {}^c V_\beta) & \text{if} \quad x \in V_\beta \\ 0 & \text{if} \quad x \notin V_\beta. \end{cases}$$

Define

$$\varphi_\beta(x) = \frac{\psi_\beta(x)}{\sum_\gamma \psi_\gamma(x)}.$$

We note that each point $x \in X$ lies in some V_γ. Therefore $\psi_\gamma(x) \neq 0$ and the sum in the denominator therefore does not vanish. Further observe that the covering $\mathcal{V}$ is locally finite, therefore, for each fixed x, the sum in the denominator is actually finite. Finally notice that, for each x, $0 \leq \varphi_\beta(x) \leq 1$.

Finally, we may check that, for each $x \in X$,

$$\sum_\beta \varphi_\beta(x) = \sum_\beta \left[\frac{\psi_\beta(x)}{\sum_\gamma \psi_\gamma(x)} \right] = \frac{\sum_\beta \psi_\beta(x)}{\sum_\gamma \psi_\gamma(x)} = 1.$$

That completes the proof. □

EXAMPLE 2.10.5 Let X be a compact metric space and let $\mathcal{U} = \{U_j\}_{j=1}^k$ be an open cover of X. Let $\delta > 0$ be the Lebesgue number of this cover (see Section 1.15). Then there is a partition of unity of X subordinate to the cover $\mathcal{U}$ such that each element of the partition φ_j has support lying in some metric ball $B(x, \delta)$. This result is immediate, since our basic theorem about the Lebesgue number says that every ball $B(x, \delta)$ will lie in one of the U_{α_j}.

2.11 An Application to Digital Imaging

Not too many years ago, most photographs were taken with what we now call "analogue cameras." An analogue camera works with film. The film is coated with light sensitive chemicals, and those chemicals record the photographic image. After exposure, the image is chemically processed to produce a *negative*.[3] Then a positive hard-copy print can be made from the negative using a special projector.

In many applications, the photographic image was run through a filter to create a *half-tone*. This was done particularly for pictures to appear in newspapers. Newspapers in those days were only in black-and-white; they needed to have a way to emulate gray-scale so that they could create the shading needed for an accurate photographic image on the page of a newspaper. The half-tone used dots of various sizes to emulate the gray-scale environment.

Now things are different. Most photography is digital, and the images are, by default, broken up into *pixels*. A pixel is a small square or rectangle that has several attributes for color, intensity, and so forth. The aggregate of the pixels makes up the image. See Figure 2.15. A great deal of modern image processing involves imposing a topology on the visual image and then studying topological attributes such as connectivity, separation axioms, and so forth. We discuss some of these ideas in the present section. The ideas presented here are found broadly in the technical literature. As an instance, the U.S. Geological Survey makes extensive use of the digital line (see [NMP]). Properties of images in cathode ray tubes are also studied with this technology.

One very simple and common operation in digital image processing is optical character recognition (OCR). The situation is that you have a printed sheet of text (words on a page) and you would like to create a computer text file with those words. The old-school, naïve way of performing the task is simply to sit down and type the words into the computer. The modern approach is to use a scanner together with OCR software to do the job for you. If the text is plain English, in a standard font like Times Roman, and the characters are all of the same size and in horizontal lines on the page (not strange fonts with the words draped across graphic images, as in an advertisement), then OCR is usually better than 95% accurate. The mathematical issues that arise for software endeavoring to recognize

---

[3]In a negative, the black and white values are reversed.

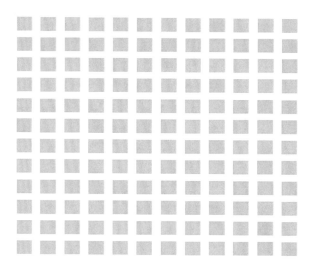

Figure 2.15: A rectangular array of pixels.

characters are largely topological ones. Examine Figure 2.16.

Now we shall define a special topology on the line, and a related topology on the plane, that mirrors the structure of a line of pixels or a rectangular array of pixels. These are called, respectively, the *digital line topology* and the *digital plane topology*.

Definition 2.11.1 Let X be the topological space consisting of the integers $\mathbb{Z}$. For each integer n, we define a basis element for the topology as follows:

$$I(n) = \{n\} \qquad \text{if } n \text{ is odd};$$

$$I(n) = \{n-1, n, n+1\} \qquad \text{if } n \text{ is even}.$$

Thus the basis element at an odd integer is a singleton, while the basis element at an even integer is a triple of integers. Figure 2.17 suggests the linear row of pixels (with their adjacencies) and the corresponding topology.

Observe that, in the digital line topology, every odd integer singleton $\{n\}$ is an open set—by the very definition of $I(n)$. By contrast, every even integer singleton $\{n\}$ is a closed set. For example,

$$\{0\} = \mathbb{Z} \setminus \bigcup_{|n| \geq 1} I(n).$$

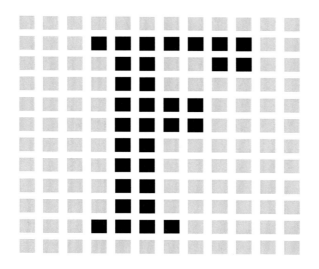

Figure 2.16: A digital rendition of the letter "F".

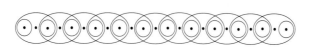

Figure 2.17: The digital line topology.

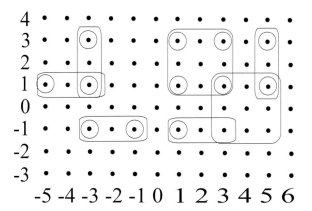

Figure 2.18: The topology of the digital plane.

It is easy to see that the digital line is connected, and in fact Figure 2.17 suggests why this is so.

On the other hand, the digital line is *not* Hausdorff. For there is no disjoint pair of open sets U, V with $2 \in U$ and $3 \in V$. It turns out that the separation axioms play an important role in digital imaging.

Now let us describe the digital plane. This is a topology on $X = \mathbb{Z} \times \mathbb{Z}$. We define basic open sets

$$J(m, n) = \{(m, n)\} \qquad \text{if } m \text{ and } n \text{ are both odd;}$$

$$J(m, n) = \{(m - 1, n), (m, n), (m + 1, n)\} \qquad \text{if } m \text{ is even and } n \text{ is odd;}$$

$$J(m, n) = \{(m, n - 1), (m, n), (m, n + 1)\} \qquad \text{if } m \text{ is odd and } n \text{ is even;}$$

$$J(m, n) = \big\{(m-1, n-1), (m, n-1), (m+1, n-1), (m-1, n), (m, n), (m+1, n),$$
$$(m-1, n+1), (m, n+1), (m+1, n+1)\big\} \qquad \text{if } m \text{ and } n \text{ are both even.}$$

This topology is a bit more difficult to apprehend; Figure 2.18 gives some guidance.

Notice how the regions in the figure correspond to the definitions: the square regions are centered at (even, even), the circles are centered at (odd, odd), the vertical rectangular regions are centered at (odd, even), and the horizontal rectangular regions are centered at (even, odd).

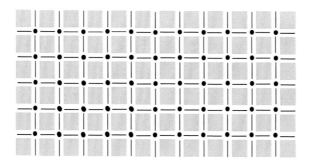

Figure 2.19: The digital plane.

Of course the single point sets $\{(m, n)\}$, with m and n both odd, are open. By contrast (exercise), the single point sets $\{(m, n)\}$, with m and n both even, are closed.

The digital plane is *not* a Hausdorff space. For the points $(1, 1)$ and $(2, 1)$ cannot be separated by open sets U, V. Figure 2.19 shows the digital plane with boundaries that our topology models.

So what do people in image processing actually do with the digital planar topology? Let us now give an indication of some of the thinking that is currently in place. Questions of particular interest are:

- How can spatial relationships and shapes and adjacencies be preserved when an image is processed or transformed?

- How can topological features of an image be preserved under various natural processes and transformations to which we might like to subject an image?

- When a digital image is stored, what steps can be taken so that the resulting computer file is not excessively large? If the image is 5 inches by 8 inches (a standard photograph size), and if the resolution is 1200 dots per inch, then this image has more than 57 million pixels. If each pixel has ten attributes (color, intensity, etc.), then this could be more than 100 million bytes of information. That is a very large file, and one that is difficult to manipulate. *Image compression* is big business these days, and that is the reason.

In the present discussion we shall concentrate on the third of these questions. First we note that compression comes up in other contexts. If one has a simple text file, it is often desirable to compress that file so that the resulting computer file will take less space. How might this work? The basic idea is to replace common letter combinations, or diphthongs, with shorter substitutes. For instance, if you replace every occurrence of "ough" with "!*" then that is a 50% savings in storage space. There is no likelihood of confusion, because "!*" never occurs in ordinary English prose. And it is easy to translate the compressed file back to its original form: just replace each occurrence of "!*" with "ough." Another example is that one could replace every occurrence of "eat" with "@@". That gives a 33% savings in storage space.

A two-dimensional visual image does not lend itself so well to tricks as described in the last paragraph. Each pixel has too many attributes, and substitutions such as we have described do not readily suggest themselves. So other ideas have been devised.

Points (m, n) in the digital plane with both m and n odd are called "open points." This simply signifies that the singleton set $\{(m, n)\}$ is open. We let $\mathcal{O}$ denote the collection of open points. The set $\mathcal{O}$ is called the *visible screen*. The visible screen is simply the mathematical model for the set of pixels in a given digital image display. This is what we actually see in the graphic. This is a dense, open set in the digital plane, and it has the discrete topology. The other points in the digital plane—the closed points and the points of mixed type—provide a non-visible infrastructure that connects up the (visible) pixels and allows us to use topological ideas to study the digital image.

Our fundamental concept for compressing a digital image is to identify regions in the plane that have the same color, or the same visible attributes. The key idea is that one can describe such a region by enumerating all the points in the region (this is the *two-dimensional approach*, and is computationally expensive) or by describing the one-dimensional "curve" that surrounds the region. Clearly the latter requires a lot less data, and will result in some compression of information.

As a simple example of this last idea, examine the set of pixels in black in Figure 2.20. We are assuming that all the pixels in this region are of the same color. This is an 8×10 pixel region, and it requires 80 pieces of data to record that this region is of one color. But we can instead examine the boundary of the region—see Figure 2.21. That boundary only has 26 pixels in it. So we compress the information by a factor of 3 using this simple device.

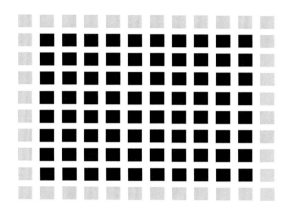

Figure 2.20: The region in the digital plane.

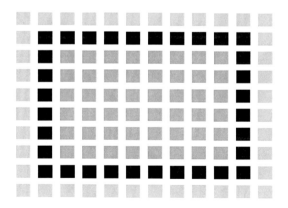

Figure 2.21: The boundary of the digital region.

The saving in data storage is dramatic in this very simple case. But the question becomes much more interesting (and more typical, in practice) when the uni-chromatic region in question is *not* a rectangle. Then one wants to know something about the nature of the boundary, and whether it can be analyzed in a similar fashion.

Definition 2.11.2 Let $\{m, m + 1, m + 2, \ldots, n\}$ be a digital *interval* in $\mathbb{Z}$. The topological space that results from identifying the endpoints m and n (in the sense of passing to a quotient) is called a *digital circle*.

Definition 2.11.3 Let X be a topological space. A *digital simple closed curve* in X is a subspace of X that is homeomorphic to a digital circle.

THEOREM 2.11.4 (Digital Jordan Curve Theorem) *Let C be a digital simple closed curve in the digital plane. Then C separates the digital plane $X = \mathbb{Z} \times \mathbb{Z}$ into two disjoint sets (or "components") $\mathcal{A}$ and $\mathcal{B}$ (that is to say, $X \setminus C = \mathcal{A} \cup \mathcal{B}$). It holds that C is the boundary of each of these sets precisely when C is closed in the digital plane.*

It is a fact that the curve C in the last theorem will be closed precisely when C does not contain any open points.

We see that the digital Jordan curve theorem would not make any sense if we did not have the aid of the points that are *not* part of the visible screen. It is the digital plane topology that we have introduced that makes this theory cohere. The paper [ROS] is one of the earliest explorations of these ideas. Later fundamental work appears in [KKM], [KOKM], and [KIS].

Exercises

1. Consider the closed intervals $[a, b]$ as a sub-basis for a topology on the real line. Describe the resulting topology.

2. Let X be the continuous, real-valued functions on the interval $[0, 1]$. For $f \in X$ and $\epsilon > 0$, consider the sets

$$\mathcal{S}_{f,\epsilon} = \{g \in X : |g(x) - f(x)| < \epsilon \ \text{ for all } x \in [0, 1]\}.$$

Using these sets as a sub-basis, describe the resulting topology.

3. If X and Y are topological spaces, then $\partial(X \times Y)$ is *not* equal to $\partial X \times \partial Y$. Explain by way of examples.

4. Let X_α be topological spaces. The *box topology* on $X = \prod_\alpha X_\alpha$ is the topology with sub-basis given by sets of the form $\prod_\alpha U_\alpha$ for U_α open in X_α. Explain the difference between the box topology and the product topology as discussed in the text. Explain why, for a product of finitely many spaces, the box topology and the product topology are the same.

5. Refer to the last exercise for terminology and notation. What is useful and canonical about the product topology is that, if Y is any topological space and if $\pi_\alpha : X \to X_\alpha$ is the standard projection, then a mapping $\varphi : Y \to X$ is continuous if and only if $\pi_\alpha \circ \varphi$ is continuous for each α. Explain why this property fails for the box topology. Explain why Tychanoff's theorem fails for the box topology.

6. Let X be the continuous, real-valued functions on the interval $[0, 1]$. For $f \in X$ and $\epsilon > 0$, consider the sets

$$\mathcal{S}_{f,\epsilon} = \{g \in X : |g(0) - f(0)| < \epsilon\}.$$

Using these sets as a sub-basis, describe the resulting topology.

7. Recall the long line from Section 2.11. Is this space first countable? Is it second countable? Why or why not?

8. Consider the space X consisting of the continuous functions on the unit interval $[0, 1]$. Equip X with the uniform topology as described in Exercise 2 above (this is the standard topology on this space). Is X first countable? Is X second countable?

9. Explain why a function $f : \mathbb{R} \to \mathbb{R}$ can be thought of as an element of the product of infinitely many copies of $\mathbb{R}$, indexed over $\mathbb{R}$. Explain why this justifies the notation $\mathbb{R}^{\mathbb{R}}$ for the space of functions from $\mathbb{R}$ to $\mathbb{R}$.

10. Recall the Cantor set C from Section 1.11. Consider the relative topology on the Cantor set, inherited from the standard topology on the real line. Give an example of an open set in C. Show that C is totally disconnected.

11. Describe the one-point compactification of $\mathbb{R}^3$. Prove that it is homeomorphic to S^3, the unit sphere in $\mathbb{R}^4$.

12. Let X be the Cartesian plane $\mathbb{R}^2$. Let Y be the lattice of points (m, n) in X with both m and n integers. Describe the quotient topology on X/Y. What topological object is X/Y?

13. Let X be the continuous functions on the unit interval $[0, 1]$, equipped with the uniform topology as described in Exercise 2. Let Y be the real numbers. Let $T : X \to Y$ be defined by

$$T(f) = \int_0^1 f(x)\, dx\,.$$

What is the quotient topology induced on Y by the mapping T?

14. Let X be the Cartesian plane $\mathbb{R}^2$. Equip X with the topology induced by the sub-basis

$$E_{x,y,\epsilon} = \{(s, t) \in X : s = x, y - \epsilon < t < y + \epsilon\}$$

for $x, y \in \mathbb{R}$, $\epsilon > 0$. Define $T : X \to \mathbb{R}$ by $T(x, y) = x$. What quotient topology is induced on $\mathbb{R}$ by this mapping?

15. Give an example of a proper mapping from $\mathbb{R}^3$ to $\mathbb{R}^2$.

16. Do an analysis, similar to our Morse theory analysis of the torus, for a torus with two handles.

17. Refer to the uniform topology on the space of continuous functions on the interval $[0, 1]$ as specified in Exercise 2. Prove that this is a uniform topology.

18. Let the unit circle in the Cartesian plane be covered by the four open sets

$$
\begin{aligned}
U_1 &= \{(x, y) : (x - 1)^2 + y^2 < 1\} \\
U_2 &= \{(x, y) : x^2 + (y - 1)^2 < 1\} \\
U_3 &= \{(x, y) : (x + 1)^2 + y^2 < 1\} \\
U_4 &= \{(x, y) : x^2 + (y + 1)^2 < 1\}
\end{aligned}
$$

Construct an explicit partition of unity subordinate to this covering.

19. Let W be an open set in the Cartesian plane. Let $\mathcal{U} = \{U_\alpha\}_{\alpha \in A}$ be a locally finite open cover of W. Let $\{\varphi_\alpha\}_{\alpha \in A}$ be a partition of unity subordinate to the cover $\mathcal{U}$. For each α, let $p_\alpha = (p_1^\alpha, p_2^\alpha) \in U_\alpha$. Define $f_\alpha(x, y) = 1/|(x, y) - p_\alpha|^2$. Use the partition of unity to patch together these singular functions and obtain a single function that is singular at all the p_α.

20. Give three examples of digital simple closed curves in the digital plane.

21. Explain why, in the digital Jordan curve theorem, one of the resulting components is bounded and one is not.

22. Construct a non-trivial proper mapping from the real line to the halfline $\mathcal{L} = \{x \in \mathbb{R} : x \geq 0\}$.

23. Show that there is no nontrivial proper mapping from the real line to the unit interval $[0, 1]$.

24. Give a topology on the collection of real-valued, continuous functions on $[0, 1]$ that is not first countable.

25. Let X be the space of continuous functions on the interval $[0, 1]$ equipped with the uniform topology. Is this space paracompact?

26. Describe a continuous, surjective mapping from the long line $\widehat{\mathbb{R}}$ to $\mathbb{R}$.

27. Describe a proper mapping from the long line $\widehat{\mathbb{R}}$ to $\mathbb{R}$.

28. Let $\mathbf{v}$ and $\mathbf{w}$ be linearly independent vectors in the plane. Consider the lattice $L = \{m\mathbf{v} + n\mathbf{w} : m, n \in \mathbb{Z}\}$. This describes a lattice, and we may consider $\mathbb{R}^2/L$. What sort of topological object is this quotient? Can you describe its topology?

29. Describe a one-point compactification for the long line.

30. Describe the Stone-Čech compactification of the open unit disc in the plane.

Chapter 3

Basic Algebraic Topology

3.1 Homotopy Theory

There is considerable interest in identifying and classifying the "holes" in a topological space. The annulus has a hole in its center, and so does the unit sphere. See Figure 3.1. But one has a sense that these two holes are different in nature. They are (perhaps) holes of different dimensions. Homotopy theory helps us to sort out these ideas.

The basis for homotopy theory is to consider the set of all loops in a given topological space X. Here, by a *loop*, we mean a continuous function $\gamma : [0, 1] \to X$ such that $\gamma(0) = \gamma(1)$. See Figure 3.2.

We want to create equivalence classes among the loops. One of the great things about topology is that it has a strong intuitive component. Certainly heuristics shape the way that we understand the ideas, and the way that we construct the proofs. The proofs are strictly rigorous, but they have a powerful pictorial component.

What we *picture* for the equivalence classes is that two curves γ_1 and γ_2 are related if one can be continuously deformed to the other—without breaking the curve, and without leaving the space X, in the process. Refer to Figure 3.3.

We need to think about what would prevent two curves from being deformable to each other. The answer—speaking strictly intuitively—is that a "hole" in X can be the source of a problem. For simplicity, imagine that X is a region in the plane. And that X has a hole in it. Figure 3.4 exhibits such an X and two curves γ_1 and γ_2 that plainly *cannot* be deformed to each other.

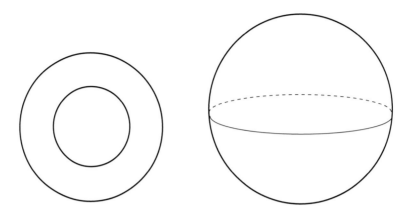

Figure 3.1: Different types of holes.

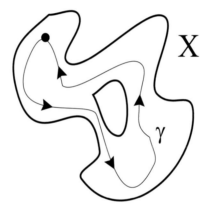

Figure 3.2: A loop.

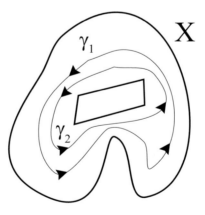

Figure 3.3: One curve can be deformed to the other.

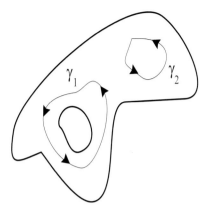

Figure 3.4: Two curves that cannot be deformed to each other.

The formal concept of homotopy makes these ideas rigorous. As you learn the new language, you should keep the intuitive picture in mind.

Definition 3.1.1 Let $(X, \mathcal{U})$ be a topological space. Fix a point $x_0 \in X$. Let $\gamma_1 : [0,1] \to X$ and $\gamma_2 : [0,1] \to X$ be two loops in X. Assume that $\gamma_1(0) = \gamma_1(1) = \gamma_2(0) = \gamma_2(1) = x_0$. We say that γ_1 and γ_2 are *fixed-point homotopic*[1] when there is a continuous function

$$\Gamma : [0,1] \times [0,1] \to X$$

with these properties:

(i) $\Gamma(0, t) = \gamma_1(t)$, $0 \le t \le 1$;

(ii) $\Gamma(1, t) = \gamma_2(t)$, $0 \le t \le 1$;

(iii) $\Gamma(s, 0) = x_0$, $0 \le s \le 1$;

(iv) $\Gamma(s, 1) = x_0$, $0 \le s \le 1$.

In practice we shall just say that loop γ_1 is homotopic to loop γ_2 (omitting the phrase "fixed-point"). The fixed point itself often only plays a tacit role in our discussions and calculations. There is also a homotopy theory that is *not* based on a fixed point. But the theory that we present here is the most standard one.

---

[1]The terminology "based homotopy" and "base-point preserving homotopy" is also used.

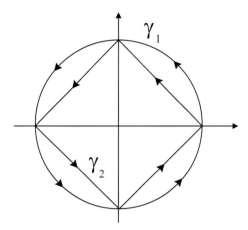

Figure 3.5: Two homotopic curves in the plane.

What we see, according to this definition, is that the two curves γ_1 and γ_2 are homotopic if they both begin and end at x_0, and if they can be continuously deformed one to the other so that all the intermediate curves lie in X and begin and end at x_0.

EXAMPLE 3.1.2 Let X be the closed unit disc in the plane. Let

$$\gamma_1(t) = (\cos 2\pi t, \sin 2\pi t)$$

and

$$\gamma_2(t) = \begin{cases} (1 - 4t, 4t) & \text{if} \quad 0 \le t \le 1/4 \\ (1 - 4t, 2 - 4t) & \text{if} \quad 1/4 < t \le 1/2 \\ (-3 + 4t, 2 - 4t) & \text{if} \quad 1/2 < t \le 3/4 \\ (-3 + 4t, -4 + 4t) & \text{if} \quad 3/4 < t \le 1. \end{cases}$$

For both curves the parameter t ranges over $[0, 1]$. Figure 3.5 illustrates the two curves.
Define
$$\Gamma(s, t) = (1 - s)\gamma_1(t) + s\gamma_2(t).$$
In this calculation the point x_0 is $(1, 0)$. Then it is easy to see that the four postulates of a homotopy are satisfied by Γ.

Now what is interesting is that Definition 3.1.1 defines an equivalence relation on the set of all loops. The reflexivity and symmetry are fairly obvious, so let us concentrate now on transitivity. Say that loop γ_1 is homotopic to γ_2

with homotopy map Γ_{12} and also that γ_2 is homotopic to γ_3 with homotopy map Γ_{23}. Now define a new homotopy map Γ_{13} by

$$\Gamma_{13}(s,t) = \begin{cases} \Gamma_{12}(2s,t) & \text{if} \quad 0 \leq s \leq 1/2 \\ \Gamma_{23}(2s-1,t) & \text{if} \quad 1/2 < s \leq 1. \end{cases}$$

It is easy to see that $\lim_{s \to 1/2+} \Gamma_{23}(2s-1,t) = \gamma_2(t) = \Gamma_{12}(1,t)$, so that the two halves of our definition mesh up. Also $\Gamma_{13}(0,t) = \Gamma_{12}(0,t) = \gamma_1(t)$ while $\Gamma_{13}(1,t) = \Gamma_{23}(1,t) = \gamma_3(t)$. So Γ_{13} is a homotopy from γ_1 to γ_3. That confirms the transivity of the relation.

We call the equivalence classes *homotopy classes* for the topological space X. The collection of homotopy classes is the *first homotopy space* (later we shall call it the *fundamental group*). We denote the first homotopy group of X, with base point x_0, by $\pi_1(X)$ or $\pi_1(X, x_0)$.

Now the profound fact about this space is that it is a group. What is the group operation? Suppose that $[\gamma]$ and $[\eta]$ are two homotopy classes. Then we set

$$[\gamma] \cdot [\eta](t) = \left[\begin{cases} \eta(2t) & \text{if} \quad 0 \leq t \leq 1/2 \\ \gamma(2t-1) & \text{if} \quad 1/2 < t \leq 1 \end{cases} \right].$$

We plainly see that the binary operation in this group is that $[\gamma]$ multiplied times $[\eta]$ is the equivalence class for the new loop formed by the loop η followed by the loop γ. See Figure 3.6. Of course one needs to check that this new operation is well defined: if we choose another representative from the equivalence class $[\gamma]$ and another representative from the equivalence class $[\eta]$, then is the resulting product the same equivalence class?

Of course this well-definedness is almost obvious. If γ' is another representative from the equivalence class $[\gamma]$ (with homotopy Γ) and η' is another representative from the equivalence class $[\eta]$ (with homotopy Φ) then

$$\Pi(s,t) = \begin{cases} \Phi(2s, 2t) & \text{if} \quad 0 \leq s \leq 1/2, 0 \leq t \leq 1/2 \\ \Gamma(2s, 2t-1) & \text{if} \quad 0 \leq s \leq 1/2, 1/2 < t \leq 1 \\ \Phi(2s-1, 2t) & \text{if} \quad 1/2 < s \leq 1, 0 \leq t \leq 1/2 \\ \Gamma(2s-1, 2t-1) & \text{if} \quad 1/2 < s \leq 1, 1/2 < t \leq 1 \end{cases}$$

is a homotopy from $\gamma \cdot \eta$ to $\gamma' \cdot \eta'$. See Figure 3.7.

Thus we have a well-defined binary group operation on the set of homotopy classes. The identity element in this group is the equivalence class

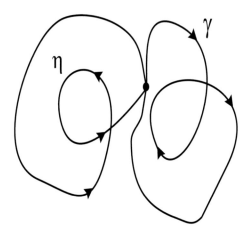

Figure 3.6: The product of two curves.

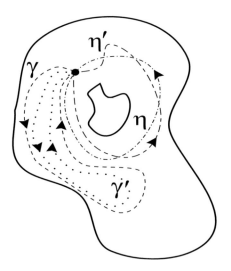

Figure 3.7: Well-definedness of the group operation.

containing the constant loop[2]

$$\gamma_0(t) \equiv x_0 \qquad \text{for all } t.$$

Also each homotopy class $[\gamma]$ has a multiplicative inverse

$$[\gamma^{-1}](t) = [\gamma(1-t)].$$

In other words, the inverse of γ is the same loop traced backwards. We leave it as an exercise to verify that $[\gamma] \cdot [\gamma^{-1}]$ is homotopic to $[\gamma_0]$ and $[\gamma^{-1}] \cdot [\gamma]$ is homotopic to $[\gamma_0]$. It is also an exercise to check that multiplication is associative.

The first homotopy group, which we have just described, is called the *fundamental group* of the topological space X. We denote it by $\pi_1(X)$. There are higher homotopy groups $\pi_k(X)$, but we shall not say much about those in this text. The important fact about π_1 is that it is *functorial*. This notion is encapsulated in the following proposition.

THEOREM 3.1.3 *Let $(X, \mathcal{U})$ and $(Y, \mathcal{V})$ be topological spaces. Let x_0 be the homotopy base point in X and y_0 be the homotopy base point in Y. Assume that $f : X \to Y$ is a mapping such that $f(x_0) = (y_0)$. Then the induced function on homotopy given by*

$$\begin{aligned} f_* : \pi_1(X) &\longrightarrow \pi_1(Y) \\ [\gamma] &\longmapsto [f \circ \gamma] \end{aligned}$$

is in fact a group homomorphism. That is to say,

$$f_*([\gamma] \cdot [\mu]) = f_*([\gamma]) \cdot f_*([\mu]).$$

Proof: Certainly the mapping is well defined, because if Γ is a homotopy between γ_1 and γ_2, then $f \circ \Gamma$ is a homotopy between $f \circ \gamma_1$ and $f \circ \gamma_2$.

For the canonical homomorphism property, note that

$$f([\gamma] \cdot [\eta]) = \left[\begin{cases} f(\gamma(2t)) & \text{if } 0 \le t \le 1/2 \\ f(\eta(2t-1)) & \text{if } 1/2 < t \le 1 \end{cases} \right] = [f \circ \gamma] \cdot [f \circ \eta].$$

---

[2]We see here why it is convenient that we are doing fixed-point homotopy.

$\square$

In fact it is not difficult to see that, if f is injective then so is f_* and if f is surjective then so is f_*. What is important to understand here is that homotopy is a topological invariant. If two spaces are topologically equivalent (i.e., homeomorphic), then their homotopy groups are isomorphic. Contrapositively, this says that if two spaces have different homotopy, then they cannot be homeomorphic.

EXAMPLE 3.1.4 Let us calculate the first homotopy group of the circle

$$S^1 = \{(x, y) \in \mathbb{R}^2 : x^2 + y^2 = 1\}.$$

This will be done in several steps.

Step 1: Let $\gamma : [0, 1] \to S^1$ be a loop. Using the Weierstrass approximation theorem, we may suppose that γ is smooth. We may also suppose that $\gamma' \neq 0$ except perhaps at finitely many points. And those exceptional points are places where γ reverses direction.

Now let t be any point of $[0, 1]$. If $\gamma'(t) \neq 0$, then either γ is proceeding counterclockwise through the point $\gamma(t)$ or clockwise through the point $\gamma(t)$. We respectively define $A(t)$ to be $+1$ or -1. Now we define the *index* of the point t to be

$$I(t) \equiv \sum_{\gamma(s)=\gamma(t)} A(s).$$

What we are counting here is all the different (finitely many) times that the curve passes through the point $\gamma(t)$, taking orientation into account.

Step 2: The index $I(t)$ is independent of the point t. This assertion is nearly obvious, for I is a continuous, integer-valued function. So it must be constant. Thus we write $I(\gamma)$ for the index of γ. If $I(\gamma) > 0$, then γ is essentially a counterclockwise rotating curve; while if $I(\gamma) < 0$, then γ is essentially a clockwise rotating curve.

Step 3: If two curves $\gamma : [0, 1] \to S^1$ and $\eta : [0, 1] \to S^1$ are homotopic, then $I(\gamma) = I(\eta)$. This is evident because if $\Gamma(s, t)$ is a homotopy of γ to η, then $I(\Gamma(s, \cdot))$ is a continuous, integer-valued function of s. So it must be constant.

Step 4: Every loop $\gamma : [0,1] \rightarrow S^1$ is homotopic to some $\rho_j(t) = e^{2\pi i j t}$. To see this, we first note that there is a continuous, real-valued function $g : [0,1] \rightarrow \mathbb{R}$ such that $\gamma(t) = (\cos g(t), \sin g(t))$. We may suppose that γ begins and ends at the point $(1,0) \in \mathbb{R}^2$. Hence $\gamma(0) = \gamma(1) = (1,0)$. It follows that $g(0) = 0$ and $g(1) = 2k\pi$ for some integer k. Now let $h(t) = 2\pi k t$. Then the graphs of g and h both begin (on the left) at $(0,0)$ and end (on the right) at $(1, 2\pi k)$. The two curves $\gamma(t) = (\cos g(t), \sin g(t))$ and $\mu(t) = \rho_k(t) = (\cos h(t), \sin h(t))$ are homotopic by the homotopy

$$\Gamma(s,t) = \left(\cos\big((1-s)g(t) + sh(t)\big), \sin\big((1-s)g(t) + sh(t)\big) \right).$$

That is what we wished to show.

Step 5: Of course the curves ρ_j, for j distinct, are homotopically distinct. This follows from **Step 3** because ρ_j has index j.

Step 6: Combining **Step 4** and **Step 5**, we see that the homotopy classes of S^1 are indexed by the loops $\{\rho_j\}_{j \in \mathbb{Z}}$. In fact these curves are representatives of the equivalence classes. We also see that

$$[\rho_j] \cdot [\rho_k] = [\rho_{j+k}].$$

In conclusion, the group $\pi_1(S^1)$ is nothing other than $\mathbb{Z}$.

Remark 3.1.5 It is an interesting fact that if

$$S^2 = \{(x, y, z) \in \mathbb{R}^3 : x^2 + y^2 + z^2 = 1\}$$

is the unit sphere in Euclidean 3-space, then $\pi_1(S^2) = \{e\}$. In other words, the first homotopy of the unit sphere is trivial. To see this, note that if $\gamma : [0,1] \rightarrow S^2$ is any closed curve, then (it can be arranged, through a small deformation, that) the image of γ lies in a subset U of the sphere that is homeomorphic to a disc in the plane. Of course the disc is homotopically trivial, so the curve can be shrunk to a point.

Higher-dimensional spheres also have trivial homotopy.

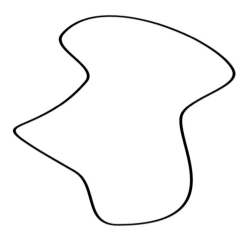

Figure 3.8: The idea of the Jordan curve theorem.

When a topological space X satisfies $\pi_1(X) = \{e\}$, then we say that X *is simply connected*. This standard terminology from topology will be used throughout the book.

We take this opportunity to mention a bellwether result of twentieth century topology. Known as the Jordan curve theorem, this result enunciates the intuitively obvious fact that a simple closed planar curve (i.e., a closed curve that does not intersect itself, as in Figure 3.8) divides the plane into two regions—a bounded region that it surrounds, and an unbounded region. See Figure 3.9, where this statement is less than obvious. Of course it should be noted that a simple closed curve (oriented counterclockwise) has index 1 with respect to any point in the interior (bounded) region—refer to Example 3.1.4, as well as Section 3.4 below, for the concept of index. That fact is critical to the proof of Jordan's result.

THEOREM 3.1.6 (Jordan curve theorem) *Let $c : [0,1] \rightarrow \mathbb{R}^2$ be a simple closed curve (i.e., a Jordan curve) in the plane. Thus $c(0) = c(1)$ and if $c(s) = c(t)$, then both s and t must be either 0 or 1. Then the complement of the image of c consists of two distinct connected components. One of these components is bounded (the interior) and the other is unbounded (the exterior). The image of c is the boundary of each component.*

The first to formulate a version of the theorem, observing that it required some proof, was Bernard Bolzano (1781–1848). The first proof was given by Camille Jordan (1838–1922), after whom the theorem is named (although

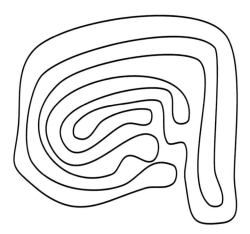

Figure 3.9: A curve for which Jordan's theorem is not obvious.

his proof was later found to be flawed). Oswald Veblen (1880–1960) finally gave a rigorous proof in 1905. Today there are a number of alternative proofs of the theorem, some of them simpler, but all still rather complicated. The reference [GAG] contains a particularly elegant proof. There are also computer verifications of the Jordan curve theorem.

Definition 3.1.7 Let X be a topological space and A a subspace. We call A a *retract* of X if there is a mapping $r : X \to A$ (called a *retraction*) such that $r(a) = a$ for every $a \in A$.

EXAMPLE 3.1.8 Let $S^1 = \{(x, y) \in \mathbb{R}^2 : x^2 + y^2 = 4\}$ be the circle of radius 2 and center the origin in the plane, and let $A = \{(x, y) \in \mathbb{R}^2 : 1 \le x^2 + y^2 \le 4\}$ be an annulus. Then of course $S^1 \subseteq A$, and we may define the mapping

$$
\begin{aligned}
r : A &\to S^1 \\
(x, y) &\mapsto \left(\frac{2x}{\sqrt{x^2 + y^2}}, \frac{2y}{\sqrt{x^2 + y^2}} \right).
\end{aligned}
$$

It is a simple matter to verify that r is a retract from A to S^1.

Lemma 3.1.9 *The fundamental group of the closed unit disc $\overline{D}$ is trivial.*

Proof: Since $\overline{D}$ can be continuously deformed to a point, it is clear that any loop in $\overline{D}$ is homotopic to any other. So the fundamental group $\pi_1(\overline{D}) = \{e\}$, the group with one element. $\square$

Proposition 3.1.10 *Let $\overline{D}$ be the closed unit disc in the plane with the usual Euclidean topology. Then the boundary circle S^1 is not a retract of $\overline{D}$.*

Proof: The lemma tells us that $\pi_1(\overline{D}) = \{e\}$, and we know that $\pi_1(S^1) = \mathbb{Z}$.

Now let $i : S^1 \to \overline{D}$ be the identity (or inclusion) map. Assume, seeking a contradiction, that $r : \overline{D} \to S^1$ is a retraction. We have the diagram

$$S^1 \xrightarrow{i} \overline{D} \xrightarrow{r} S^1 \, .$$

According to Theorem 3.1.3, this induces homomorphisms

$$\pi_1(S^1) \xrightarrow{i_*} \pi_1(\overline{D}) \xrightarrow{r_*} \pi_1(S^1)$$

or

$$\mathbb{Z} \xrightarrow{i_*} \{e\} \xrightarrow{r_*} \mathbb{Z} \, .$$

Since $r \circ i$ is the identity on S^1, the composition $r_* \circ i_*$ is supposed to be a group isomorphism on $\mathbb{Z}$. But it is impossible to factor such an isomorphism through the one-point-group $\{e\}$. That is a contradiction.   □

THEOREM 3.1.11 (Brouwer) *Let $\overline{D}$ be the closed unit disc in the plane and $f : \overline{D} \to \overline{D}$ a mapping. Then there must be some point $P = (x, y) \in \overline{D}$ such that $f(P) = P$.*

Proof: Suppose not. Then $f(P) \neq P$ for every $P \in \overline{D}$. We define a mapping $r : \overline{D} \to S^1$ as follows. If $P \in \overline{D}$ then consider the line segment that begins at $f(P)$, passes through P, and ends on S^1. That terminal point is $r(P)$. We see that $P \mapsto r(P)$ is a retraction of $\overline{D}$ to S^1. By the preceding proposition, we know that this is impossible. Thus f must have a fixed point.   □

Theorem 3.1.11 is a bellwether of twentieth century mathematics. It established the Dutchman L. E. J. Brouwer as a major figure in modern topology. Interestingly, he soon rejected his theorem and its proof because they were *nonconstructive*. He subsequently founded the school of intuitionism, which asserted that existence results should be established constructively (*not* by contradiction).

Let us say that two mappings $F : X \to Y$ and $G : X \to Y$ are *homotopic* if there is a mapping $\Lambda : X \times [0,1] \to Y$ such that

$$\begin{aligned} \Lambda(x,0) &= F(x) \\ \Lambda(x,1) &= G(x) \, . \end{aligned}$$

Two spaces X and Y are *homotopy equivalent* if there are mappings $f : X \to Y$ and $g : Y \to X$ such that $g \circ f$ is homotopic to the identity map id_X on X and $f \circ g$ is homotopic to the identity map id_Y on Y.

EXAMPLE 3.1.12 Let X be the closed unit disc in the plane and Y be the single point $\{0\}$ in the plane. Obviously these two spaces are *not* homeomorphic since there does not even exist a set-theoretic equivalence between them (one set has uncountably many elements while the other set has just one element). But the two spaces *are* homotopy equivalent.

To see this, let

$$\begin{aligned} f : X &\to Y \\ (x,y) &\mapsto (0,0) \, . \end{aligned}$$

And let

$$\begin{aligned} g : Y &\to X \\ (x,y) &\mapsto (x,y) \, . \end{aligned}$$

Then $g \circ f(x,y) \equiv (0,0)$, and this map is homotopic to the identity by way of the homotopy $\Gamma((x,y),t) = (tx,ty)$. A similar analysis applies to $g \circ f$.

In fact an argument as in the last example shows that Euclidean cells of any dimensions are homotopy equivalent. Here a cell in Euclidean space is just the set of points having distance less than or equal to 1 from the origin. A Euclidean 0-cell is a point, a Euclidean 1-cell is an interval, a Euclidean 2-cell is a disc, and so forth. All these are homotopy equivalent. For each is homotopy equivalent to a point by the argument just given, and the property of homotopy equivalence is transitive. We say that a space X is *contractible* if the identity map $\mathrm{id}_X : X \to X$ is homotopic to the trivial map of X to a point $x \in X$. The last example shows that the closed disc in the plane is contractible. It is easy to see that the fundamental group of a contractible space is trivial.

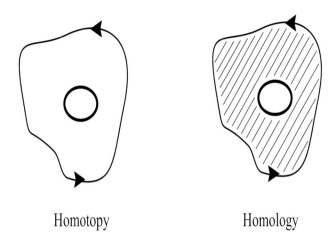

Homotopy   Homology

Figure 3.10: Homotopy vs. homology of $\mathbb{R}^2$ minus a disc.

EXAMPLE 3.1.13 The circle S^1 in the plane is *not* contractible, since it has nontrivial fundamental group. In fact no sphere of any dimension is contractible for a similar reason (i.e., any sphere has nontrivial homotopy of *some* dimension).

3.2   Homology Theory

Homology theory is another method for detecting "holes" in topological spaces. Whereas the fundamental obstruction in homotopy theory is whether a loop can be contracted to a point, the fundamental obstruction in homology theory is whether a loop bounds a cell or region. See Figure 3.10.

The two theories have a variety of interesting relationships, and they also contrast in notable ways. The homology group is always abelian, but the homotopy group in general is not. In fact we shall prove below that the first homology group is actually the abelianization of the first homotopy group. In many situations homology is easier to calculate than homotopy, and that fact is important in practice.

Homology theory was originally developed, in the 1920s, on simplices. The version of the theory that we present here (inspired by the exposition in [GRH]) is based on the idea of simplices, but in fact has a more general context.

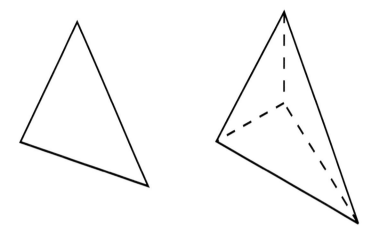

Figure 3.11: Simplices in dimensions 2 and 3.

3.2.1 Fundamentals

The entire theory of the homology of a space X is based on maps of classical Euclidean simplices into X. So we had better begin by defining Euclidean simplices and their related artifacts.

The canonical home for a simplex is real Euclidean space. Let $P_0, \ldots, P_m$ be points in $\mathbb{R}^N$ such that the vectors $\{\overrightarrow{P_0 P_j}\}_{j=1}^m$ are linearly independent. We say that the points $P_0, \ldots, P_m$ are *independent*. Then the set of points

$$S = \sum_{j=0}^m a_j P_j \,,$$

with all $a_j \geq 0$ and $\sum_j a_j = 1$ is called the *geometric simplex* spanned by $P_0, \ldots, P_m$. We call $(a_0, a_1, \ldots, a_m)$ the *barycentric coordinates* of the point S (with respect to the base points $P_0, \ldots, P_m$). We note that the geometric simplex so defined is also the *closed convex hull* of the points $P_0, \ldots, P_m$ (i.e., the intersection of all closed, convex sets containing these points). Figure 3.11 illustrates simplices in dimensions 2 and 3.

The natural morphisms on simplices are affine maps. A function or mapping $f : \mathbb{R}^M \to \mathbb{R}^N$ is called an *affine map* if

$$f(tP + (1-t)Q) = tf(P) + (1-t)f(Q)$$

for any points $P, Q \in \mathbb{R}^M$ and any real number t. An affine map sends lines to lines, and more generally affine objects to affine objects.[3]

In the space $\mathbb{R}^N$, consider the points

$$
\begin{aligned}
E_0 &= (0, 0, \ldots, 0) \\
E_1 &= (0, 1, \ldots, 0) \\
E_2 &= (0, 0, 1, \ldots, 0) \\
&\quad \cdots \\
E_{N-1} &= (0, 0, \ldots, 1, 0) \\
E_N &= (0, 0, \ldots, 1)
\end{aligned}
$$

The *standard (geometric) N-simplex* $\triangle_N$ is the geometric simplex spanned by $E_0, E_1, \ldots, E_N$.

Finally, if $P_0, \ldots, P_M$ are points in some Euclidean space $\mathcal{E}$, then $(P_0, \ldots, P_M)$ will denote the restriction to $\triangle_M$ of the unique affine map $\mathbb{R}^M \to \mathcal{E}$ that takes E_0 to P_0, E_1 to P_1, $\ldots$, E_M to P_M. In particular, $(E_0, E_1, \ldots, E_M)$ denotes the identity map on $\triangle_M$. We denote this last map by i_M.

3.2.2   Singular Homology

Now let X be a topological space. A *singular N-simplex* in X is a continuous map $\varphi : \triangle_N \to X$. We see that

- For $N = 0$, the singular N-simplex is a point in X.

- For $N = 1$, the singular N-simplex is the image of a line segment in X.

- For $N = 2$, the singular N-simplex is the image of a Euclidean triangle (and its interior) in X.

- Etc.

Our aim is to set up a formal calculus on the singular N-simplices. Let R be a commutative ring with unit (for us this ring will usually be either $\mathbb{Z}$ or $\mathbb{R}$). Our formal calculus will entail (formal) addition and (formal) scalar multiplication by elements of R.

_____

[3]For those who know some linear algebra, an affine map is very much like a linear map—except that we no longer require that the origin go to the origin.

We will be dealing here with a mathematical construct called a *module*. A module is like a vector space, except that the coefficients come from a ring rather than from a field (see Appendix 5). All the other (familiar) properties are analogous to what you have seen in your study of vector spaces in linear algebra.

Definition 3.2.1 Let $S_N(X)$ be the free R-module generated by all the singular N-simplices in X. We call the elements of S_N *chains*. A typical element of $S_N(X)$ is a formal (finite) linear combination

$$\sum_\sigma \nu_\sigma \sigma\,, \qquad\qquad (3.2.1.1)$$

where σ is an index over the set of all N-simplices and the coefficients ν_σ come from the ring R.

We stress that this is a *formal* sum. The only way that such a sum can vanish is if $\nu_\sigma = 0$ for every σ. We call such a sum a *singular N-chain*. Chains are the basic units of singular homology theory.

We add two formal sums in the obvious way by

$$\sum_\sigma \nu_\sigma^1 \sigma + \sum_\sigma \nu_\sigma^2 \sigma = \sum_\sigma (\nu_\sigma^1 + \nu_\sigma^2)\sigma$$

and we perform scalar multiplication by

$$c \cdot \left(\sum_\sigma \nu_\sigma \sigma \right) = \sum_\sigma (c\nu_\sigma)\sigma\,.$$

Definition 3.2.2 If $N \geq 1$ and $0 \leq j < N$, then we define $F_N^j : \triangle_{N-1} \to \triangle_N$ to be the affine map that sends

$$(E_0, E_1, \ldots, E_{N-1}) \longmapsto (E_0, E_1, \ldots, E_{j-1}, E_{j+1}, \ldots, E_N)\,.$$

[One must learn to read this correctly: We are sending E_0 to E_0, E_1 to E_1, ..., E_{j-1} to E_{j-1}, E_j to E_{j+1}, E_{j+1} to E_{j+2}, ..., E_{N-1} to E_N.]

EXAMPLE 3.2.3 Let $N = 2$ and let us calculate F_N^0, F_N^1, and F_N^2. We see that

$$F_2^0 : (E_0, E_1) \longmapsto (E_1, E_2)\,,$$

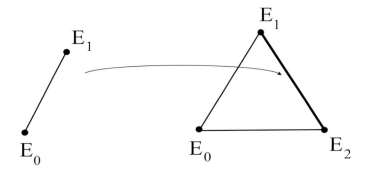

Figure 3.12: The idea of F_2^0.

$$F_2^1 : (E_0, E_1) \longmapsto (E_0, E_2),$$
$$F_2^2 : (E_0, E_1) \longmapsto (E_0, E_1).$$

It is important to see the geometric interpretation of these mappings. The affine map F_2^0 sends the singular 1-simplex (a line segment) to the edge of the singular 2-simplex (a triangle) that is opposite to E_0 (Figure 3.12). The affine map F_2^1 sends the singular 1-simplex (a line segment) to the edge of the singular 2-simplex (a triangle) that is opposite to E_1. The affine map F_2^2 sends the singular 1-simplex (a line segment) to the edge of the singular 2-simplex (a triangle) that is opposite to E_2.

Now if σ is any singular N-simplex in the topological space X, then we define the jth *face* $\sigma^{(j)}$ of σ to be the singular $(N-1)$-simplex $\sigma \circ F_N^j$. Thus, in particular, F_N^j is the jth face of i_N. In the special case that $\sigma = (P_0, \ldots, P_N)$ is a simplex in Euclidean space then we have that

$$\sigma^{(j)} = (P_0, P_1, \ldots, P_{j-1}, P_{j+1}, \ldots, P_N).$$

We may summarize these ideas by noting that F_N^j maps $\triangle_{N-1}$ homeomorphically and affinely onto the face of $\triangle_N$ that is opposite to the vertex E_j. Again see Figure 3.12.

Now the essential operation in any homology (or cohomology) theory is the boundary operator. That is what we now define. Some intuition is helpful here. A region (in the plane) that has no holes will have a simple boundary consisting of a single curve. A planar region with holes will have a boundary with several components or pieces. We will need to take care to orient these components properly so that the algebraic aspects of our theory work out as they should.

Let σ be a singular N-simplex and set

$$\partial(\sigma) = \sum_{j=0}^{N} (-1)^j \sigma^{(j)}.$$

This is the *boundary* of the simplex σ. Notice that this boundary is the formal sum of simplices of one lower dimension (which is what we would expect). For instance, the boundary of a 2-simplex (which is a topological triangle together with its interior) should be the union of segments (which are 1-simplices). We extend ∂ to a module homomorphism by linearity:

$$\partial : S_N(X) \;\rightarrow\; S_{N-1}$$
$$\sum \nu_\sigma \sigma \;\mapsto\; \sum \nu_\sigma \partial(\sigma).$$

When $N = 0$, the boundary of a 0-chain is defined to be 0.

We notice that the components of the boundary of a simplex are defined with *signs*, or orientations. Good sense can be made of this convention by looking at an example:

EXAMPLE 3.2.4 Consider the standard 2-simplex $\sigma = \triangle_2$. Then

$$\partial(\sigma) = \sum_{j=0}^{2} (-1)^j \sigma^{(j)} = \sum_{j=0}^{2} (-1)^j \sigma \circ F_2^j = (E_1, E_2) - (E_0, E_2) + (E_0, E_1).$$

Figure 3.13 shows how the signs guarantee that the algebraically specified boundary traverses the geometric boundary of the figure counterclockwise (i.e., $-(E_0, E_2)$ is nothing other than (E_2, E_0), etc.).

Proposition 3.2.5 *We have that $\partial \circ \partial = 0$.*

Proof: By linearity and naturality, it suffices to prove that

$$\partial(\partial(c)) = 0$$

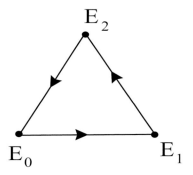

Figure 3.13: The boundary of $\triangle_2$.

when c is a singular N-simplex. Now

$$\partial(\partial(c)) = \sum_{j=0}^{N}(-1)^j\partial(c^{(j)})$$

$$= \sum_{j=0}^{N}(-1)^j\sum_{\ell=0}^{N-1}(-1)^\ell(c \circ F_N^j) \circ F_{N-1}^\ell$$

$$= \sum_{\ell<j=1}(-1)^{j+\ell}c \circ \left(F_N^\ell F_{N-1}^{j-1}\right) + \sum_{0=j\leq\ell}(-1)^{j+\ell}c \circ \left(F_N^j F_{N-1}^\ell\right).$$

Now we must note that if we set $j' = \ell$ and $\ell' = j - 1$ in the first sum, then it becomes

$$\sum_{j'\leq\ell'=0}^{N-1}(-1)^{j'+\ell'+1}c \circ \left(F_N^{j'} F_{N-1}^{\ell'}\right).$$

Plainly this cancels with the second sum on the far right of the above display. So the final result is 0, as required.   □

Now we have all the elements in place so that we can define homology. This will require three definitions in sequence.

Definition 3.2.6 Any singular N-chain c such that $\partial(c) = 0$ is called a *cycle*. The collection of all cycles is denoted by Z_N, and these form a module.

Definition 3.2.7 If c is a cycle and $c = \partial(c')$ for some other $(N + 1)$-chain c', then c is called a *boundary*. We denote the collection of boundaries by B_N, and B_N forms a module.

Notice that, by the last proposition, any boundary is a cycle. In fact the set of boundaries B_N forms a *submodule* of the cycles Z_N.

Definition 3.2.8 The quotient Z_N/B_N is called the Nth *singular homology module* $H_N(X; R)$ of the space X.

We note in passing that two cycles whose difference is a boundary are called *homologous*. Clearly these are two chains which are equal in the quotient.

Notice that the homology module depends on the index N, the topological space X, and the ring of coefficients R. Often, in context, the ring of coefficients is understood. So we shall write simply $H_N(X)$.

EXAMPLE 3.2.9 Let X be the topological space consisting of a single point. Then, for each dimension N, there is just one singular N-simplex σ_N. Also

$$\partial(\sigma_N) = \begin{cases} \sigma_{N-1} & \text{if} \quad N \text{ even} \\ 0 & \text{if} \quad N \text{ odd} . \end{cases}$$

Therefore

$$Z_N = B_N = \begin{cases} 0 & \text{if} \quad N > 0 \text{ even} \\ S_N & \text{if} \quad N \text{ odd} . \end{cases}$$

Thus $H_N = 0$ for all $N > 0$.

Note, however, that $Z_0 = S_0$ (the collection of all the 0-chains) while $B_0 = 0$. We conclude that $H_0 = R$; the isomorphism here is $\nu\sigma_0 \mapsto \nu$.

EXERCISE FOR THE READER 3.2.10 Calculate the homology of X the closed unit disc in the plane. The answers should be the same as in the last example, but the calculations a bit more sophisticated.

EXAMPLE 3.2.11 Consider the chain X shown in Figure 3.14. Let us consider the homology ring $H_1(X)$ with coefficients in $\mathbb{Z}$. There are two types of cycles in X, and these are illustrated in Figure 3.15. Note that neither of these 1-chains has itself a boundary, and that is why they are cycles. But the first cycle (on the left) is itself a boundary—it bounds a 2-chain in X. So it is a boundary. But the second cycle (on the right) does not bound a chain in X. It is *not* a boundary.

Thus the collection Z_1 of cycles has two generators, and the collection B_1 of boundaries has one generator. The quotient then has a single generator

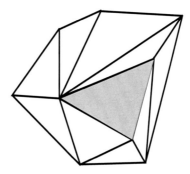

Figure 3.14: A chain with hole shaded in grey.

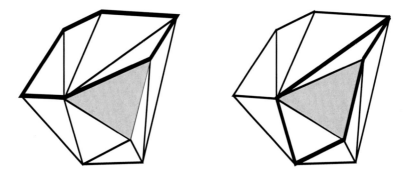

Figure 3.15: Two cycles in the chain (left is a boundary, right is not).

σ^* (the image of the second cycle). Since the coefficient ring is $\mathbb{Z}$, we then see that the quotient $H_1 = Z_1/B_1$ is simply the collection of all $\nu\sigma^*$ with $\nu \in \mathbb{Z}$, and this collection is canonically isomorphic to $\mathbb{Z}$ itself.

EXERCISE FOR THE READER 3.2.12 Calculate H_2 for the chain in the last example.

If X is a topological space, then we may define an equivalence relation on X by the condition that $x_1 \sim x_2$ if there is a path connecting x_1 to x_2. The resulting equivalence classes are called *path components* of X.

In fact H_0 of a topological space counts the number of path components in the space. We shall formulate this assertion more rigorously in the discussion that follows.

EXAMPLE 3.2.13 Let X be the circle S^1. We shall omit many details in the discussion that follows, and let the interested reader fill them in as desired.

For Z_1, there is the path σ_1 that circles S^1 once counterclockwise. And then there is the path σ_2 that circles twice counterclockwise; and, more generally, the path σ_k that circles S^1 a total of k times counterclockwise. [We classified these in our discussion of the homotopy of S^1 in the last section.] The only one of these that can be a boundary is σ_1. Of course there are also the paths that circle clockwise, but they are just the negatives of the σ_j. Thus Z_1 identifies naturally with $\mathbb{Z}$, and B_1 is just the group with one element. So H_1, the quotient of these, is isomorphic to $\mathbb{Z}$.

It is not difficult to see that any element of $S_2(X)$ or, more generally, $S_j(X)$ for $j \geq 2$, will be topologically trivial—deformable to a point. We conclude that $H_j = 0$ for $j \geq 2$. As usual, $H_0 = R$ since S^1 has just one path component.

Proposition 3.2.14 *Let $\{X_k\}$ be the path components of the topological space X. Then, for any $j \geq 0$,*

$$H_j(X) \equiv \oplus_k H_j(X_k).$$

This proposition tells us that, in order to calculate the homology of a space, we may examine each path component separately.

Proof of the Proposition: There is an obvious isomorphism

$$S_j(X) \equiv \oplus_k S_j(X_k)$$

for any $j \geq 0$. And the boundary operator ∂ acts component-by-component.

Observe that, since $\triangle_j$ is path-connected, any singular j-simplex σ maps $\triangle_j$ into some one of the path components X_k. Hence each j-chain c decomposes naturally and uniquely into a sum

$$c = \sum_k c_k \,,$$

where c_k is a singular j-chain in X_k.   □

The next result validates an assertion that we made, informally, earlier.

Proposition 3.2.15 *The module $H_0(X)$ is a free R-module on as many generators as there are path-components of the space X.*

Proof: We may, by the last result, assume that X is path-connected. Let $x_0 \in X$ be a base point. For any $x \in X$, let σ_x be a path from x_0 to x. Therefore $\partial(\sigma_x) = x - x_0$.

Now consider a 0-chain

$$c = \sum_x \nu_x x \,.$$

We assert that c is a boundary if and only if the sum of its coefficients is equal to 0. In case the sum is 0, then

$$c = \sum_x \nu_x x - \left(\sum_x \nu_x\right)x_0 = \partial\left(\sum_x \nu_x \sigma_x\right).$$

So c is a boundary. The converse statement is immediate.

Now we see that every 0-chain is a cycle. The map that sends c onto the sum of its coefficients is a homomorphism of S_0 onto R with kernel B_0. Therefore

$$H_0(X) \equiv R \,.$$ □

Definition 3.2.16 Now we define a slight modification of the homology construct. The purpose of this variant is to address certain anomalies that occur at the 0-level.

Define a new boundary operator on 0-chains by

$$\partial^{\#}\left(\sum_{x}\nu_{x}x\right)=\sum_{x}\nu_{x}\,.$$

Then certainly $\partial^{\#}\partial=0$. We set

$$H_0^{\#}(x)=\ker(\partial^{\#})/B_0\,.$$

Note that, if X is path-connected, then $H_0^{\#}(X)=0$. On the other hand, if X has r path components, $r>1$, then $H_0^{\#}(X)$ is a free module on $(r-1)$ generators.

For $j>0$ we set $H_j^{\#}(X)=H_j(X)$. This reinforces our earlier statement that the $H^{\#}$ homology theory is a modification only at the 0-level.

Now it is time to discuss the functorial properties of homology. This is what makes the idea important. Let $f:X\to X'$ be a continuous mapping. Let σ be a singular j-simplex in X. Then $f\circ\sigma$ is a singular j-simplex in X'. Thus we have a homomorphism

$$\begin{array}{ccc}S_j(f):S_j(X)&\to&S_j(X')\\[4pt]\displaystyle\sum_{\sigma}\nu_{\sigma}\sigma&\longmapsto&\displaystyle\sum\nu_{\sigma}(f\circ\sigma)\,.\end{array}$$

We see that (in case $g:X'\to X''$ is another mapping)

(i) $S_j(\mathrm{id})=\mathrm{id}$,

(ii) $S_j(g\circ f)=S_j(g)\circ S_j(f)$.

Of course it follows that, in case f is a homeomorphism, we have

$$S_j(f^{-1})=[S_j(f)]^{-1}\,.$$

Thus a homeomorphism f of topological spaces induces an isomorphism $S_j(f)$ of modules.

Lemma 3.2.17 *We have that*

$$\partial S_j(f) = S_{j-1}(f)\partial.$$

Proof: The result follows immediately from the simple identity

$$(f \circ \sigma) \circ F_j^{\ell} = f \circ (\sigma \circ F_j^{\ell}).$$ □

In sum, if z is a j-cycle on the topological space X and if $\overline{z}$ is its homology class, then we have the homomorphism

$$H_j(f) : H_j(X) \to H_j(X')$$

defined by

$$H_j(f)(\overline{z}) = \overline{S_j(f)(z)}.$$

Since **(i)** and **(ii)** above also hold for $H_j(f)$, we conclude that H_j takes topological spaces to R-modules, and does so in a natural way. In particular, H_j *is a topological invariant.*

We note in passing that $S_j(\sigma)(i_j) = \sigma$, and the reader may verify this assertion as an exercise.

Since we formulate all the ideas here in terms of chains, it is natural to wonder when a space can be triangulated—that is, broken up in a natural way into "triangles." This is a subtle matter—especially in higher dimensions. We refer the reader to [ARM2] for some of the details of this matter.

3.2.3   Relation to Homotopy

In some sense homotopy is a stronger invariant than homology. The next result suggests what this means in a concrete sense.

THEOREM 3.2.18 *Let $f : X \to X'$ and $g : X \to X'$ be homotopic maps. Then, for every $j \geq 0$, the induced homomorphisms $H_j(f)$ and $H_j(g)$ on the homology modules are equal.*

Proof: The proof will employ an interesting and useful new tool called the chain homotopy.

Begin by defining $\lambda_t : X \to X \times I$ by $\lambda_t(x) = (x, t)$. Now λ_t gives a homotopy between λ_0 and λ_1. Let $\Gamma : X \times I \to X'$ be the homotopy from f

to g. So $\Gamma \circ \lambda_0 = f$ and $\Gamma \circ \lambda_1 = g$. If we can prove the result for λ_0 and λ_1, then the general result follows; for

$$H_j(f) = H_j(\Gamma \circ \lambda_0) = H_j(\Gamma)H_j(\lambda_0) = H_j(\Gamma)H_j(\lambda_1) = H_j(g) \,.$$

So we concentrate on λ_0 and λ_1. We construct a homomorphism

$$P : S_j(X) \to S_{j+1}(X \times I)$$

so that

$$\partial P + P\partial = S_j(\lambda_1) - S_j(\lambda_0) \,.$$

This last relationship is called a *chain homotopy* on $S_j(\lambda_0)$ and $S_j(\lambda_1)$. Now, for any j-cycle z, $P\partial z = 0$. Thus $S_j(\lambda_1)(z)$ and $S_j(\lambda_0)(z)$ differ by ∂Pz; that is to say, they are homologous.

It remains for us to construct P. We call P the *prism operator*. We want P to have this property: For any mapping $h : Y \to X$ (for some space Y), we ask that this diagram be commutative:

$$
\begin{array}{ccc}
S_j(Y) & \xrightarrow{\ P\ } & S_{j+1}(Y \times I) \\
{\scriptstyle S_j(h)} \downarrow & & \downarrow {\scriptstyle S_{j+1}(h) \times \mathrm{id}} \\
S_j(X) & \xrightarrow{\ P\ } & S_{j+1}(X \times I)
\end{array}
$$

If σ is a singular j-simplex in X, then we have

$$P(\sigma) = S_{j+1}(\sigma \times \mathrm{id})P(i_j) \,. \tag{3.2.18.1}$$

Conversely, this last equation implies the commutative diagram above. So it all comes down to the need to define $P(i_j)$.

Observe that $\triangle_j \times I$ has vertices $A_0 = (E_0, 0)$, $A_1 = (E_1, 0)$, ..., $A_j = (E_j, 0)$ and $B_0 = (E_0, 1)$, $B_1 = (E_1, 1)$, ... $B_j = (E_j, 1)$. Set

$$P(i_j) = \sum_{\ell=0}^{j} (-1)^\ell (A_0 \cdots A_\ell B_\ell \cdots B_j) \,.$$

Now we must apply both sides of the chain homotopy to i_j. On the right-hand side we obtain

$$(B_0 \cdots B_j) - (A_0 \cdots A_j) \,.$$

For the left-hand side we have

$$\partial(PI_j) \;=\; \sum_{\ell=0}^{j} (-1)^{\ell} \partial(A_0 \cdots A_\ell B_\ell \cdots B_j)$$

$$=\; \sum_{m\le\ell=0}^{j} (-1)^{\ell+m}(A_0 \cdots A_{m-1}A_{m+1} \cdots A_\ell B_\ell \cdots B_j)$$

$$+\; \sum_{0\le\ell\le m} (-1)^{\ell+m+1}(A_0 \cdots A_\ell B_\ell \cdots B_{m-1}B_{m+1} \cdots B_j)\,.$$

Notice that the terms with $\ell = m$ all cancel except for $(B_0 \cdots B_j) - (A_0 \cdots A_j)$, so we are left with

$$\sum_{m<\ell=1}^{\ell=j} (-1)^{\ell+m}(A_0 \cdots A_{m-1}A_{m+1} \cdots A_\ell B_\ell \cdots B_j)$$

$$+\; \sum_{\ell<m=1}^{m=j} (-1)^{\ell+m+1}(A_0 \cdots A_\ell B_\ell \cdots B_{m-1}B_{m+1} \cdot B_j)\,. \qquad (3.2.18.2)$$

Now we have

$$P(\partial i_j) = \sum_{p=0}^{j} (-1)^p P(F_j^p)\,.$$

Using (3.2.18.1) with $\sigma = F_j^p$, we get the *negative* of (3.2.18.2), since

$$S_j(F_j^p \times \mathrm{id})(A_0 \cdots A_k B_k \cdots B_{j-1})$$

$$= \begin{cases} (A_0 \cdots A_k B_k \cdots B_{p-1}B_{p+1} \cdots B_j) & \text{if} \quad p > k \\ (A_0 \cdots A_{p-1}A_{p+1} \cdots A_{k+1}B_{k+1} \cdots B_j) & \text{if} \quad p \le k\,. \end{cases}$$

That completes the proof. $\square$

EXERCISE FOR THE READER 3.2.19 Suppose that σ is a loop with $\sigma(0) = \sigma(1) = x_0$. Sketch a picture to show $P\sigma$.

Now our key result to relate homology and homotopy is the following. We specialize down to the ring R being the integers $\mathbb{Z}$. For simplicity, we shall write $H_1(X)$ for $H_1(X;\mathbb{Z})$. In this context it is convenient to think of $H_1(X)$ as a group under addition.

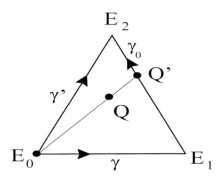

Figure 3.16: Construction of a singular 2-simplex.

THEOREM 3.2.20 *Let X be a topological space and $x_0 \in X$ a fixed point. There is a homomorphism $\chi : \pi_1(X, x_0) \to H_1(X)$ such that the homotopy class of a loop γ is sent into the homology class of the singular 1-simplex γ. If X is path-connected, then χ is surjective; also the kernel of the homomorphism is the commutator subgroup of $\pi_1(X, x_0)$.*

Corollary 3.2.21 *If X is path-connected, then χ is an isomorphism if and only if the fundamental group of X is commutative.*

Remark 3.2.22 Of course it is immediate from the theorem that if C is the kernel of the homomorphism (the commutator subgroup), then $\pi_1/C \cong H_1$. It is in this sense that the first homology group is the abelianization of the first homotopy group.

Proof of Theorem 3.2.20: Let γ, γ' be loops at x_0 and suppose that F is a homotopy of γ to γ'. Define σ a singular 2-simplex in X as follows:

- First set $\sigma(E_0) = x_0$.

- If $Q \in \triangle_2$ is any point other than E_0, then consider the line through E_0 and Q. That line meets the edge of $\triangle_2$ opposite E_0 in a unique point Q'. See Figure 3.16.

- We write

$$Q' = tE_2 + (1-t)E_1 \qquad \text{and} \qquad Q = sQ' + (1-s)E_0 \,. \qquad (3.2.20.1)$$

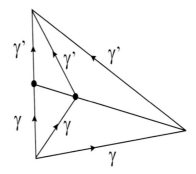

Figure 3.17: The mapping χ is a homomorphism.

- Now set $\sigma(Q) = F(s, t)$ for the choice of s and t indicated in equation (3.2.20.1).

Now σ is continuous because $\triangle_2$ is the quotient space of $I^2 \equiv I \times I$ under the mapping

$$(s, t) \mapsto \begin{cases} Q & \text{if} \quad s \neq 0 \\ E_0 & \text{if} \quad s = 0. \end{cases}$$

Also, if γ_0 is the constant loop at x_0, we have

$$\partial(\sigma) = \gamma - \gamma' + \gamma_0.$$

Of course γ_0 is the boundary of the trivial 2-simplex at x_0, so $\gamma \sim \gamma'$ and χ (as defined in the statement of the theorem) is well defined.

To see that the mapping χ is indeed a homomorphism, we examine the diagram given in Figure 3.17.

Define a singular 2-simplex σ such that

$$\partial(\sigma) = \gamma + \gamma' - \gamma\gamma'.$$

Here $\gamma\gamma'$ denotes the usual product of loops.

This formula makes sense any time $\gamma\gamma'$ is defined. If now X is path-connected, then let $z = \sum_j \nu_j(\alpha_j)$ be a 1-cycle, so that

$$0 = \sum \nu_j(\alpha_j(1) - \alpha_j(0)) = \partial(z).$$

We see, after collecting terms, that all coefficients are equal to 0. Choose paths η_{j0} from x_0 to $\alpha_j(0)$ and η_{j1} from x_0 to $\alpha_j(1)$ so that the paths depend only on the vertices, not on the indexing. Collecting terms, we see that

$$0 = \sum_j \nu_j(\eta_{j1} - \eta_{j0}).$$

If we set $\beta_j = \eta_{j0} + \alpha_j - \eta_{j1}$, we find that

$$z = \sum_j \nu_j \beta_j.$$

If γ_j is the loop $\eta_{j0}\alpha_j\eta_{j1}^{-1}$, we find that

$$\chi\left[\prod_j \gamma_j^{\nu_j}\right] = \overline{z}.$$

Thus χ is surjective.

Now let γ be any loop that is homologous to 0. We may write

$$\gamma = \partial\left(\sum_j \nu_j \sigma_j\right).$$

Write $\partial(\sigma_j) = \alpha_{j0} - \alpha_{j1} + \alpha_{j2}$. After collecting terms in the sum

$$\sum_j(\alpha_{j0} - \alpha_{j1} + \alpha_{j2}),$$

we see that γ occurs with coefficient 1. All other paths have coefficient 0. Again now choose paths $\eta_{j\ell}$ with $\ell = 0, 1, 2$ from x_0 to $\alpha_{j2}(0)$, $\alpha_{j0}(0)$, $\alpha_{j1}(1)$, respectively, so that they depend only on the vertices and not on the indexing. Also select the constant path to the vertex x_0. See Figure 3.18.

Now consider these loops based at x_0:

$$\begin{aligned}
\beta_{j0} &= \eta_{j1}\alpha_{j0}\eta_{j2}^{-1} \\
\beta_{j1} &= \eta_{j0}\alpha_{j1}\eta_{j2}^{-1} \\
\beta_{j2} &= \eta_{j0}\alpha_{j2}\eta_{j1}^{-1}
\end{aligned}$$

We have

$$\beta_j = \beta_{j0}\beta_{j1}^{-1}\beta_{j2} \approx \eta_{j1}\alpha_{j0}\alpha_{j1}^{-1}\alpha_{j2}\eta_{j1}^{-1} \approx x_0 \ \operatorname{rel}(0,1),$$

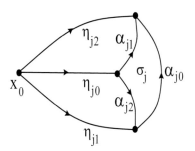

Figure 3.18: Properties of the mapping χ.

where the "rel" notation simply means that we have a homotopy from $t = 0$ to $t = 1$. Thus

$$\prod_j [\beta_j]^{\nu_j} = 1 \,.$$

Now let $\prod_j'$ be the quotient group of $\prod_j(X, x_0)$ by its commutator subgroup. Let $\widetilde{\beta}_j$ be the coset of $[\beta_j]$ in this quotient. Then we have

$$\prod_j \widetilde{\beta}_j^{\nu_j} = 1 \,.$$

Since $\prod_1'$ is commutative, we can collect terms as before and find that

$$\prod_j \widetilde{\beta}_j^{\nu_j} = \widetilde{\gamma} \,.$$

Hence $[\gamma]$ belongs to the commutator subgroup.

Finally, since $H_1(X)$ is commutative, the kernel of χ must contain the commutator subgroup. That completes the proof.   □

EXERCISE FOR THE READER 3.2.23 Let X be the figure 8, illustrated in Figure 3.19. Verify that its fundamental group is the free group on 2 generators. In particular, *this fundamental group is not abelian.* Check that our theorem tells us that $H_1(X)$ must be the free *abelian group* on 2 generators— in other words, it must be $\mathbb{Z} \times \mathbb{Z}$.

Remark 3.2.24 It is difficult to say anything useful about the homomorphism χ when $j > 1$.

Figure 3.19: The figure 8 space.

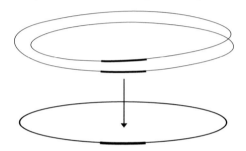

Figure 3.20: The idea of a covering space.

3.3 Covering Spaces

The idea of a covering space is this: a topological space X is given. We produce another space $\widehat{X}$ and a map

$$\pi : \widehat{X} \longrightarrow X$$

which is locally trivial in a certain sense. Refer to Figure 3.20. Often the space $\widehat{X}$ is topologically simpler than is X. The upshot of this construction is that one can come to understand X by instead studying π and $\widehat{X}$.

Definition 3.3.1 A mapping $\pi : \widehat{X} \to X$ is called a *covering map* and $\widehat{X}$ is said to be a *covering space* if this condition holds:

> For each $x \in X$ there is a neighborhood V of x so that $\pi^{-1}(V)$ decomposes into pairwise disjoint open sets $\{U_\alpha\}_{\alpha \in A}$ such that the restriction of π to each U_α is a homeomorphism of U_α onto V.

If $x \in X$, then it is common to refer to the set $\pi^{-1}(X)$ as the *fiber* over X.

EXAMPLE 3.3.2 Let $X = S^1$ be the unit circle in the plane as usual and let $\widehat{X} = \mathbb{R}$, the real line. We define the mapping

$$\pi : \widehat{X} \longrightarrow X$$
$$t \longmapsto (\cos 2\pi t, \sin 2\pi t).$$

Then $\widehat{X}$ is a covering space over X and π is a covering map. To see this, pick a point $x \in S^1$. We may as well assume, after applying a rotation, that $x = (1, 0)$. Let $V = \{(\cos s, \sin s) : s \in (-\pi/2, \pi/2)\}$. Then

$$\pi^{-1}(V) = \bigcup_{j=-\infty}^{+\infty} (-1/4 + j, 1/4 + j).$$

So the inverse image of V under π decomposes into the pairwise disjoint open sets $U_j = (-1/4 + j, 1/4 + j)$. And π maps each U_j homeomorphically onto V as required.

It is sometimes convenient to think of S^1 as living in the complex plane, and then to write the covering mapping as

$$t \longmapsto e^{2\pi i t}.$$

[Of course $e^{2\pi i t} = (\cos 2\pi t, \sin 2\pi t)$.] We shall do so in what follows without comment.

Now we are going to learn about the relationship between the homotopy of X and the homotopy of $\widehat{X}$. This is where the power of covering space theory comes from. Choose a base point $x \in X$ and corresponding $\widehat{x} \in \widehat{X}$ so that $\pi(\widehat{x}) = x$ (of course the choice of $\widehat{x}$ is *not* unique). Write $G = \pi_1(X, x)$ and $H = \pi_1(\widehat{X}, \widehat{x})$. Our goal, naturally, is to relate G to H.

Lemma 3.3.3 *If γ is a path in X which begins (but does not necessarily end) at x, then there is a unique path $\widehat{\gamma}$ in $\widehat{X}$ which begins at $\widehat{x}$ and satisfies $\pi \circ \widehat{\gamma} = \gamma$. We call $\widehat{\gamma}$ a lift of γ.*

Proof: We prove the result by a connectivity argument. Let

$$S = \{t \in [0, 1] : \text{ the path } \gamma \text{ restricted to the interval}$$
$$[0, t] \text{ can be lifted to a path in } \widehat{X}\}.$$

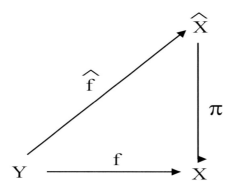

Figure 3.21: The lift of a map.

We shall show that S is nonempty, open, and closed.

Of course S is nonempty, for $0 \in S$ (the trivial curve $\gamma(0) \equiv x$ can be lifted to the trivial curve $\widehat{\gamma}(0) \equiv \widehat{x}$). To see that S is open, let $t_0 \in S$. Define $p_0 = \gamma(t_0)$. Then, by the definition of covering map, there is an open neighborhood V of p_0 so that $\pi^{-1}(V)$ is a pairwise disjoint union of open sets U_α and π is a homeomorphism of each U_α onto V. Choose that U_{α_0} that contains the point of the curve $\widehat{\gamma}$ that projects under π to p_0. Let g be the inverse of π from V to U_{α_0}. There is a segment $[t_0, t_1]$ so that $\gamma([t_0, t_1])$ lies in V. Hence $g \circ \gamma$ makes sense on $[t_0, t_1]$ and extends the lifted curve to the interval $[0, t_1]$. We conclude that S is open.

We verify that S is closed by showing that its complement is open. So suppose that $s_0 \in [0, 1] \setminus S$. Certainly $\gamma(s_0)$ lies in some neighborhood V with the property that $\pi^{-1}(V)$ is a pairwise disjoint union of open sets U_α with π a homeomorphism of each U_α onto V. Surely there will be an $\epsilon > 0$ so that the image of the interval $(s_0 - \epsilon, s_0 + \epsilon)$ under γ lies in V. But then it follows, just as in the preceding paragraph, that none of the points in that interval can lie in S (for if they did then the preceding argument would show that s_0 lies in S). Thus the complement of S is open.

We conclude by connectivity that $S = [0, 1]$ and the entire curve γ can be lifted. $\square$

Given $f : Y \to X$, a new map $\widehat{f} : Y \to \widehat{X}$ with the property that $\pi \circ \widehat{f} = f$ is called a *lift* of f. See Figure 3.21, which suggests the idea of a lift. The fact that we can lift paths and homotopies up to the covering space is of central significance, as we shall now see.

Lemma 3.3.4 *If* $\Gamma : I \times I \to X$ *is a map so that* $\Gamma(0,t) = \Gamma(1,t) = x$ *for* $0 \leq t \leq 1$, *then there is a unique* $\widehat{\Gamma} : I \times I \to \widehat{X}$ *such that* $\pi \circ \widehat{\Gamma} = \Gamma$ *and* $\widehat{\Gamma}(0,t) = \widehat{x}$ *for all* $0 \leq t \leq 1$.

Proof: The proof is similar to that of Lemma 3.3.3. For each fixed s, use a connectivity argument in the t variable to see that the homotopy can be lifted for all $0 \leq t \leq 1$. Now use a connectivity argument in s. $\square$

THEOREM 3.3.5 *The induced homomorphism* $\pi_* : H \to G$ *is one-to-one.*

Proof: Let $\widehat{\alpha}$ be a loop in $\widehat{X}$ which is based at $\widehat{x}$ and so that $\alpha \equiv \pi \circ \widehat{\alpha}$ is null homotopic in X. Select a specific homotopy Γ from the constant loop at x to α and apply the homotopy lifting Lemma 3.3.3 to find a $\widehat{\Gamma} : I \times I \to \widehat{X}$ such that $\pi \circ \widehat{\Gamma} = \Gamma$ and $\widehat{\Gamma}(0,t) = \widehat{x}$ for all $0 \leq t \leq 1$.

Let $\mathcal{P}$ denote the union of the left- and right-hand edges of $I \times I$ together with the bottom edge. Then Γ sends all of $\mathcal{P}$ to x. But

- $\pi \circ \widehat{\Gamma} = \Gamma$;

- the set $\pi^{-1}(x)$ is a discrete set of points;

- the set $\mathcal{P}$ is connected.

Thus $\widehat{\Gamma}$ maps all of $\mathcal{P}$ to $\widehat{x}$.

In addition, the path in $\widehat{X}$ defined by $\widehat{\Gamma}(s,1)$ is a lift of α which begins at $\widehat{x}$. It must therefore, by the uniqueness part of 3.3.3, be $\widehat{\alpha}$. We conclude that $\widehat{\Gamma}$ is a homotopy from the constant loop at $\widehat{x}$ to $\widehat{\alpha}$ as we wished to prove. $\square$

The following result gives a direct link between the topology of the base space X and the topology of the total space $\widehat{X}$.

THEOREM 3.3.6 *A loop* $\alpha \in X$ *based at* x *lifts to a loop* $\widehat{\alpha}$ *in* $\widehat{X}$ *based at* $\widehat{x}$ *if and only if the homotopy class of* α *lies in* $\pi_*(H)$.

Proof: Certainly one implication is obvious: If $\widehat{\alpha}$ is a loop upstairs, then (using brackets $\langle \ \rangle$ to denote the homotopy class) $\langle \alpha \rangle = \langle \pi \circ \widehat{\alpha} \rangle \in \pi_*(H)$.

For the converse direction, suppose that $\langle \alpha \rangle \in \pi_*(H)$. Then there is a loop β based at $\widehat{x}$ in $\widehat{X}$ so that $\alpha \approx \pi \circ \beta$. Choose a particular homotopy

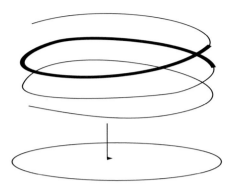

Figure 3.22: Lifts of a loop.

between α and $\pi \circ \beta$ and lift that homotopy to $\widehat{X}$ using Lemma 3.3.4. An argument similar to that in the proof of Theorem 3.3.5 now shows that β and $\widehat{\alpha}$ must have the same endpoint; hence $\widehat{\alpha}$ is a loop based at $\widehat{x}$. $\qquad \square$

It is worth noting that a loop in X could have one lift which is another loop and a second lift which is a path (not a loop) with distinct endpoints. See Figure 3.22.

THEOREM 3.3.7 *For any point $x \in X$, the cardinality of the set $\pi^{-1}(x)$ is equal to the index of $\pi_*(H)$ in G.*

Proof: Fix a point $x \in X$. Let $y \in X$ be any other point. Then $\pi^{-1}(x)$ and $\pi^{-1}(y)$ have the same cardinality (as the statement of the theorem certainly suggests). For let γ_y be a path in X that joins x to y. We lift γ_y to a path $\widehat{\gamma}_y$ in $\widehat{X}$ which begins at $\widehat{x}$ over x. This gives a function from the set $\pi^{-1}(x)$ to the set $\pi^{-1}(y)$ which is defined by $\widehat{x} \mapsto \widehat{\gamma}_y(1)$. This function is perforce one-to-one and onto, since we can explicitly produce its inverse using γ_y^{-1}. This confirms the statement about cardinality.

Now examine the set $\pi^{-1}(x)$. Let α be a loop in X which is based at x. We lift α to a path $\widehat{\alpha}$ in $\widehat{X}$ which begins at $\widehat{x}$; notice that $\widehat{\alpha}(1)$ is a point of $\pi^{-1}(x)$. Conversely, if $\widehat{t} \in \pi^{-1}(x)$ then projecting a path joining $\widehat{x}$ to $\widehat{t}$ down into X gives a loop based at x; so every point of $\pi^{-1}(x)$ arises in the manner described in the preceding paragraph.

Now two loops α and β give rise to the same point $\pi^{-1}(x)$ if and only if $\alpha\beta^{-1}$ lifts to a loop based at $\widehat{x}$; thus the preceding theorem tells us that this can happen if and only if the respective homotopy classes of α and β

determine the same right coset of $\pi_*(H)$ in G. Thus there is a one-to-one correspondence from the right cosets of $\pi_*(H)$ in G to the set $\pi^{-1}(x)$. That proves the result.   □

Now some terminology is in order. If the inverse image of each point under the covering map π contains a finite number of points, say n points (and we know that the number is the same for each pre-image), then we say that $\widehat{X}$ is an *n-sheeted covering of X* (or sometimes an *n-fold covering*).

EXAMPLE 3.3.8 Let $X = \mathbb{C}\setminus\{0\}$ and $\widehat{X} = \mathbb{C}\setminus\{0\}$. Let $\pi : \widehat{X} \to X$ be given by $z \mapsto z^n$ (where this is the ordinary power operation in the complex number field). Then π is an *n*-fold covering. It "winds" the punctured complex plane n times on itself.

Notice that, in this case, we have $H = G = \pi_1(\mathbb{C}\setminus\{0\}) \equiv \mathbb{Z}$. Also $\pi_*(H) = n\mathbb{Z} \subseteq \mathbb{Z}$. Hence the index of $\pi_*(H)$ in G is indeed n (as it should be).

THEOREM 3.3.9 *The groups $\pi_*(\pi_1(\widehat{X},\widehat{x}))$, for $\widehat{x} \in \pi^{-1}(x)$, form a conjugacy class of subgroups in G.*

Proof: The conjugacy class that has been described here is nothing other than the conjugacy class determined by $\pi_*(H)$. Of course $\pi : \widehat{x} \mapsto x$. Suppose now that $\widehat{\widehat{x}} \in \pi^{-1}(x)$ (here $\widehat{\widehat{x}}$ could be $\widehat{x}$ or some other point in the fiber). Then joint $\widehat{x}$ to $\widehat{\widehat{x}}$ with a path $\widehat{\gamma}$ in $\widehat{X}$. Let $\gamma \equiv \pi \circ \widehat{\gamma}$. Then it is easy to verify that the following diagram commutes:

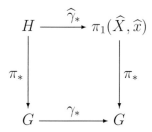

As a result, the inner automorphism[4] of G defined by γ_* sends $\pi_*(H)$ to $\pi_*(\pi_1(\widehat{X},\widehat{x}))$. See Appendix 5.

---

[4]Recall here that an *inner automorphism* on a group G is one defined by a mapping of the form $g \mapsto aga^{-1}$ for some fixed element $a \in G$.

For the converse direction, suppose that $K = \langle \alpha \rangle^{-1} \pi_*(H) \langle \alpha \rangle$, where $\langle \alpha \rangle \in G$ is the homotopy equivalence class containing the loop α. Now lift α to a path $\widehat{\alpha}$ in $\widehat{X}$ which begins at $\widehat{x}$, and set $\widehat{\widehat{x}} = \widehat{\alpha}(1)$. Then $\widehat{\widehat{x}} \in \pi^{-1}(x)$ and $K = \pi_*(\pi_1(\widehat{X}, \widehat{\widehat{x}}))$. $\qquad\qquad\qquad\qquad\qquad\qquad\qquad\qquad\qquad\qquad\qquad\qquad \square$

In order to make any further progress, we need a more general map-lifting theorem. In what follows, let Y be a path-connected and locally path-connected space with base point y. Assume that $f : Y \to X$ is a mapping which takes y to the base point x in X. We take it, as usual, that we have a covering map $\pi : \widehat{X} \to X$ and that the base point $\widehat{x}$ in $\widehat{X}$ is mapped by π to the base point $x \in X$. As usual H is the fundamental group of $(\widehat{X}, \widehat{x})$.

THEOREM 3.3.10 *There is a lift of f which takes y to $\widehat{x}$ if and only if $f_*(\pi_1(Y,y)) \subseteq \pi_*(H)$; the lift is unique.*

Proof: Necessity in this theorem is clear since a lift $\widehat{f}$ gives a commutative diagram

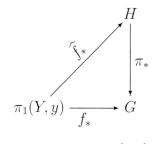

Uniqueness is also clear. Suppose that $\widehat{f}_1, \widehat{f}_2$ both lift f and send y to $\widehat{x}$. If $t \in Y$, joint y to t by a path γ. Then $\widehat{f}_1 \circ \gamma$ and $\widehat{f}_2 \circ \gamma$ are both lifts of $f \circ \gamma$ which begin at $\widehat{x}$, so they must agree. In particular, the two lifts have the same terminal point. We may conclude then that $\widehat{f}_1(t) = \widehat{f}_2(t)$. And that is the uniqueness that we seek.

For the converse direction, suppose that $f_*(\pi_1(Y,y)) \subseteq \pi_*(H)$. Then we can construct a lift $\widehat{f} : Y \to \widehat{X}$ in the following manner. Given $t \in Y$, join y to t by a path γ. Now lift the path $\alpha = f \circ \gamma$ to a path $\widehat{\alpha}$ in $\widehat{X}$ which begins at $\widehat{x}$; set $\widehat{f}(t) = \widehat{\alpha}(1)$. This construction is independent of the choice of the curve γ, for if γ' is another such curve, then set $\beta = f \circ \gamma'$. Thus $\alpha\beta^{-1}$ is a loop in X based at x. Also the homotopy equivalence class $\langle \alpha\beta^{-1} \rangle$ lies in $f_*(\pi_1(Y,y))$ and hence in $\pi_*(H)$. Therefore Theorem 3.3.4 tells us that $\alpha\beta^{-1}$

lifts to a loop based at $\widehat{x}$ in $\widehat{X}$. For this to occur, it must be that $\widehat{\alpha}$ and $\widehat{\beta}$ have the same terminal point.

The final task is to check that $\widehat{f}$ is continuous. Assume that $\widehat{f}(t) = \widehat{x}$ and also that $\pi(\widehat{x}) = x$. Let N be a neighborhood of $\widehat{x}$ in $\widehat{X}$. Following the standard properties of a covering space, let V be a neighborhood of $x \in X$ and U a neighborhood of $\widehat{x}$ in $\widehat{X}$ so that the restriction of π to U, $\pi : U \to V$, is a homeomorphism. Then $f^{-1} \circ \pi(N \cap U)$ is a neighborhood of t in Y. Since Y is locally path-connected, we may choose a path-connected neighborhood W of t in $f^{-1} \circ \pi(N \cap U)$. We claim that $\widehat{f}(W) \subseteq N$. If we can prove this assertion then we are done, because this shows that $f^{-1}(N)$ is open.

Let $s \in W$, and joint t to s in W (by local path-connectedness) by a path σ. To find $\widehat{f}(s)$, we lift the path $f \circ (\gamma\sigma) = (f \circ \gamma)(f \circ \sigma)$ to a path which begins at $\widehat{x}$ in $\widehat{X}$; now look at the terminal point of this path. We see that $f \circ \sigma$ lies inside $\pi(N \cap U)$ and its lift must begin at the terminal point of the lift of $f \circ \gamma$; this is the point $\widehat{x}$. But $\pi\big|_{N \cap U}$ is a homeomorphism, hence the endpoint of this lift lies in $N \cap U$. Thus the endpoint lies in N, as required. $\square$

Our ultimate goal here is to construct the "universal covering space" for a suitable topological space X. Heuristically, the universal covering space is a covering space above X which sits above all other covering spaces. A key property of the universal covering space is that it is simply connected. Of course the results that we have proved thus far (and there will be more evidence below) suggest why this must be so.

Now let $\Pi_1 : \widehat{X}_1 \to X$ and $\Pi_2 : \widehat{X}_2 \to X$ be covering maps. Let $\widehat{x}_1 \in \widehat{X}_1$ and $\widehat{x}_2 \in \widehat{X}_2$ be base points. Hence $\Pi_1(\widehat{x}_1) = \Pi_2(\widehat{x}_2) = x$. We write $H_1 = \pi_1(\widehat{X}_1, \widehat{x}_1)$ and $H_2 = \pi_1(\widehat{X}_2, \widehat{x}_2)$ (these are the fundamental groups).

THEOREM 3.3.11 *If $\Pi_{2*}(H_2) \subseteq \Pi_{1*}(H_1)$, then there is a covering map $\Pi : \widehat{X}_2 \to \widehat{X}_1$ which sends $\widehat{x}_2$ to $\widehat{x}_1$ and satisfies $\Pi_1 \circ \Pi = \Pi_2$.*

Proof: Apply the previous theorem to lift the map $\Pi_2 : \widehat{X}_2 \to X$ to a map $\Pi : \widehat{X}_2 \to \widehat{X}_1$ which sends $\widehat{x}_2$ to $\widehat{x}_1$. This lifted map is certainly a covering map. $\square$

Remark 3.3.12 In the special circumstance that $\Pi_{1*}(H_1) = \Pi_{2*}(H_2)$, we can run the preceding argument in the reverse direction. Thus there will be maps $g : \widehat{X}_1 \to \widehat{X}_2$ and $h : \widehat{X}_2 \to \widehat{X}_1$ which satisfy $\Pi_2 \circ g = \Pi_1$ and $\Pi_1 \circ h = \Pi_2$. Now $\Pi_1 \circ h \circ g = \Pi_1$, hence $h \circ g$ and $\mathrm{id}_{\widehat{X}_1}$ both lift the mapping $\Pi_1 : \widehat{X}_1 \to X$ to a map from $\widehat{X}_1$ to $\widehat{X}_1$ which sends $\widehat{x}_1$ to $\widehat{x}_1$. By the uniqueness part of the last theorem, we may conclude that $h \circ g = \mathrm{id}_{\widehat{X}_1}$. In like manner, $g \circ h = \mathrm{id}_{\widehat{X}_2}$ and we may conclude that $g : \widehat{X}_1 \to \widehat{X}_2$ (and also $h : \widehat{X}_2 \to \widehat{X}_1$) is a homeomorphism.

Definition 3.3.13 Let us say that two covering spaces $\widehat{X}_1$, $\widehat{X}_2$ are *equivalent* if there is a homeomorphism $h : \widehat{X}_1 \to \widehat{X}_2$ such that $\Pi_2 \circ h = \Pi_1$.

Combining this (and the preceding) discussion with Theorem 3.3.9, we may conclude that two covering spaces of X are equivalent if and only if they determine the same conjugacy class of subgroups of the fundamental group of X.

EXAMPLE 3.3.14 Let X be the unit circle in the plane as usual. Let $\widehat{X}$ also be the circle, and define

$$\pi_k : \widehat{X} \longrightarrow X$$

by

$$\pi_k(\cos t, \sin t) = (\cos kt, \sin kt).$$

Of course each π_k is a covering, and these coverings are distinct (i.e., inequivalent) for distinct k.

One can convince oneself of this last assertion by using the original definition of equivalence/inequivalence of coverings. But it is even easier to use the remark immediately preceding this example—just compare the conjugacy classes (i.e., compare $\mathbb{Z}_k$ for different k) to see that the coverings are inequivalent.

EXAMPLE 3.3.15 Let X be the unit circle in the plane and let $\widehat{X}$ be the real line. Define

$$\pi : \widehat{X} \longrightarrow X$$

by

$$\pi(t) = (\cos 2\pi t, \sin 2\pi t).$$

Then one may check that this covering is inequivalent to the coverings described in the last example. Certainly this $\widehat{X}$ is not homeomorphic to the $\widehat{X}$ in that example.

Definition 3.3.16 Let $\pi : \widehat{X} \to X$ be a covering map. Define a *covering transformation* of $\widehat{X}$ to be a homeomorphism $h : \widehat{X} \to \widehat{X}$ such that $\pi \circ h = \pi$. The collection of all covering transformations forms a group K under composition of mappings.

Proposition 3.3.17 *The group of covering transformations acts freely. This means that if $g \in K$ and $\widehat{x} \in \widehat{X}$ and $g\widehat{x} = \widehat{x}$ then $g = e$ (the group identity).*

Proof: We leave the verification of this assertion to the reader. [This is obviously just the uniqueness of liftings.] $\qquad\qquad\square$

EXAMPLE 3.3.18 Let π be the covering in the last example. Then a mapping

$$h_k(t) = t + k \, ,$$

for any integer k, will be a covering transformation. You can verify directly that the collection of covering transformations forms a group.

Notice that we may consider the quotient space $\widehat{X}/K$, where two elements $\widehat{x}_1$, $\widehat{x}_2$ of $\widehat{X}$ are considered to be related if there is a $g \in K$ such that $g\widehat{x}_1 = \widehat{x}_2$. The set of resulting equivalence classes is, by definition, the quotient space. This quotient tells us a good deal about the nature of the covering.

THEOREM 3.3.19 *If $\pi_*(H)$ is a normal subgroup of G, then X is homeomorphic to the orbit space $\widehat{X}/K$. Also K is group-theoretically isomorphic to the factor group $G/\pi_*(H)$.*

Proof: Both the covering map $\pi : \widehat{X} \to X$ and the "natural projection" $\widehat{X} \mapsto \widehat{X}/K$ are identification maps. Our job now is to check that the inverse image of a point under the first map is the same as the inverse image of a point under the second map. Put in other words, we must prove that the inverse image of a point $x \in X$ under π is also an orbit of K.

Given $x \in X$, we know that each $g \in K$ permutes the points of $\pi^{-1}(x)$— just because g is a covering transformation. Also, if $\widehat{x}_1$ and $\widehat{x}_2$ are elements of $\pi^{-1}(x)$, then $\pi_*(\pi_1(\widehat{X}, \widehat{x}_1)) = \pi_*(H) = \pi_*(\pi_1(\widehat{X}, \widehat{x}_2))$. Here we use Theorem 3.3.6 and the fact that $\pi_*(H)$ is normal in G. Thus we can find a covering transformation $h \in K$ that send $\widehat{x}_1$ to $\widehat{x}_2$.

It remains to show that K is isomorphic to $G/\pi_*(H)$. Let α be a loop based at $x \in X$. Lift α to a path $\widehat{\alpha}$ which begins at $\widehat{x} \in \widehat{X}$; and let k_α be the unique covering transformation which sends $\widehat{x}$ to $\widehat{\alpha}(1)$. Certainly every element of K can be produced in this fashion. Also Theorem 3.3.4 shows that two loops α and β give the same element of K if and only if $\langle \alpha\beta^{-1} \rangle \in \pi_*(H)$. Thus the correspondence $\alpha \mapsto k_\alpha$ induces a one-to-one function from $G/\pi_*(H)$ to K. To check that this function is in fact a homomorphism, we note that if α, β are loops based at x, then the lift of $\alpha\beta$ which begins at $\widehat{x}$ is the path $\widehat{\alpha}(k_\alpha \circ \widehat{\beta})$; also the endpoint of this last loop is $k_\alpha(k_\beta(\widehat{x}))$. In other words, $\alpha\beta$ corresponds to $k_\alpha \circ k_\beta$. $\qquad\square$

Definition 3.3.20 In the case that $\pi_*(H)$ is a normal subgroup of G, we say that $\widehat{X}$ is a *regular* covering space.

In the case that a covering is not regular, there may not be enough covering transformations to map all the elements of $\pi^{-1}(x)$, $x \in X$, to each other.

In the case that $\widehat{X}$ is a simply connected covering space of the base space X, then it is unique up to homeomorphism—for any two such spaces must be equivalent. Also this simply connected covering space must in fact be a regular covering of any other covering space of X—by Theorem 3.3.11. Thus we shall call such an $\widehat{X}$ the *universal covering space* of X.

EXAMPLE 3.3.21 The universal covering space of the unit circle in the plane is the line.

EXAMPLE 3.3.22 The universal covering space of the torus is the plane. In fact if we think of the torus as $S^1 \times S^1$, then the covering map is

$$\pi(x, y) = \big((\cos 2\pi x, \sin 2\pi x), (\cos 2\pi y, \sin 2\pi y)\big).$$

The main point of our discussion below will be to give conditions under which the universal covering space exists. But first some preliminary discussion. If $\widehat{X}$ is a universal covering space for X, then the covering transformation group K is isomorphic to the fundamental group of X. Given any subgroup L of $\pi_1(X)$, we see that L acts on $\widehat{X}$; the associated orbit space $\widehat{X}/L$ is a covering space of X whose fundamental group is isomorphic to L. Thus, if X has a universal covering space, then it has a covering space corresponding to any subgroup of its fundamental group.

Definition 3.3.23 Let X be a topological space. We say that X is *semi-locally simply connected* if each point of X has a neighborhood U such that each loop in U is null homotopic in X.

THEOREM 3.3.24 *Let X be a topological space which is path-connected, locally path-connected, and semi-locally simply connected. Then X has a universal covering space.*

Proof: The basic idea is that $\widehat{X}$ is the path space in X. We shall outline the key ideas here, and refer the reader to an authoritative source like [FUL] or [SPA] for all the details.

Now fix a base point $x \in X$. A point of $\widehat{X}$ will be an equivalence class of paths in X which begin at x and so that two paths α and β are equivalent precisely when **(i)** they have the same endpoint and **(ii)** the loop $\alpha\beta^{-1}$ is homotopic to the constant loop at x. We define $\pi : \widehat{X} \to X$ in this fashion: if $\widehat{x} \in \widehat{X}$, then $\widehat{x}$ is represented by a path α in X; now let $\pi(\widehat{x})$ be the endpoint $\alpha(1)$.

Now we construct a basis for the topology on $\widehat{X}$. Begin with a path-connected open set V in X so that any loop in V is null homotopic in X (this is the property of semi-local simple-connectivity). Also choose a point $\widehat{x} \in \widehat{X} \in \pi^{-1}(V)$. Represent this point $\widehat{x}$ by a path α in X which joins x to $\pi(\widehat{x})$. Define a neighborhood $V_{\widehat{x}}$ to be the subset of $\widehat{X}$ determined by paths in X of the form $\alpha\beta$, where β lies in V. These sets $V_{\widehat{x}}$ are the basic open sets in $\widehat{X}$.

One may check (we omit the details), that $\pi : \widehat{X} \to X$ is a covering map and that $\widehat{X}$ is path-connected and simply connected. The semi-local simple-connectivity guarantees that the map $\pi\big|_{V_{\widehat{x}}} : V_{\widehat{x}} \to V$ is one-to-one. $\square$

3.4   The Concept of Index

Implicit in Example 3.1.4—where we calculated the fundamental group of the circle—is the concept of index. Here we formulate the idea and develop some of its applications.

Let S^1 be the unit circle in the plane and let $\gamma : [0, 1] \to S^1$ be a loop. So γ is a continuous function with $\gamma(0) = \gamma(1)$. By Lemma 3.3.3, there is a lift of γ to a map

$$G : [0, 1] \to \mathbb{R}$$

so that (refer to our discussion of this covering map in Example 3.2.2)

$$e^{2\pi i G(t)} = \gamma(t) , \qquad 0 \le t \le 1$$

and

$$G(0) = 0 .$$

It follows that the terminal point $G(1)$ must be an integer (so that $e^{2\pi i G(1)} = 1$). We call that integer the *index* of γ and denote it by ind(γ).

We saw the index playing a role in Example 3.1.4: a loop that wraps k times counterclockwise around S^1 has index k, and a loop that wraps m times clockwise around S^1 has index $-m$.

Proposition 3.4.1 *Two loops γ_1 and γ_2 in S^1, both based at 1, are in the same homotopy class if and only if they have the same index. The mapping*

$$[\gamma] \mapsto ind(\gamma)$$

is an isomorphism of $\pi_1(S^1)$ with the group of integers $\mathbb{Z}$.

Proof: The proof of this result is contained in the discussion of Example 3.1.4. □

Let B^N denote the closed Euclidean unit ball in $\mathbb{R}^N$. In case $N = 2$ we shall identify $\mathbb{R}^2$ with the complex plane and think of B^2 as $\{z \in \mathbb{C} : |z| \le 1\}$.

Lemma 3.4.2 *Let*

$$f : B^2 \to S^1$$

be a mapping with $f(1) = 1$. Then the loop

$$\gamma(t) = f(e^{2\pi i t})$$

(i.e., the image of the boundary circle under f) has index 0 in S^1.

Proof: Let

$$G(t) = e^{2\pi i t}$$

be the boundary loop in B^2. Then of course $\gamma = f \circ G$.

Let $c(s)$ be the constant loop at 1. Since B^2 is convex, the homotopy

$$\Gamma(s, t) = (1 - t)\gamma(s) + tc(s)$$

shows that γ is homotopic to c. It follows then from Proposition 3.4.1 that γ has index 0. □

Corollary 3.4.3 *There is no map* $g : B^2 \to S^1$ *so that* $g(z) = z$ *for all* $z \in S^1$.

Proof: If there were such a map, then the loop $s \to e^{2\pi i s}$ in the circle S^1 would have index 0 (by the lemma). But obviously the index of this loop is 1. This is a contradiction. □

We can now give an alternative proof of the Brouwer fixed-point theorem (Theorem 3.1.11). The ideas here are not fundamentally different from the homotopy concepts that we used in our original proof. But the language of index allows for a particularly elegant presentation.

THEOREM 3.4.4 *Let* $F : B^2 \to B^2$ *be a mapping. Then* F *has a fixed point.*

Proof: Suppose to the contrary that there is no fixed point. For each $z \in B^2$, consider the segment beginning at $F(z)$, passing through z, and terminating at a point $r(z)$ in the circle $S^1 = \partial B$. Then the mapping $z \mapsto r(z)$ fixes points of S^1, so we have contradicted the corollary. □

Next we present a version of the celebrated Borsuk-Ulam theorem. In what follows, S^2 will be the unit sphere in $\mathbb{R}^3$. Two points of S^2 are called *antipodal* if they are at opposite ends of a diameter of S^2.

THEOREM 3.4.5 (Borsuk-Ulam) *Let* $\varphi : S^2 \to \mathbb{R}^2$ *be a mapping. Then there are antipodal points* P *and* $-P$ *in* S^2 *so that* $\varphi(P) = \varphi(-P)$.

Proof: We set

$$F(p) = \varphi(p) - \varphi(-p)$$

for $p \in S^2$. Our job is to show that F vanishes at some point of the sphere. Notice that F is odd, in the sense that

$$F(-p) = -F(p).$$

This will be important in the reasoning that follows.

Now if $z = (x, y) \in B^2$, we set

$$G(z) = G(x, y) = F(x, y, +\sqrt{1 - x^2 - y^2}).$$

Of course it now follows that

$$G(-z) = -G(z) \qquad (3.4.5.1)$$

when $z \in S^1$ (for then $\sqrt{1 - x^2 - y^2} = \sqrt{1 - |z|^2} = 0$). We shall show that any mapping $G : B^2 \to \mathbb{R}^2$ satisfying (3.4.5.1) must vanish at some point (x, y) of B^2. This will in turn tell us that F vanishes at $(x, y, \sqrt{1 - x^2 - y^2})$, and that gives the result.

Suppose to the contrary that a G such as we have described does *not* vanish at any point of B^2. Set

$$\psi(z) = \frac{G(z)}{|G(z)|} \cdot \frac{|G(1)|}{G(1)}.$$

Then ψ is a mapping from B^2 to S^1 satisfying

(3.4.5.2) $\psi(-z) = -\psi(z)$, for all $z \in S^1$,

(3.4.5.3) $\psi(1) = 1$.

Now Lemma 3.4.2 tells us that the mapping

$$\beta(s) = \psi(e^{2\pi i s}), \qquad 0 \le s \le 1,$$

has index 0. We shall show that in fact this index must be odd, and that will be a contradiction.

Define a function $\mu : [0, 1] \to \mathbb{R}$ so that

$$\psi(e^{2\pi i s}) = e^{2\pi i \mu(s)}, \qquad 0 \le s \le 1,$$

and

$$\mu(0) = 0.$$

It follows that $\text{ind}(\beta) = \mu(1)$. Then line (3.4.5.2) tells us that

$$\begin{aligned}
\exp[2\pi i \mu(s + 1/2)] &= -\exp[2\pi i \mu(s)] \\
&= \exp[2\pi i (\mu(s) + 1/2)], \qquad 0 \le s \le 1/2.
\end{aligned}$$

It follows that, for each $s \in [0, 1/2]$, the number

$$\mu(s + 1/2) - \mu(s) - 1/2 \qquad (3.4.5.4)$$

is an integer. But the function defined in (3.4.5.4) is continuous and has range in a discrete space. It follows that this function must be constant. Take the value to be $m \in \mathbb{Z}$. Hence

$$\mu(s + 1/2) - \mu(s) = m + \frac{1}{2}, \qquad 0 \leq s \leq 1/2.$$

But then we may calculate that

$$\begin{aligned} \operatorname{ind}(\beta) &= \mu(1) \\ &= \mu(1) - \mu(1/2) + \mu(1/2) - \mu(0) \\ &= m + 1/2 + m + 1/2 \\ &= 2m + 1. \end{aligned}$$

This shows that the index of β is an odd integer, and that is the desired contradiction. $\qquad\qquad\square$

3.5   Mathematical Economics

Today economics is a very mathematical discipline. Of course the subject of economics is very large and complex, and one of the main goals is to use mathematical/analytical tools to bring the various vectors in the subject under control. In the present section we illustrate one means of attacking this program.

One fairly simple view of an economy is that it is a mechanism for trading commodities. So we use those commodities as the basis of our analysis. To keep things fairly simple, let us suppose that we are working in an economy with *three* commodities: Fords, computers, and wristwatches. We denote these three commodities by F, C, and W. The total supply of each, measured in units, is S_F, S_C, and S_W, respectively. Each person participating in this economy will own a certain number of each commodity. So we think of that person as an ordered triple of numbers: $\mathbf{X}^j = (x_F^j, x_C^j, x_W^j)$. Here is how to read this:

> The person under consideration is person j (where $j = 1, 2, \dots$). The nonnegative integer x_F^j denotes the number of Fords that person j owns; the nonnegative integer x_C^j denotes the number of computers that person j owns; the nonnegative integer x_W^j denotes the number of wristwatches that person j owns.

Observe that if we calculate the sum

$$\mathbf{S} = \sum_{j} \mathbf{X}^j = \sum_{j} (x_F^j, x_C^j, x_W^j),$$

then this will give us an ordered triple of the total number of Fords, the total number of computers, and the total number of wristwatches in the entire economy. Note that every unit of every commodity is owned by someone— there are no commodities just sitting around unclaimed.

There are various simplifying assumptions that must be made in our analysis. First of all, we take it that commodities are not consumed, disabled, or destroyed. So the number $\mathbf{S}$ remains constant. Second, we adopt the assumption (that goes back to Adam Smith (1723–1790)) that the price of a commodity is determined by supply and demand. To wit, if there is more demand for a product than there is supply, then the price will go up. If, instead, there is more supply than demand, then the price will go down.

With these premises in mind, we let supply and demand drive the market. Of course the *lingua franca* for this process is *price*. Let the current price per unit for Fords, computers, and wristwatches be p_F, p_C, and p_W respectively. We take it that all these prices are nonnegative (meaning that each commodity has at least some value for any consumer). The price of an item *could* be 0. Following the model set above, we use a single vector $\mathbf{p} = (p_F, p_C, p_W)$ to denote price.

And in fact, given that the economy we are describing has three commodities and three commodities only, it is not the *actual* price of an item that matters. It is instead the relative price. Put in other words, the fact that a wristwatch costs \$50 is of no interest. What *is* important is that a wristwatch costs one-twentieth of what a computer costs. Thus we define the *relative* or *normalized* prices of the three commodities to be

$$p_F' = \frac{p_F}{p_F + p_C + p_W},$$
$$p_C' = \frac{p_C}{p_F + p_C + p_W},$$
$$p_W' = \frac{p_W}{p_F + p_C + p_W}.$$

Notice that $p_F' + p_C' + p_W' = 1$.

Given the setup that we have, it is possible to graph the collection of all possible price vectors in 3-dimensional space. Thus we are graphing the set

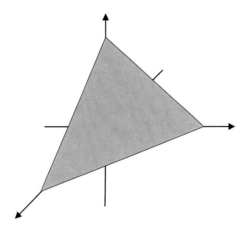

Figure 3.23: The space $\mathcal{T}$ of all price vectors.

of all $\mathbf{p}' = (p'_F, p'_C, p'_W)$ with $p'_F \geq 0$, $p'_C \geq 0$, $p'_W \geq 0$ and $p'_F + p'_C + p'_W = 1$. This is the portion of a plane that lies in the first octant. See Figure 3.23. Since the region is shaped like a triangle (or, to use our earlier terminology, it is a simplex together with its interior), we denote it by $\mathcal{T}$.

For the jth consumer, we can measure that person's "wealth" by calculating the value of all his/her goods. This wealth quantity will depend on the price $\mathbf{p}$. The wealth will be the number

$$w^j(\mathbf{p}') = p'_F x^j_F + p'_C x^j_C + p'_W x^j_W = \mathbf{p}' \cdot \mathbf{X}^j .$$

Here, of course, the dot $\cdot$ is the ordinary "dot product" from vector analysis. When the price vector $\mathbf{p}'$ varies, so the wealth of any given individual will vary; so the wealth of individual j is a function of $\mathbf{p}'$.

Now a matter of particular interest—something that economists would like to quantify—is the level of a consumer's satisfaction with his/her current possessions, and whether or not that consumer would like to reconfigure his/her possessions to a different state. In other words, would consumer j prefer to own fewer wristwatches and more computers? Or perhaps fewer computers and more Fords? We have let $\mathbf{X}^j$ denote the current holdings of consumer j. Let us let $\mathbf{Y}^j(\mathbf{p}')$ denote the "optimal" or "desired" holdings for consumer j, given the price structure $\mathbf{p}'$. It is understood in this model that the only thing that consumer j can do, to move from $\mathbf{X}^j$ to $\mathbf{Y}^j$, is to trade. There will be no outside influx of money to change the position of j in the economy. In other words, the total wealth of consumer j will remain fixed. Thus

$$w^j(\mathbf{p}') = \mathbf{p}' \cdot \mathbf{X}^j = \mathbf{p}' \cdot \mathbf{Y}^j(\mathbf{p}') \,. \tag{3.5.1}$$

We call $\mathbf{Y}^j(\mathbf{p}')$ the *demand vector*. It is important to understand that the demand vector is constrained by Equation (3.5.1).

A simple example illustrates the concepts discussed thus far:

EXAMPLE 3.5.2 Imagine that Ataturk has 3 Fords, 8 computers, and 4 wristwatches. So his vector of commodities is $\mathbf{X}^A = (3, 8, 4)$. Here the boldface $\mathbf{X}$ is our usual notation and the A stands for "Ataturk." Suppose further that the price vector is $\mathbf{p}' = (0.86, 0.1, 0.04)$. Notice that each of the entries here is nonnegative and the three numbers sum to 1. Also the entry for Fords is notably larger than that for computers, and that for computers larger than that for wristwatches. This reflects the relative value of these items.[5] Thus Ataturk's net worth is

$$\mathbf{p}' \cdot \mathbf{X}^A = 0.86 \cdot 3 + 0.1 \cdot 8 + 0.04 \cdot 4 = 3.54.$$

It is natural for Ataturk to ponder the price vector and perhaps therefore to formulate an optimal or desired distribution of commodities. He decides that Fords are a particularly attractive buy these days, so he sets $\mathbf{Y}^A(\mathbf{p}') = (8, 1, 1)$. But this will not work! It is not allowed, because Equation (3.5.1) dictates that

$$\mathbf{p}' \cdot \mathbf{X}^A = \mathbf{p}' \cdot \mathbf{Y}^A(\mathbf{p}') \,.$$

We already calculated the left-hand side, and it equals 3.54. But the right-hand side equals 7.02. In short, Ataturk cannot trade up to 8 Fords. So Ataturk will re-assess his resources and define a new demand vector to be $(4, 1, 0)$. Now we may calculate that

$$\mathbf{p}' \cdot \mathbf{Y}^A(\mathbf{p}') = 0.86 \cdot 4 + 0.1 \cdot 1 + 0.04 \cdot 0 = 3.54 \,.$$

Now all is right with the world, and Ataturk's new demand vector is consistent (according to Equation (3.5.1)) with his original vector of commodities (i.e., his original resources).

Another consumer, named Bärbel, has commodities vector $(1, 3, 9)$. She has a need for a good many wristwatches, but only needs one car. Of course

---

[5]Of course there exist wristwatches that cost more than any car. And there exist computers that cost more than any car or wristwatch. The relative values that we assign here are for typical household items.

the price vector is the same for her as for any other consumer (namely, $(0.86, 0.1, 0.04)$). So her net worth is

$$\mathbf{p}' \cdot \mathbf{X}^B = 0.86 \cdot 1 + 0.1 \cdot 3 + 0.04 \cdot 9 = 1.52\,.$$

We see that her net worth is considerably less than Ataturk's, because she owns fewer Fords and fewer computers. She ponders her status in the economy and formulates the demand vector $(1, 5, 4)$. Bärbel is every mindful of the constraint imposed by Equation (3.5.1), so when we calculate

$$\mathbf{p}' \cdot \mathbf{Y}^A(\mathbf{p}') = 0.86 \cdot 1 + 0.1 \cdot 5 + 0.04 \cdot 4 = 1.52\,.$$

we obtain the right answer. Bärbel's choices reflect her lack of interest in cars and her greater valuing of computers and wristwatches.

If the prices change, then demand is likely to change as well. If cars suddenly become very cheap, then some people will want to stockpile cars while others will perceive that a car is no longer a prestige item and will lose interest in cars. The view of computers is probably rather different. A computer is a utility item that most everyone needs for strictly practical purposes. If computers become cheaper, then people will stock up—perhaps buy an extra one for the vacation home or for the kitchen. But if computers become very expensive, then most everyone will still buy one because it is essential for managing modern life.

Suppose now that the price vector changes to $(0.92, 0.06, 0.02)$. So cars are more expensive, and computers and wristwatches less so (this may be a reflection of market relations with Japan). Now Ataturk's net worth becomes

$$\mathbf{p}' \cdot \mathbf{X}^A = 0.92 \cdot 3 + 0.06 \cdot 8 + 0.02 \cdot 4 = 3.32.$$

Observing that wristwatches have become very cheap, he may wish to buy several to give as gifts. So his demand vector may become $(3, 6, 10)$. Again, we may verify that Equation (3.5.1) is satisfied:

$$\mathbf{p}' \cdot \mathbf{Y}^A(\mathbf{p}') = 0.92 \cdot 3 + 0.06 \cdot 6 + 0.02 \cdot 10 = 3.32\,.$$

Notice once again that, in the economic model we have set up, a person's net worth does not change. He or she can shift it around among various commodities; but the aggregate value of those commodities does not change.

We know from the above discussion that each consumer j has a demand function $\mathbf{Y}^j(\mathbf{p}')$. The variable (or domain) is the normalized price vector, and the values (or range) are triples of nonnegative numbers. Following our standard paradigm, we may sum all the demand functions to obtain an aggregate demand vector function $\mathcal{Y}(\mathbf{p}')$ for the economy as whole at the value $\mathbf{p}'$:

$$\mathcal{Y}(\mathbf{p}') = \sum_j \mathbf{Y}^j(\mathbf{p}').$$

We will make the very natural, and plausible, assumption that $\mathcal{Y}$ depends *continuously* on the variable vector $\mathbf{p}'$—that is to say, small variations in $\mathbf{p}'$ result in small variations of $\mathcal{Y}$.

We have already noted that $\mathbf{p}' \cdot \mathbf{Y}^j(\mathbf{p}')$ is the desired worth of consumer j, given the normalized price vector $\mathbf{p}'$. As a result, the quantity $\mathbf{p}' \cdot \mathcal{Y}(\mathbf{p}')$ is the total desired wealth of the entire population of the economy, given the normalized price vector $\mathbf{p}'$. We may now use Equation (3.5.1) to calculate that

$$\mathbf{p}' \cdot \mathcal{Y}(\mathbf{p}') = \sum_j \mathbf{p}' \cdot \mathbf{Y}^j(\mathbf{p}') = \sum_j \mathbf{p}' \cdot \mathbf{X}^j = \mathbf{p}' \cdot \mathbf{S}.$$

We have derived *Walras's Law* (Leon Walras (1829–1910)):

$$\mathbf{p}' \cdot \mathcal{Y}(\mathbf{p}') = \mathbf{p}' \cdot \mathbf{S}.$$

Leon Walras was one of the first mathematical economists. His law says that the component of price in the direction of the aggregate demand equals the component of price in the direction of aggregate holdings. This is a statement of conservation of resources.

As an instance of the philosophy of Walras's law, if the coordinate of $\mathcal{Y}(\mathbf{p}')$ corresponding to computers is greater than the coordinate of $\mathbf{S}$ corresponding to computers, then we see that this price yields more demand for computers than there is supply. One consequence will be, according to Adam Smith, that the price of computers will rise. By contrast, if the current array of prices—of Fords, computers, and wristwatches—results in everyone being able to obtain their optimal or desired selection of commodities, so that nobody is tempted to trade, or to change their status in the economy, then the economy is in equilibrium. It is of great interests to economists to be able to detect, and particularly to be able to predict, such a state.

Definition 3.5.3 We say that a normalized price vector $\mathbf{p}'$ is an *equilib-rium price vector* if the jth component of $\mathcal{Y}$ is less than or equal to the jth component of $\mathbf{S}$ for $j = 1, 2, 3$. In practice we shall write $\mathcal{Y}(\mathbf{p}') \leq \mathbf{S}(\mathbf{p}')$.

In what follows, if $\mathbf{V}$ is a vector, then we will let V_j (or sometimes $\mathbf{V}_j$) denote its components for $j = 1, 2, 3$. Thus the concept of equilibrium price vector could be defined by $Y_j(\mathbf{p}') \leq S_j$.

Definition 3.5.4 We define

$$\mathbf{E}(\mathbf{p}') = \mathcal{Y}(\mathbf{p}') - \mathbf{S}(\mathbf{p}')\,.$$

This is the *excess demand vector*.

These new definitions bear some consideration. If $\mathbf{p}'$ is an equilibrium price vector, then the demand for any particular item is not greater than the supply of that item (here we think of $\mathbf{S}$ as the supply and $\mathcal{Y}$ as the demand). Obviously $\mathbf{p}'$ is an equilibrium price vector precisely when $\mathbf{E}(\mathbf{p}') \leq \mathbf{0}$ (where $\mathbf{0}$ is the 3-vector of 0s). It is naturally a matter of some interest to be able to see that an equilibrium price vector always exists. We shall in fact prove this assertion below using the Brouwer fixed-point theorem.

Remark 3.5.5 Using our new notation, Walras's Law may be rewritten as

$$\mathbf{p}' \cdot \mathbf{E}(\mathbf{p}') = 0\,.$$

Notice that the vector $\mathbf{E}(\mathbf{p}')$ points in the direction that it is anticipated that prices will move from $\mathbf{p}'$.

Our goal now is to define a function to which we may apply the Brouwer fixed-point theorem. We first note that the region $\mathcal{T}$ illustrated in Figure 3.23 is homeomorphic to the closed unit disc in the plane. Thus the Brouwer theorem would apply to a continuous function $F : \mathcal{T} \to \mathcal{T}$. Now we need one additional technical definition before we may proceed:

Definition 3.5.6 Let X be a topological space and $g : X \to \mathbb{R}$ a function. We define g^+ to be the function

$$g^+(x) = \begin{cases} g(x) & \text{if} \quad g(x) \geq 0 \\ 0 & \text{if} \quad g(x) < 0\,. \end{cases}$$

Thus g^+ is the modified function that replaces negative values of g by 0.

Now here is our point of view: The function to which we wish to apply the Brouwer fixed-point theorem is $h(\mathbf{p'}) = \mathbf{p'} + \mathbf{E}(\mathbf{p'})$. This function is continuous, and it makes sense on the domain $\mathcal{T}$. That is the good news. However, it does not (necessarily) map elements of $\mathcal{T}$ to elements of $\mathcal{T}$. For one thing, the entries of $h(\mathbf{p'})$ could be negative. For another thing, they might not sum to 1.

But a good mathematician knows how to make adjustments so that the function under study will fit the needed hypotheses.

Definition 3.5.7 We define a *price change function* with domain $\mathcal{T}$ and range $\mathcal{T}$ by

$$f(\mathbf{p'}) = \frac{(\mathbf{p'} + \mathbf{E}(\mathbf{p'}))^+}{\sum_{j=1}^{3}(\mathbf{p'} + \mathbf{E}(\mathbf{p'}))_j^+}. \qquad (3.5.7.1)$$

We note that the entries of $f(\mathbf{p'})$ are certainly nonnegative. And, because we have normalized them in the usual fashion (by dividing by the sum), they certainly sum to 1. Thus $f : \mathcal{T} \to \mathcal{T}$ as desired. It will be useful in what follows to write $\mathbf{B}(\mathbf{p'}) = \sum_{j=1}^{3}(\mathbf{p'} + \mathbf{E}(\mathbf{p'}))_j^+$.

Remark 3.5.8 The quantity $\mathbf{p'} + \mathbf{E}(\mathbf{p'})$ measures how prices move because of excess supply or demand. We note that

$$\mathbf{p'} \cdot (\mathbf{p'} + \mathbf{E}(\mathbf{p'})) = \mathbf{p'} \cdot \mathbf{p'} + \mathbf{p'} \cdot \mathbf{E}(\mathbf{p'}) \overset{\text{Walras}}{=\!=} \|\mathbf{p'}\|^2 = 1 > 0.$$

Therefore it cannot be that all the entries of $\mathbf{p'} + \mathbf{E}(\mathbf{p'})$ are ≤ 0 (since the entries of $\mathbf{p'}$ are all ≥ 0). This will be useful information in what follows. And we have already made use of the information, because we need to know that the denominator in (3.5.7.1) is nonvanishing.

THEOREM 3.5.9 *In the economic system that we have set up, there will always exist an equilibrium price vector.*

This result is a triumph for the methods of topology, and also for mathematical economics. It really says something about the economic system we have been discussing.

In order to prove the theorem, we first need a technical lemma:

Lemma 3.5.10 *If $\widetilde{\mathbf{p}}$ is a fixed point for f, then $\mathbf{B}(\widetilde{\mathbf{p}}) = 1$.*

Proof: To simplify the argument, we shall assume at first that the entries of $\widetilde{\mathbf{p}} + \mathbf{E}(\widetilde{\mathbf{p}})$ are all positive. We shall comment on the more general situation at the end of the proof.

Our initial hypothesis means then that

$$(\widetilde{\mathbf{p}} + \mathbf{E}(\widetilde{\mathbf{p}}))^+ = \widetilde{\mathbf{p}} + \mathbf{E}(\widetilde{\mathbf{p}}) \,.$$

Armed with this information, we know that

$$f(\widetilde{\mathbf{p}}) = \widetilde{\mathbf{p}}$$

so

$$\widetilde{\mathbf{p}} + \mathbf{E}(\widetilde{\mathbf{p}}) = \mathbf{B}(\widetilde{\mathbf{p}})\widetilde{\mathbf{p}} \,.$$

Taking the dot product of both sides with $\widetilde{\mathbf{p}}$ gives

$$\widetilde{\mathbf{p}} \cdot \widetilde{\mathbf{p}} + \mathbf{E}(\widetilde{\mathbf{p}}) \cdot \widetilde{\mathbf{p}} = \mathbf{B}(\widetilde{\mathbf{p}})\widetilde{\mathbf{p}} \cdot \widetilde{\mathbf{p}} \,. \qquad (3.5.10.1)$$

This simplifies, using Walras' Law, to

$$\widetilde{\mathbf{p}} \cdot \widetilde{\mathbf{p}} = \mathbf{B}(\widetilde{\mathbf{p}})\widetilde{\mathbf{p}} \cdot \widetilde{\mathbf{p}} \,.$$

Rewriting this as

$$[\widetilde{\mathbf{p}} \cdot \widetilde{\mathbf{p}}](\mathbf{B}(\widetilde{\mathbf{p}}) - 1) = 0 \,,$$

and remembering that $\widetilde{\mathbf{p}} \cdot \widetilde{\mathbf{p}} = 1$, we conclude that

$$\mathbf{B}(\widetilde{\mathbf{p}}) = 1$$

as required.

In the more general case that $(\widetilde{\mathbf{p}} + \mathbf{E}(\widetilde{\mathbf{p}}))^+$ has some zero entries, we simply notice that (3.5.10.1) trivializes to $0 = 0$, so the argument still goes through.   □

Proof of Theorem 3.5.9: As in the lemma, $\widetilde{\mathbf{p}}$ is a fixed point for f (which of course exists by Brouwer's theorem). Since, by the lemma, $\mathbf{B}(\widetilde{\mathbf{p}}) = 1$, we may simplify the equation $f(\widetilde{\mathbf{p}}) = \widetilde{\mathbf{p}}$ to

$$(\widetilde{\mathbf{p}} + \mathbf{E}(\widetilde{\mathbf{p}}))^+ = \widetilde{\mathbf{p}} \,. \qquad (3.5.9.1)$$

Fix an index $j = 1, 2, 3$. Now there are two cases:

The case $(\widetilde{\mathbf{p}} + \mathbf{E}(\widetilde{\mathbf{p}}))_j^+ > 0$. Then we have from (3.5.9.1) that

$$\widetilde{\mathbf{p}}_j + \mathbf{E}_j(\widetilde{\mathbf{p}}) = \widetilde{\mathbf{p}}_j$$

or

$$\mathbf{E}_j(\widetilde{\mathbf{p}}) = 0. \qquad (3.5.9.2)$$

The case $(\widetilde{\mathbf{p}} + \mathbf{E}(\widetilde{\mathbf{p}}))_j^+ = 0$. Then we have from (3.5.9.1) that

$$(\widetilde{\mathbf{p}} + \mathbf{E}(\widetilde{\mathbf{p}}))^+ = 0$$

so that

$$\mathbf{E}_j(\widetilde{\mathbf{p}})^+ = 0.$$

In particular, $\mathbf{E}_j(\widetilde{\mathbf{p}}) \leq 0$.

The two cases taken together tell us that $\mathbf{E}_j(\widetilde{\mathbf{p}}) \leq 0$ for all j. Therefore $\mathcal{Y}_j(\widetilde{\mathbf{p}}) \leq \mathbf{S}_j(\widetilde{\mathbf{p}})$ for all j. Hence $\widetilde{\mathbf{p}}$ is an equilibrium price vector. $\qquad \square$

Of course it goes without saying that our analysis applies just as well to an economy with M commodities for any integer $M \geq 3$. And for any finite number of consumers. One needs to apply a higher-dimensional version of the Brouwer fixed point theorem; other than that, the reasoning is just the same.

Exercises

1. Calculate the fundamental group of the unit sphere S^2 in $\mathbb{R}^3$.

2. Calculate the fundamental group of the unit sphere S^{N-1} in $\mathbb{R}^N$.

3. Explain why the Brouwer fixed-point theorem fails for the closed annulus

$$\overline{A} = \{(x, y) \in \mathbb{R}^2 : 1 \leq x^2 + y^2 \leq 4\}.$$

4. Explain why the annulus

$$A = \{(x, y) \in \mathbb{R}^2 : 1 < x^2 + y^2 < 4\}$$

and the circle
$$S^1 = \{(x, y) \in \mathbb{R}^2 : x^2 + y^2 = 1\}$$

are homotopy equivalent.

5. Explain why the sphere

$$S^2 = \{(x, y, z) \in \mathbb{R}^3 : x^2 + y^2 + z^2 = 1\}$$

and the circle
$$S^1 = \{(x, y) \in \mathbb{R}^2 : x^2 + y^2 = 1\}$$

are *not* homotopy equivalent.

6. Calculate the first and second homology of the sphere

$$S^2 = \{(x, y, z) \in \mathbb{R}^3 : x^2 + y^2 + z^2 = 1\}.$$

7. Fix a sphere
$$S^N = \{x \in \mathbb{R}^{N+1} : \|x\| = 1\}.$$

Explain why, for k sufficiently large, $H_k(S^N, \mathbb{Z}) = 0$.

8. Calculate the first and second homology of the torus.

9. Give an example of two subsets X and Y of the plane that have the same fundamental group but are not homeomorphic.

10. Let U be a bounded, open subset of the Cartesian plane with the property that its fundamental group is equal to $\mathbb{Z}$. Prove that U is homeomorphic to an annulus.

11. The alternating group on three letters cannot be the first homology group of any planar domain. Explain why.

12. Let U be an open set in the Cartesian plane that is finitely connected—that is to say, its complement has just finitely many connected components. Prove that U is homotopically equivalent to a curve in the plane.

13. Prove that if an open set U in the plane is bounded and simply connected, then it is homotopically equivalent to the unit disc.

14. Give a concrete example of a consumer in the model economy that we discussed in Section 3.5 who owns at least 6 of each commodity. Formulate a suitable demand vector and verify Walras's Law for the data you have created.

15. The Borsuk-Ulam theorem tells us that there are two diametrically opposite points on the surface of the earth at which both the wind speed and the temperature are the same. Explain why this is so.

16. Complete this outline to prove the famous "generalized ham sandwich theorem." The result says that a sandwich made of some quantity of bread (in *any* shape), some quantity of ham (in *any* shape), and some quantity of cheese (in *any* shape) can be cut with a single knife-cut so that the bread is bisected (by volume), the ham is bisected (by volume), and the cheese is bisected (by volume). First formulate a precise mathematical version of the result. Now complete these steps for the proof:

(a) If the point $P \in S^2$, then consider the line ℓ_P through the origin and through P. There will be a unique point X_P on ℓ_P so that the plane perpendicular to ℓ_P and passing through X_P will bisect the bread. Set $G_B(P) = X_P$.

(b) Reasoning as in part **(a)**, we may define a point $G_H(P)$ on the line ℓ_P so that the plane through $G_H(P)$ and orthogonal to ℓ_P will bisect the ham. Likewise, there is a point $G_C(P)$ that gives rise to a plane bisecting the cheese.

(c) Observe that $G_B(-p) = -G_B(p)$, $G_H(-p) = -G_H(p)$, and $G_C(-p) = -G_C(p)$ for each point $p \in S^2$.

(d) Define a mapping

$$F(p) = \bigl(F_B(p) - F_H(p), F_B(p) - F_C(p)\bigr) , \qquad p \in S^2 .$$

(e) Use the Borsuk-Ulam theorem to argue that F vanishes at some point of S^2.

(f) Explain why the result now follows.

17. What can you say about the universal covering space of the annulus

$$A = \{(x, y) \in \mathbb{R}^2 : 1 < x^2 + y^2 < 4\}$$

equipped with the usual Euclidean topology?

18. What can you say about the universal covering space of the "hollow ball"

$$C = \{(x, y, z) \in \mathbb{R}^3 : 1 < x^2 + y^2 + z^2 < 4\}$$

equipped with the usual Euclidean topology?

19. If X is simply connected, then its universal covering space is X itself. Prove this assertion.

20. Let X be a topological space with universal covering space $\widehat{X}$. Let Y be simply connected. What can you say about the universal covering space of $X \times Y$? [*Hint:* Begin with X the circle and Y the unit interval.]

21. Is there an analogue of the Borsuk-Ulam theorem for the circle in the plane? Is there an analogue for the three-dimensional sphere S^3? Why or why not?

22. Let X be the planar topological space depicted in Figure 3.24. Can you describe its universal covering space?

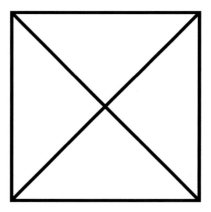

Figure 3.24: The topological space X.

23. Give an example of an economic system with three commodities such that there are two equilibrium price vectors.

24. Give an example of two simply connected spaces that are not homeo-morphic.

25. Let X be a sphere with two handles. What is the first homology group of X?

26. Let X and Y be topological spaces. How can you relate the first ho-motopy group of $X \times Y$ with that of X and Y?

27. Define the planar topological space X by

$$X = \{(x, y) \in \mathbb{R}^2 : 1 < x^2 + y^2 < 2\} \cup \{(x, y) \in \mathbb{R}^2 : 3 < x^2 + y^2 < 4\}.$$

What can you say about the universal covering space of X?

28. Prove that the product of two simply connected spaces is simply con-nected.

29. Let U_j be simply connected open sets in the plane, and assume that $U_1 \subseteq U_2 \subseteq U_3 \subseteq \cdots$. Define $\mathcal{U} = \cup_j U_j$. Is $\mathcal{U}$ simply connected? Why or why not?

30. If X and Y are simply connected topological spaces, does it then follow that $X \cap Y$ is simply connected?

31. If X and Y are simply connected topological spaces, does it then follow that $X \cup Y$ is simply connected?

Chapter 4

Manifold Theory

4.1 Basic Concepts

The concept of a 2-dimensional surface in 3-dimensional space is intuitively appealing. See Figure 4.1.

The surface is a 2-dimensional geometric object. It *could be* a hyperplane, as in Figure 4.2. Or it could be something more general (Figure 4.1). The key fact about a 2-dimensional surface is that it is *locally like Euclidean 2-space*. What does this mean? If p is a point of the surface, then there is a neighborhood of p that is homeomorphic to an open set in $\mathbb{R}^2$: reference Figure 4.3.

The idea of manifold grew, in the late nineteenth century and early twentieth century, out of these considerations. There was a need for a for-

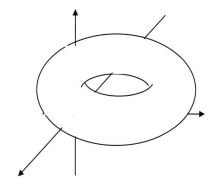

Figure 4.1: A 2-dimensional surface in 3-dimensional space.

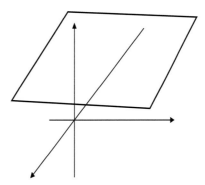

Figure 4.2: A hyperplane in 3-dimensional space.

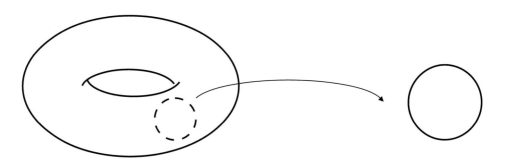

Figure 4.3: Geometric characterization of a 2-dimensional surface.

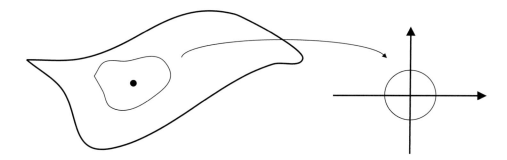

Figure 4.4: The idea of a manifold.

malization of the idea of Riemann surface, and also a need to make rigorous such constructs as the Klein bottle. The Klein bottle (to be discussed below) is a legitimate and intuitively appealing 2-dimensional surface, but it cannot be rigorously realized as a surface in 3-dimensional Euclidean space (in fact it requires 4-dimensional space for an embedding). In the present chapter we shall study how the rigorous concept of "manifold" can be presented.

4.2 The Definition

We begin with the basic definition of manifold:

Definition 4.2.1 Let $(X, \mathcal{U})$ be a second countable Hausdorff space. We say that X is a *manifold of dimension N* if each point $x \in X$ has a neighborhood U_x and a corresponding open set $V_x \subseteq \mathbb{R}^N$ so that U_x and V_x are homeomorphic via a *coordinate mapping* φ_x. We call X an *N-manifold*. Refer to Figure 4.4.

In this last definition, we call U_x a *coordinate patch* or *coordinate chart*. The mapping φ_x is a *coordinate map*. The collection of $\{U_x\}$ is called an *atlas* for the manifold structure.

EXAMPLE 4.2.2 Fix a positive integer N and consider the Euclidean space $X = \mathbb{R}^N$. Certainly X is a manifold. If $x \in X$ then the ball $B(x, 1) = \{t \in \mathbb{R}^N : |x - t| < 1\}$ will serve as a coordinate chart and the identity can be the coordinate mapping.

EXAMPLE 4.2.3 If M is any manifold as defined above and $\mathcal{W} \subseteq M$ is an open set, then $\mathcal{W}$ is a manifold. For if U is any coordinate chart of M and

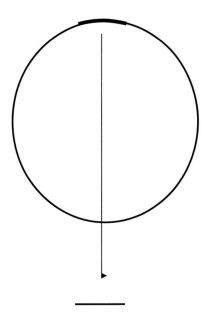

Figure 4.5: The circle S^1 is a 1-manifold.

$U \cap W \neq \emptyset$ then $U \cap W$ is a coordinate chart for W. And the corresponding coordinate map ϕ restricted to $U \cap W$ will serve nicely as the coordinate map.

EXAMPLE 4.2.4 Let S^1 be the unit circle in the Euclidean plane. Then S^1 is a 1-manifold. To see this, note that the point $(0, 1) \in S^1$ has a neighborhood $\{(x, y) \in \mathbb{R}^2 : x^2 + y^2 = 1, |x| < 1/10\}$ that is homeomorphic to the open line segment $(-1/10, 1/10) \subseteq \mathbb{R}$ by way of the mapping

$$(x, y) \mapsto (x, 0).$$

See Figure 4.5. Any other point in the circle S^1 has a similar neighborhood, with a similar mapping, as one can see by rotating that point to $(0, 1)$.

EXAMPLE 4.2.5 The unit sphere $\Sigma = \{(x, y, z) \in \mathbb{R}^3 : x^2 + y^2 + z^2 = 1\}$ in $\mathbb{R}^3$ is a 2-manifold. To see this, note that the point $(0, 0, 1) \in \Sigma$ has a neighborhood $\{(x, y, z) \in \mathbb{R}^3 : x^2 + y^2 + z^2 = 1, x^2 + y^2 < 1/10\}$ that is homeomorphic to the disc $d = \{(x, y) \in \mathbb{R}^2 : x^2 + y^2 < 1/100\}$ by way of the mapping

$$(x, y, z) \mapsto (x, y).$$

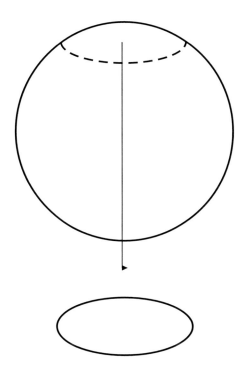

Figure 4.6: The sphere Σ is a 2-manifold.

See Figure 4.6. Any other point of the sphere Σ has a similar neighborhood, with a similar mapping, as one can see by rotating that point to $(0, 0, 1)$.

Remark 4.2.6 Similar reasoning shows that the sphere S^N in $\mathbb{R}^{N+1}$ is an N-manifold.

EXAMPLE 4.2.7 If M is an m-manifold and N is an n-manifold, then $M \times N$ is an $m \cdot n$ manifold. For if $(x, y) \in M \times N$ is any point then x has a coordinate neighborhood U_x in M with coordinate map φ_x to V_x and likewise y has a coordinate neighborhood U'_y in N with coordinate map φ'_y to V'_y. Then $U_x \times U'_y$ is a coordinate chart for (x, y). The coordinate mapping is $\varphi \times \varphi'$ which maps to $V_x \times V'_y$.

EXAMPLE 4.2.8 Consider the torus

$$S^1 \times S^1.$$

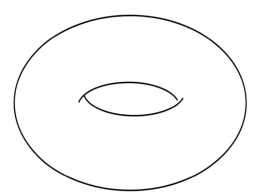

Figure 4.7: The torus is a 2-manifold.

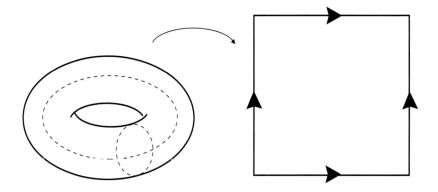

Figure 4.8: Cutting open the torus.

This is a 2-manifold. Refer to Figure 4.7. But it is awkward to write down an embedding of this surface into $\mathbb{R}^3$, and even more awkward to explicitly write down the coordinate mappings. Thus we shall now introduce another way to think about the matter.

Imagine taking a pair of scissors and cutting the torus around a circumference of one of the second circles. And then cutting again around a circumference of one of the first circles. See Figure 4.8.

We see that we have reduced the torus to a rectangle, with the understanding that the upper and lower edges are identified and the left and right edges are identified (with orientations indicated by the arrows). Now it is clear that if P is any point of the torus that corresponds to an interior point of the rectangle, then there is certainly a coordinate chart at P (for the rectangle is already Euclidean). If P is instead a point of the torus that cor-

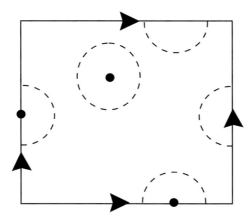

Figure 4.9: Coordinate charts on the torus.

responds to a boundary point of the rectangle, then it also has a coordinate chart just because the upper and lower edges of the rectangle are identified and the left and right edges are identified. See Figure 4.9.

EXAMPLE 4.2.9 Now we will use the device from the last example to examine the Klein bottle. Look at Figure 4.10. It is similar to Figure 4.8, except that there is a twist when the left and right edges are joined. The result is the surface that is suggested (only poetically, because this surface *cannot* be embedded into Euclidean 3-space) in Figure 4.11. This is the classic Klein bottle. It is the primordial example of what we call a *non-orientable* surface.

EXERCISE FOR THE READER 4.2.10 The Möbius strip is obtained from the rectangle shown in Figure 4.12. Explain why the Klein bottle is obtained from Möbius strip by "closing it up."

EXAMPLE 4.2.11 Let $0 < k < n$ and consider the collection $G(k, n)$ of all k-dimensional affine spaces in $\mathbb{R}^n$. In the case that $k = 1$ and $n = 2$ this would be the collection of all lines in the plane. In the case that $k = 2$ and $n = 3$ this would be the collection of all 2-planes in 3-dimensional space. It turns out that $G(k, n)$ can be given the structure of a manifold called the *Grassmannian*.

Let us examine carefully the case $k = 1$, $n = 2$. If the line $y = ax + b$ is a line ℓ that is an element of $G(1, 2)$, then we may identify this line naturally

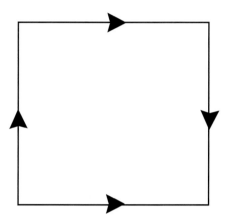

Figure 4.10: Rectangular model for the Klein bottle.

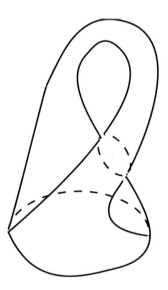

Figure 4.11: Poetic depiction of the Klein bottle.

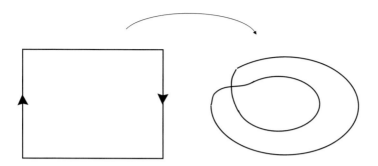

Figure 4.12: The Klein bottle is a "closed up" Möbius strip.

with the ordered pair $p_\ell = (a, b) \in \mathbb{R}^2$ and then a neighborhood of p_ℓ is just an open Euclidean disc. That gives a neighborhood of ℓ in $G(1, 2)$. If instead the line ℓ' has the form $x = \alpha y + \beta$, then we may identify the line naturally with the point $q_{\ell'} = (\alpha, \beta) \in \mathbb{R}^2$. And a neighborhood of $q_{\ell'}$ is an open disc. That defines a corresponding neighborhood of ℓ' in $G(1, 2)$. Thus we have set up a coordinate chart around any element of $G(1, 2)$, and we see thereby that $G(1, 2)$ is a manifold.

EXAMPLE 4.2.12 Consider projective space $\mathbb{P}^N$, which was introduced in Example 2.6.5. Let λ be an element of that space. This is a line ℓ (less the origin) through 0. There is a pair of points p_ℓ and $-p_\ell$ that lie in this line and have Euclidean norm 1. We may identify the pair $(p_\ell, -p_\ell)$ with a point in $S^N \times S^N$ (the product of spheres). By Example 4.2.5, and the subsequent Remark 4.3.6, S^N is a manifold. It is easy to see (see Example 4.2.7) that $S^N \times S^N$ is therefore a manifold. It follows then that $\mathbb{P}^N$ is a manifold.

EXAMPLE 4.2.13 The unit sphere S^2 in $\mathbb{R}^3$ is not homeomorphic to the torus T. This is an interesting assertion. Each of these is a compact 2-manifold. Our intuition tells us that these surfaces are quite different. While they each have a hole, we think that the holes have a different character. But it is conceivable that there is some devilishly clever mapping between the two that is one-to-one, onto, and bicontinuous. It turns out that there is not.

   The way to see this is to observe that the torus is the product of S^1 and S^1. It follows (exercise) that the fundamental group of the torus is $\mathbb{Z} * \mathbb{Z}$, the free product of two copies of $\mathbb{Z}$ (exercise). And any loop in the sphere can be shrunk to a point (exercise). So the fundamental group of the sphere is the singleton $\{e\}$. If there were a homeomorphism $\phi : S^2 \to T$, then that would induce an isomorphism of fundamental groups. But obviously $\mathbb{Z} * \mathbb{Z}$ is not isomorphic to $\{e\}$. So the surfaces cannot be homeomorphic.

Exercises

1. Consider the unit sphere

$$S^2 = \{(x, y, z) \in \mathbb{R}^3 : x^2 + y^2 + z^2 = 1\}$$

   in $\mathbb{R}^3$. Let us identify two points of the sphere if they are antipodal— that is, if they lie at opposite ends of the same diameter. The resulting space is projective space. Prove that this new space is a 2-dimensional manifold.

2. Consider $\mathbb{R}^3$ equipped with the relation

$$(x, y, z) \sim (x', y', z')$$

   if and only if $x - x'$, $y - y'$, and $z - z'$ are all integers. This is an equivalence relation. Show that the collection of equivalence classes is a 3-dimensional manifold. Can you describe this manifold?

3. If M and N are both manifolds, then $M \times N$ is a manifold in a natural way. Explain.

4. The Klein bottle can be defined by examining Figure 4.13. We identify the upper and lower edges with the same orientation (as the parallel arrows indicate), but we identify the left and right edges with reversed orientations. Explain why the result is a manifold.

   Figure 4.14 shows an imaginative rendition of the Klein bottle (it is in fact impossible to embed the Klein bottle into Euclidean 3-space).

5. A *cross cap* is a surface that is topologically equivalent to a Möbius strip (or sometimes a Möbius strip with a disc attached). Figure 4.15 shows two renditions of the resulting surface. Part of the classification by Jordan and Möbius of 2-manifolds is that any non-orientable 2-manifold (such as the Klein bottle) can be obtained by adding handles and cross caps to a sphere. We already know how to add a handle. We add a cross cap by cutting a hole in the cross cap and cutting a hole

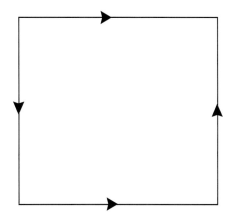

Figure 4.13: Formation of the Klein bottle.

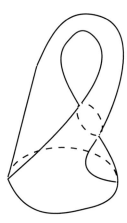

Figure 4.14: Artist's rendition of a Klein bottle.

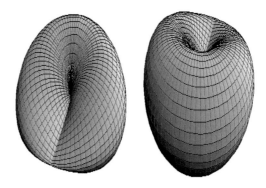

Figure 4.15: Two views of a crosscap.

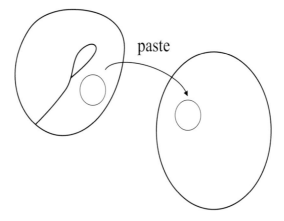

Figure 4.16: Pasting a crosscap to a sphere.

in the sphere and then pasting the two holes together. Refer to Figure 4.16. Explain why the Klein bottle is a sphere with two cross-caps attached.

6. The set of all lines in the plane can be turned into a manifold (this is an old idea of Hermann Grassman (1809–1877)). To see this, notice that every line in the plane can be parametrized by **(i)** where it crosses the y-axis and **(ii)** what angle it makes with the x-axis. [Of course there are exceptions to this description—the vertical lines—but you will see how to handle these after a moment's thought.] It is easy to turn $\mathbb{R} \times (0, \pi)$ into a manifold. Carry out the details of this program.

7. Refer to the last exercise for inspiration. Explain why the set of 2-planes in $\mathbb{R}^3$ can be realized as a manifold.

8. We know that the torus is a 2-dimensional manifold. Now remove a closed disc from the torus. Is the result still a manifold? Why or why not?

9. Consider a smooth, simple, closed curve in $\mathbb{R}^3$. This is a mapping $\gamma : [0,1] \to \mathbb{R}^3$ such that

   - γ is C^2;
   - γ' is never 0;
   - $\gamma(0) = \gamma(1)$;
   - $\gamma(s) \neq \gamma(t)$ if $s \neq t$ and $\{s,t\} \neq \{0,1\}$;
   - the two derivatives of γ match up at 0 and 1.

   Let S be the union of all the tangent lines to all the points of γ. Show that this "tangent bundle" S is a manifold in a natural way.

10. Let $(X, \mathcal{U})$ be a second countable Hausdorff space. We say that X is a C^k *manifold of dimension* N if each point $x \in X$ has a neighborhood U_x and a corresponding open set $V_x \subseteq \mathbb{R}^N$ and a coordinate homeomorphism $\varphi_x : U_x \to V_x$ such that, whenever $U_x \cap U_y \neq \emptyset$, the map $\varphi_x \circ \varphi_y^{-1} : V_x \cap V_y \to V_x \cap V_y$ is a C^k diffeomorphism. [**Note:** We call the map $varphi_x \circ \varphi_y^{-1}$ a *transition map*.]

    Prove that the torus is a C^k manifold of dimension 2 for every k.

11. Refer to the last exercise for terminology. Show that the unit sphere in $\mathbb{R}^3$ is a C^k manifold for every k.

12. Refer to Exercise 10 for terminology. Give an example of a C^1 manifold that is not obviously a C^2 manifold.

13. Let $f : \mathbb{R} \to \mathbb{R}$ be a C^k function. Show that the graph of f is a C^k manifold.

14. (*For those who know a little complex analysis.*) Modify the definition in Exercise 10 to define a (real) 2-dimensional [or complex 1-dimensional] complex manifold. That is, replace $\mathbb{R}^N$ by $\mathbb{C}$ and replace "C^k" by

"holomorphic." Think of the Riemann sphere from complex variable theory. Show that the Riemann sphere is a complex manifold. You may find the stereographic projection to be useful here.

15. Let M be an m-dimensional C^k manifold and let N be an n-dimensional C^k manifold with $0 < n < m$. Suppose that $N \subseteq M$. We say that N is a *regularly embedded submanifold* of M if, for each point $p \in N$, there is a neighborhood U of p in M, an open set $V \subseteq \mathbb{R}^m$, and a C^k diffeomorphism $\varphi_p : U \to V$ so that the image of N is a n-dimensional affine subspace of V. Show that a great circle is a 1-dimensional regularly embedded submanifold of the unit sphere in $\mathbb{R}^3$.

16. Refer to the last exercise for terminology. Show that the torus has two obvious regularly embedded circular submanifolds.

17. Let $f : \mathbb{R} \to \mathbb{R}$ be a C^k function. Show that the graph of f is a regularly embedded C^k submanifold of $\mathbb{R}^2$.

18. Formulate and prove a result similar to that in the last exercise for functions from $\mathbb{R}^2$ to $\mathbb{R}$.

19. Let $F : \mathbb{R}^2 \to \mathbb{R}$ be a continuously differentiable function, and set $M = \{(x,y) \in \mathbb{R}^2 : F(x,y) = 0\}$. Assume that if $(x,y) \in M$ then $\nabla F(x,y) \neq (0,0)$. Prove that M is a 1-dimensional manifold.

20. Formulate and prove a result like that in the last exercise for mappings $F : \mathbb{R}^M \to \mathbb{R}^N$.

21. Let M and N be C^k, 1-dimensional manifolds (see Exercise 10) for terminology. Let $f : M \to N$ be a mapping. If $P \in M$ and $Q \in N$ and U, V are coordinate charts for P and Q respectively and $\varphi : U \to \widetilde{U}$ and $\psi : V \to \widetilde{V}$ the respective coordinate mappings then we demand that $\psi \circ f \circ \phi^{-1}$ be C^k. Under this circumstance we say that f is a C^k mapping of manifolds. Give an example of a C^2 mapping of the unit circle to the ellipse $x^2 + 2y^2 = 4$.

22. If X_j are manifolds and $X_1 \subseteq X_2 \subseteq X_3 \subseteq \cdots$ then define $X = \cup_j X_j$. Is X a manifold? Why or why not?

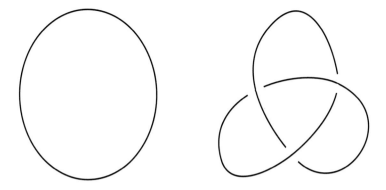

Figure 4.17: The circle and the trefoil.

23. Let $X \subseteq \mathbb{R}^2$ be a 1-manifold and let $Y \subseteq \mathbb{R}^2$ be a 1-manifold. Does it follow that $X \cup Y$ is a 1-manifold? Why or why not? How about $X \cap Y$?

24. Let $\pi : \widehat{X} \to X$ be a covering mapping. Assume that $\widehat{X}$ is a manifold of dimension k. Show that therefore X is a manifold of dimension k.

25. Let $\pi : \widehat{X} \to X$ be a covering mapping. Assume that X is a manifold of dimension k. Show that therefore $\widehat{X}$ is a manifold of dimension k.

26. Describe all the manifolds of dimension 1.

27. Figure 4.17 shows two topological spaces. One is simply the circle, and the other is a knot known as the trefoil. Both are 1-manifolds. Explain why they are homeomorphic.

28. Let S^2 be the standard unit sphere in $\mathbb{R}^3$. It is of course a 2-manifold. Now create a new topological space by identifying antipodal points of S^2. What space results? Is it a manifold? Why or why not?

29. Let $p(x, y)$ be a polynomial of two real variables and let V be the zero set of p. Then p is not necessarily a 1-manifold. Explain, by way of an example, why not. But there is an exceptional, 0-dimensional set X such that $V \setminus X$ *is* a manifold. Explain why that is the case.

30. Let

$$Q = \left\{ (x, y, z) \in \mathbb{R}^3 : -1 \le x \le 1, -1 \le y \le 1, -1 \le z \le 1 \right\}.$$

This is of course the unit cube in $\mathbb{R}^3$. Now form a new space by identifying opposite faces of Q. The result is a 3-dimensional manifold. Prove that assertion. What manifold is it?

Chapter 5

Moore-Smith Convergence and Nets

5.1   Introductory Remarks

One of the nice features of the metric space setting is that all topological notions can be formulated in terms of sequences. Such is not the case in an arbitrary topological space. In that more general setting we must use the theory of nets, and the associated ideas of Moore-Smith convergence. That is the topic of the present chapter.

Whereas a sequence is modeled on the natural numbers, a net is modeled on a more general object called a directed set. The general feel of the subject is similar to that for sequences, but it is rather more abstract. We shall find good use for nets later in the book, especially in the chapter on function spaces.

5.2   Nets

We first define directed sets, and then nets.

Definition 5.2.1 Let D be a nonempty set and $\geq$ a binary relation on D. We say that $\geq$ *directs* D provided that:

  (a) If m, n, p are members of D such that $m \geq n$ and $n \geq p$, then $m \geq p$;

  (b) If $m \in D$, then $m \geq m$;

(c) If m, n are elements of D, then there exists $p \in D$ such that $p \geq m$ and $p \geq n$.

We note that **(a)** is a *transitivity* property and **(b)** is a *reflexivity* property. Condition **(c)** might be called the *Archimedean property*. We call D or $(D, \geq)$ a *directed set*.

EXAMPLE 5.2.2 The set $\mathbb{R}$, the real numbers, equipped with the usual ordering $\geq$, is a directed set. The nonnegative integers $\mathbb{N}$, together with the ordering $\geq$, is a directed set.

EXAMPLE 5.2.3 Let $(X, \mathcal{U})$ be a topological space. Let $x \in X$ and let $\mathcal{E}$ be the family of all neighborhoods of x. Then $\mathcal{E}$ is directed by the relation $\subseteq$.

EXAMPLE 5.2.4 Let X be any infinite set and let $\mathcal{E}$ be the collection of all finite subsets of X. Then $\mathcal{E}$ is directed by $\subseteq$.

Definition 5.2.5 A *net* is a pair $(f, \geq)$, where $f : X \to Y$ is a function and $\geq$ directs X. It is sometimes convenient to write a net as $(S_n, n \in D, \geq)$, where the directed set D is the domain of the function S (here $n \in D$ is playing the role of $x \in X$ and S takes n to $S(n) = S_n$). When context is understood, we shall often write the net as $(S, \geq)$.

Definition 5.2.6 We say that the net $(S_n, n \in D, \geq)$ is *in* the set A if $S_n \in A$ for all $n \in D$. We say that $(S_n, n \in D, \geq)$ is *eventually in* A if there is an element $m \in D$ such that, if $n \in D$ and $n \geq m$, then $S_n \in A$. Finally, the net is *frequently in* A if, for each $m \in D$, there is an $n \in D$ such that $n \geq m$ and $S_n \in A$.

Remark 5.2.7 If $(S_n, n \in D, \geq)$ is frequently in A, then the set E of all members $n \in D$ such that $S_n \in A$ has the property that, for each $m \in D$ there is a $p \in E$ such that $p \geq m$. The reader may verify this assertion as an exercise. The subset E is called *cofinal*.

EXERCISE FOR THE READER 5.2.8 Every confinal subset E of D is also directed by $\geq$. Also prove the following equivalence: A net $(S_n, n \in D, \geq)$ is frequently in a set A if and only if a cofinal subset of D maps into the set A, and this is so if and only if the net is not eventually in the complement of A.

Definition 5.2.9 A net $(S, \geq)$ in a topological space $(X, \mathcal{U})$ *converges* to s relative to $\mathcal{U}$ if and only if it is eventually in every $\mathcal{U}$-neighborhood of s.

It is not difficult to build on what has been presented thus far and describe the accumulation points of a set, the closure of a set, and indeed the topology of a space in terms of convergence of nets.

THEOREM 5.2.10 *Let $(X, \mathcal{U})$ be a topological space. Then*

(a) *A point s is an accumulation point of a subset A of X if and only if there is a net in $A \setminus \{s\}$ which converges to s.*

(b) *A point s belongs to the closure of a subset A of X if and only if there is a net in A converging to s.*

(c) *A subset A of X is closed if and only if no net in A converges to a point of $X \setminus A$.*

Proof: If s is an accumulation point of A, then for each neighborhood U of s there is a point t_U of A which belongs to $U \setminus \{s\}$. The family $\mathcal{U}$ of all neighborhoods U of s is directed by $\subseteq$. If U and V are neighborhoods of s such that $V \subseteq U$, then $t_V \in V \subseteq U$. The net $(t_U, U \in \mathcal{U}, \subseteq)$ therefore converges to s. For the converse, if a net in $A \setminus \{s\}$ converges to s, then this net has values in every neighborhood of s and $A \setminus \{s\}$ certainly intersects every neighborhood of s. This proves **(a)**.

To prove **(b)**, we recall that the closure of a set A consists of A together with all the accumulation points of A. For each accumulation point a of A, there is (by our discussion above) a net in A converging to a. For each point s of A, any net whose value at every element of its domain is s certainly converges to s. Therefore each point of the closure of A has a net in A converging to it. Conversely, if there is a net in A converging to s, then every neighborhood of s intersects A and s belongs to the closure of A.

Part **(c)** is now immediate. $\qquad\qquad\square$

EXAMPLE 5.2.11 A net in a topological space may converge to several different points. As an example, consider the space of integers equipped with the topology consisting of sets whose complements are finite. Let S be the net $S_j = j$, with the domain $\mathbb{N}$ directed by the usual ordering $\geq$. Then it is easy to see that S is eventually in every open set. So S converges to every element of our topological space.

THEOREM 5.2.12 *A topological space $(X, \mathcal{U})$ is Hausdorff if and only if each net in the space converges to at most one point.*

Proof: Let X be Hausdorff and let $s, t \in X$ be distinct points. Then there are disjoint neighborhoods U, V of s, t respectively. Since a net cannot eventually be in each of two disjoint neighborhoods, it is clear that no net in X converges to both points s and t.

For the converse, assume that X is *not* a Hausdorff space. Let s, t be distinct points of X so that every neighborhood of s intersects every neighborhood of t. If $\mathcal{U}_s$ is the collection of all neighborhoods of s and $\mathcal{U}_t$ is the collection of all neighborhoods of t, then each of these collections is directed by $\subseteq$. Let us order the Cartesian product $\mathcal{U}_s \times \mathcal{U}_t$ by saying that $(T, U) \geq (V, W)$ provides that $T \subseteq V$ and $U \subseteq W$. Then the Cartesian product so described is directed by $\geq$. If now $(T, U) \in \mathcal{U}_s \times \mathcal{U}_t$, then the intersection $T \cap U$ is nonempty so we may select a point $p_{T,U} \in T \cap U$. If $(V, W) \geq (T, U)$ then $p_{V,W} \in V \cap W \subseteq T \cap U$. As a result, the net $(p_{T,U}, (T, U) \in \mathcal{U}_s \times \mathcal{U}_t, \geq)$ converges to both s and t. $\qquad\square$

The idea of nets, and the associated concept of convergence, is referred to in the literature as Moore-Smith convergence.

Exercises

1. Let X be the space of C^∞ functions on $\mathbb{R}$. For each compact set $K \subseteq \mathbb{R}$, each $f \in X$, each $k \in \{0, 1, 2, \dots\}$, and each $\epsilon > 0$ we define

$$\mathcal{U}_{f,k,K,\epsilon} = \left\{ g \in X : \sup_{x \in K} \left| \frac{d^j}{dx^j}(f - g)(x) \right| < \epsilon, 0 \leq j \leq k \right\}.$$

   We let the $U_{f,k,K,\epsilon}$ be a sub-basis for a topology $\mathcal{U}$ on X. Describe what Moore-Smith convergence means in the context of this space.

2. Refer to Exercise 1 for terminology and notation. Let T be the collection of polynomial functions in X. What is the closure of T in the topology $\mathcal{U}$? [*Hint:* The Weierstrass approximation theorem may prove useful here.]

3. Refer to Exercise 1 for terminology and notation. Verify directly that X is a Hausdorff space.

4. Use the language of nets to define a boundary point of a set.

5. Use the language of nets to define continuous function. Prove that your new definition is equivalent to the one formulate in the language of the inverse image of an open set.

6. Use the language of nets to define the concept of $\lim_{x \to +\infty} f(x) = \ell$ for a real-valued function f on the real line.

7. Let X be the real line equipped with the topology consisting only of the empty set and the entire space X. What are the convergent nets in X?

8. Formulate the concept of completeness of a topological space in the language of nets.

9. When we discuss sequences, we often speak of subsequences. Now formulate a concept of subnet. A set K in a metric space is compact if and only if every sequence in K has a convergent subsequence. Is there an analogous characterization of compact sets in an arbitrary topological space using the language of nets?

10. Let X be the space of continuous functions on $\mathbb{R}$. For each $f \in X$, each compact $K \subseteq \mathbb{R}$, and each $\epsilon > 0$, define

$$V_{f,K,\epsilon} = \left\{ g \in X : \sup_{x \in K} |f(x) - g(x)| < \epsilon \right\}.$$

We let the $V_{f,K,\epsilon}$ be a sub-basis for a topology $\mathcal{V}$. Verify by hand that the set of all f in X that are strictly bounded by 1 is open.

11. Refer to Exercise 10 for terminology and notation. Check that X is a T_1 space. Is it also T_2?

12. Refer to Exercise 10 for terminology and notation. Construct a net of bounded, continuous functions that converges to e^{x^2}.

13. Refer to Exercise 10 for terminology and notation. Find a countable dense subset for X in the topology $\mathcal{V}$.

14. Refer to Exercise 10 for terminology and notation. Find a net of unbounded, continuous functions that converges to the identically 0 function.

15. Refer to Exercise 10 for terminology and notation. Let $f_j(x) = \sin jx$, $j = 1, 2, \ldots$. Discuss convergence and limiting properties of this net.

16. Can you formulate a definition of "open set" using the language of nets?

17. Let X be the real numbers equipped with the trivial topology such that every set is open. What do the nets look like in this space?

Chapter 6

Function Spaces

6.1 Preliminary Ideas

Many interesting examples of topological spaces arise in the context of function spaces. Function spaces are very natural artifacts of analysis, and they are interesting because they are usually infinite dimensional. What does this mean?

If V is a vector space over the field $\mathbb{R}$, then V will have a basis.[1] If the basis has finitely many elements $\mathbf{v}_1, \ldots, \mathbf{v}_k$, then any other basis for V will also have k elements. We call k the *dimension* of V. There is also a notion of dimension for a topological space that is not necessarily a vector space. We shall not provide the details here, but refer the interested reader to [HUW] and also to our brief discussion of the idea in Section 1.10.

EXAMPLE 6.1.1 Let X be the linear space of all continuous functions on the real line. This is certainly a vector space—closed under addition and scalar multiplication. Define

$$\varphi_j(x) = \begin{cases} 0 & \text{if} & x \le j \\ x - j & \text{if} & j < x \le j + 1/2 \\ (j+1) - x & \text{if} & j + 1/2 < x \le j + 1 \\ 0 & \text{if} & j + 1 < x. \end{cases}$$

Then it is plain that the functions φ_j lie in X and are linearly independent.

_____

[1]This statement is actually a nontrivial theorem that requires the Axiom of Choice for proof.

Also there are infinitely many of these functions. Thus X does *not* have a finite basis. We say that X is an *infinite-dimensional* vector space.

Most any space of functions having infinitely many elements, and so that every function in the space has the same domain, will be infinite dimensional. One result is that there will be a great many different topologies on such a space (finite-dimensional spaces tend to have few topologies). We shall study some of them here.

6.2   The Topology of Pointwise Convergence

Let S be any set and let $\mathcal{F}(S)$ be the space of real-valued functions with domain S. We also denote this space as $\mathbb{R}^S$, and we think of this as a product space (see Section 2.2). The topology of pointwise convergence is nothing other than the product topology on $\mathbb{R}^S$. A sub-basis for the topology is the collection of sets

$$\mathcal{E}_{s,U} = \{f \in \mathcal{F}(S) : f(s) \in U\}$$

for $s \in S$ a fixed point and $U \subseteq \mathbb{R}$ an open set.

In plain language, a sequence of functions f_j on S converges pointwise if, for each $s \in S$, $\lim_{j \to \infty} f_j(s)$ exists. Then we define a limit function by $f_0(s) = \lim_{j \to \infty} f_j(s)$.

Definition 6.2.1 Let $\mathcal{E}$ be a family of functions from the set S to the real numbers $\mathbb{R}$. We say that $\mathcal{E}$ is *pointwise closed* if it is closed as a subset of the space $\mathbb{R}^S$.

Proposition 6.2.2 *Let $\mathcal{E}$ be a family of functions from a set S to $\mathbb{R}$. Then $\mathcal{E}$ is compact with respect to the topology of pointwise convergence provided that*

(a) *$\mathcal{E}$ is pointwise closed in $\mathbb{R}^S$;*

(b) *for each point $s \in S$, the set $\mathcal{E}(s) \equiv \{f(s) : f \in \mathcal{E}\}$ has compact closure in $\mathbb{R}$.*

*The conditions **(a)** and **(b)** are also necessary for $\mathcal{E}$ to be compact in the topology of pointwise convergence.*

Proof: Of course the family $\mathcal{E}$ is a subfamily of $\mathbb{R}^S$. But it is also contained in $\times_{s \in S} \overline{\mathcal{E}}(s)$. If condition **(b)** holds, then this latter product is a compact subset of $\mathbb{R}^S$ by Tychanoff's theorem. If $\mathcal{E}$ is pointwise closed, then $\mathcal{E}$ is compact. This proves the sufficiency of **(a)** and **(b)** for compactness of $\mathcal{E}$ in the topology of pointwise convergence.

For the converse, suppose that $\mathcal{E}$ is compact in the topology of pointwise convergence. Then of course $\mathcal{E}$ is closed. The set $\mathcal{E}(s)$ is compact for each $s \in S$ and it is closed because the point evaluation map $e_s : \mathcal{E} \to \mathbb{R}$ given by $e_s(f) = f(s)$ is continuous. That proves the result. $\qquad \square$

Definition 6.2.3 Let $\mathcal{E}$ be a family of functions with common domain S. Let $A \subseteq S$. We say that the set A *distinguishes members of* $\mathcal{E}$ provided that, if f and g are distinct members of $\mathcal{E}$, then there is a point $a \in A$ such that $f(a) \neq g(a)$.

Proposition 6.2.4 *Let $\mathcal{E}$ be a family of functions on the set S, with values in $\mathbb{R}$. Let $A \subseteq S$. The family $\mathcal{E}$ with the topology of pointwise convergence on A is a Hausdorff space if and only if A distinguishes members of $\mathcal{E}$.*

Proof: The product space $\mathbb{R}^A$ is Hausdorff. Of course $\mathcal{E}$ with the topology of pointwise convergence on A is Hausdorff if and only if the map that takes $f \in \mathcal{E}$ to its restriction to the domain A is one-to-one. This will hold if and only if A distinguishes members of $\mathcal{E}$. $\qquad \square$

6.3 The Compact-Open Topology

The compact-open topology arises naturally in complex variable theory, in the study of isometry groups, and in many other contexts as well. One motivating question for this topology is the following: Let $\mathcal{E}$ be a family of functions from S to $\mathbb{R}$. Under what circumstances is the mapping

$$\begin{aligned} i : \mathcal{E} \times S &\to \mathbb{R} \\ (f, s) &\mapsto f(s) \end{aligned}$$

continuous? When it is continuous we say that the topology is *jointly continuous*.

Definition 6.3.1 Let $(X,\mathcal{U})$ and $(Y,\mathcal{V})$ be topological spaces. Let $\mathcal{E}$ be a family of functions from X to Y. If $K \subseteq X$ and $U \subseteq Y$, we let $\mathcal{W}(K,U)$ denote those $f \in \mathcal{E}$ such that $f(K) \subseteq U$. The family of subsets $\mathcal{W}(K,U)$ when K is compact in X and U is open in Y forms a sub-basis for a topology $\mathcal{C}$ on $\mathcal{E}$ called the *compact-open topology*.

Because each singleton set is compact, it is now simple to compare the compact-open topology with the topology of pointwise convergence.

Proposition 6.3.2 *The compact-open topology $\mathcal{C}$ contains the topology $\mathcal{P}$ of pointwise convergence. The space $(\mathcal{E},\mathcal{C})$ is a Hausdorff space provided that the range space Y is Hausdorff.*

Proof: For each $x \in X$ and each open subset $U \subseteq Y$, the set

$$\mathcal{W}(\{x\},Y) = \{f : f(x) \in U\} \qquad (6.3.2.1)$$

belongs to $\mathcal{C}$ because $\{x\}$ is compact. Therefore $\mathcal{P} \subseteq \mathcal{C}$ because the family of all sets of the form (6.3.2.1) is a sub-basis for the pointwise topology $\mathcal{P}$.

If Y is a Hausdorff space, then $(\mathcal{E},\mathcal{P})$ is a Hausdorff space, just because the product of Hausdorff spaces is Hausdorff. If U, V are disjoint $\mathcal{P}$-neighborhoods of distinct members f, g of $\mathcal{E}$, then they are also $\mathcal{C}$-neighborhoods. So $(\mathcal{E},\mathcal{C})$ is Hausdorff. $\qquad\square$

6.4 Uniform Convergence

Here we study the concept of uniformity for a family $\mathcal{E}$ of functions from a set X to a uniform space $(Y,\mathcal{V})$. Note that our results do not depend on, and do not require, any topological structure on the set X. However, in the case that X *does* have a topology, we will be able to consider the question of whether the uniform limit of continuous functions is continuous.

Definition 6.4.1 Let $\mathcal{E}$ be a family of functions from a set X to a uniform space $(Y,\mathcal{V})$. For each $V \in \mathcal{V}$, let $W(V)$ be the set of all pairs of functions (f,g) such that $(f(x),g(x)) \in V$ for each $x \in X$.[2] Now let $W(V)[f]$ be the set of all g such that $g(x) \in V[f(x)]$ for every $x \in X$. Then

---

[2]Remember here that $\mathcal{V}$ is a uniformity, so it consists of pairs. Refer back to Section 2.7 for our discussion of uniformities.

- $W(V^{-1}) = \left(W(V)\right)^{-1}$;

- $W(U \cap V) = W(U) \cap W(V)$;

- $W(U \circ V) \supseteq W(U) \circ W(V)$

for all $U, V \in \mathcal{V}$. Thus the family of sets $W(V)$ for $V \in \mathcal{V}$ is a basis for a new uniformity $\mathcal{U}$ for $\mathcal{E}$. We call $\mathcal{U}$ the *uniformity of uniform convergence*. The topology induced by $\mathcal{U}$ is the *topology of uniform convergence*.

EXERCISE FOR THE READER 6.4.2 Prove that the uniformity $\mathcal{U}$ in the last definition is larger than the uniformity of pointwise convergence. Also uniform convergence implies pointwise convergence.

THEOREM 6.4.3 *Let $\mathcal{E}$ be the family of all functions from a set X to a uniform space $(Y, \mathcal{V})$. Let $\mathcal{U}$ be the uniformity of uniform convergence on Y. Then*

(a) *The uniformity $\mathcal{U}$ is generated by the family of all pseudometrics of the form $d^*(f, g) = \sup\{d(f(x), g(x)) : x \in X\}$, where d is a bounded member of the gage of $(Y, \mathcal{V})$ (see Section 1.12 for the concept of pseudometric and Section 2.7 for the concept of gage).*

(b) *A net $\{f_j : j \in D\}$ converges uniformly to g if and only if it is a Cauchy net relative to $\mathcal{U}$ and $\{f_j(x), j \in D\}$ converges to $g(x)$ for each $x \in X$.*

(c) *If $(Y, \mathcal{V})$ is complete, then so is the uniform space $(\mathcal{E}, \mathcal{U})$.*

Proof: For part **(a)**, note that if d is a bounded member of the gage of $\mathcal{V}$, then the family of all sets of the form $\{(y, z) : d(y, z) \leq r\}$, $r > 0$, is a basis for $\mathcal{V}$. This is so because if e is a pseudometric, then the pseudometric $d^* = \min\{1, e\}$ is bounded and has the same uniformity. But

$$\begin{aligned}
\{(f, g) : d^*(f, g) \leq r\} &= \{(f, g) : d^*(f, g) \leq r \text{ for each } x \in X\} \\
&= W(\{(y, z) : d(y, z) \leq r\}),
\end{aligned}$$

where W is the correspondence used above to define the uniformly continuous uniformity. In conclusion, d^* belongs to the gage of $\mathcal{U}$ and the pseudometrics of this form generate the gage. That establishes **(a)**.

The "only if" part of **(b)** is obvious. For the "if" part, suppose that a Cauchy net $\{f_j : j \in D\}$ converges pointwise to g; we must then show that it converges uniformly to g. Let V be an arbitrary, closed, symmetric member of $\mathcal{V}$. Select $m \in D$ so that, if $j \geq m$ and $k \geq m$, then $f_k(x) \in V[f_j(x)]$ for each $x \in X$. We may do this because the net is supposed to be Cauchy relative to $\mathcal{U}$. Since $V[f_j(x)]$ is closed and $f_k(x)$ converges to $g(x)$, it follows that $g(x) \in V[f_j(x)]$ hence $f_j(x) \in V[g(x)]$ for each $j \geq m$ and all $x \in X$. So **(b)** is proved.

We note that **(c)** is immediate by **(b)** and the fact that the product of complete spaces is complete.   □

Now the main properties of the uniformity for uniform convergence are capsulized in the following theorem.

THEOREM 6.4.4 *Let $\mathcal{E}$ be the family of all continuous functions from a topological space X to a uniform space $(Y, \mathcal{V})$. Let $\mathcal{U}$ be the uniformity of uniform convergence. Then*

   (a) *The family $\mathcal{E}$ is closed in the space of all functions from X to Y. As a result, $(\mathcal{E}, \mathcal{U})$ is complete if $(Y, \mathcal{V})$ is complete.*

   (b) *The topology of uniform convergence is jointly continuous.*

Proof: We prove part **(a)** indirectly by showing that the set of all non-continuous functions is an open subset of the space $\mathcal{G}$ of all functions from X to Y. If in fact f is not continuous at a point $x \in X$, then there is a member $V \in \mathcal{V}$ such that $f^{-1}[V[f(x)]]$ is not a neighborhood of x. Select a symmetric member W of $\mathcal{V}$ such that $W \circ W \circ W \subseteq V$. We shall show that if g is a function such that $(g(y), f(y)) \in W$ for each y, then $g \subseteq W \circ f$ and $f^{-1} \subseteq f^{-1} \circ W^{-1} = f^{-1} \circ W$ and therefore $g^{-1} \circ W \circ g \subseteq f^{-1} \circ W \circ W \circ W \circ f \subseteq f^{-1} \circ V \circ f$. As a result, $g^{-1}[W[g(x)]]$ is a subset of $f^{-1}[V[f(x)]]$ and is thus not a neighborhood of x. Thus **(a)** is proved.

For **(b)**, we need to demonstrate the continuity of the map of $\mathcal{E} \times X$ into Y at a point (f, x). It is necessary to verify that, for $V \in \mathcal{V}$, if $y \in f^{-1}[V[f(x)]]$ and $g(z) \in V[f(z)]$ for all z, then $g(y) \in V[f(y)] \subseteq V \circ V[f(x)]$. That completes the argument.   □

It is a useful device to consider uniform convergence on each member of a family $\mathcal{A}$ of subsets of the domain space X. For example, in complex variable theory we commonly consider the topology of uniform convergence on compact sets. In detail, if $\mathcal{E}$ is a family of functions from a set X to a uniform space $(Y, \mathcal{V})$ and if $\mathcal{A}$ is a family of subsets of X, then the uniformity of *uniform convergence on members of* $\mathcal{A}$, abbreviated $\mathcal{U}|\mathcal{A}$, has for a subbasis the family of all sets of the form

$$\{(f, g) : (f(x), g(x)) \in V \text{ for all } x \in A\}$$

where $V \in \mathcal{V}$ and $A \in \mathcal{A}$.

EXERCISE FOR THE READER 6.4.5 Prove that the topology of uniform convergence on compact sets is just the same as the compact-open topology.

6.5 Equicontinuity and the Ascoli-Arzela Theorem

Equicontinuity is a notion of uniform continuity over a family of functions. It is useful when we want to prove compactness theorems for families of functions.

Definition 6.5.1 Let $\mathcal{E}$ be a family of mappings from a topological space X into a uniform space $(Y, \mathcal{V})$. The family $\mathcal{E}$ is said to be *equicontinuous at a point* x if, for each $V \in \mathcal{V}$, there is a neighborhood U of x such that $f[U] \subseteq V[f(x)]$ for every $f \in \mathcal{E}$.

Remark 6.5.2 In case (X, d), (Y, e) are metric spaces then there is a particularly elegant and compelling formulation of equicontinuity. We say that a family of functions $f_\alpha : X \to Y$ is equicontinuous at x if, for any $\epsilon > 0$, there is a $\delta > 0$ such that if $d(x, t) < \delta$ then $e(f_\alpha(x), f_\alpha(t)) < \epsilon$ for all α.

Proposition 6.5.3 *If the family $\mathcal{E}$ of functions from X to Y is equicontinuous at x, then the closure of $\mathcal{E}$ relative to the topology $\mathcal{P}$ of pointwise convergence is also equicontinuous at x.*

Proof: Let $x \in X$ and U a neighborhood of x. Let V be a member of the uniformity of Y that is a closed set. Then the class of all functions f which satisfy

$f[U] \subseteq V[f(x)]$ is closed relative to the topology $\mathcal{P}$ of pointwise convergence because the set is just the same as $\bigcap\{\{f : (f(y), f(x)) \in V\} : y \in U\}$. It follows that the pointwise closure of $\mathcal{E}$ is equicontinuous.   □

Definition 6.5.4 A family $\mathcal{E}$ of functions is said to be *equicontinuous* if it is equicontinuous at each point.

It follows that the closure of an equicontinuous family in the topology of pointwise convergence is also equicontinuous.

Proposition 6.5.5 *Let $\mathcal{E}$ be an equicontinuous family of functions. Then the topology of pointwise convergence on $\mathcal{E}$ is jointly continuous and therefore coincides with the topology of uniform convergence on compact sets.*

Proof: We want to show that the map of $\mathcal{E} \times X \to Y$ given by $(f, x) \mapsto f(x)$ is continuous at a point (f, x). So let V be a member of the uniformity of Y and let U be a neighborhood of x such that $g[U] \subseteq V[g(x)]$ for all $g \in \mathcal{E}$. If g is a member of the $\mathcal{P}$-neighborhood $\{h : h(x) \in V[f(x)]\}$ of f and $y \in U$, then $g(y) \in V[g(x)]$ and $g(x) \in V[f(x)]$. As a result, $g(y) \in V \circ V[f(x)]$. Joint continuity follows. One may verify that each jointly continuous topology is larger than the compact-open topology, and the compact-open topology coincides with that of uniform convergence on compact sets.   □

One interpretation of this last result is that an equicontinuous family of functions is compact relative to the topology of uniform convergence on compact sets if it is compact relative to pointwise convergence. Also the Tychanoff theorem gives suffcent criteria for compactness in the pointwise topology. Thus we see that equicontinuity plus some other conditions gives compactness for a family of functions. The next result is a sort of converse.

THEOREM 6.5.6 *Let $\mathcal{E}$ be a family of functions from a topological space X to a uniform space $(Y, \mathcal{V})$. If this family is compact relative to a jointly continuous topology, then $\mathcal{E}$ is equicontinuous.*

Proof: Let $x \in X$ be a fixed point and V a symmetric member of $\mathcal{V}$. If we can show that there is a neighborhood U of x such that $g[U] \subseteq V \circ V[g(x)]$ for each g in $\mathcal{E}$, then the result will follow.

Because the topology on $\mathcal{E}$ is jointly continuous, there is for each $f \in \mathcal{E}$ a neighborhood $\mathcal{G}$ of f and a neighborhood W of x such that $\mathcal{G} \times W$ maps into $V[f(x)]$. If $g \in \mathcal{G}$ and $w \in W$, then $g(x)$ and $g(w)$ both belong to $V[f(x)]$ and hence $g(w) \in V \circ V[g(x)]$. That is to say, $g[W] \subseteq V \circ V[g(x)]$ for each $g \in \mathcal{G}$. Since $\mathcal{E}$ is compact, there is a finite family of open sets $\mathcal{G}_1, \mathcal{G}_2, \ldots, \mathcal{G}_k$ covering $\mathcal{E}$ and corresponding neighborhoods $W_1, \ldots, W_k$ of x such that $g[W_j] \subseteq V \circ V[g(x)]$ for each $g \in \mathcal{G}_j$. If we let U be the intersection of the neighborhoods W_j of x, then it is clear that $g[U] \subseteq V \circ V[g(x)]$ for every $g \in \mathcal{G}$. $\qquad\square$

Now we have the celebrated Ascoli-Arzela theorem.

THEOREM 6.5.7 *Let $\mathcal{C}$ be the family of all continuous functions from a regular, locally compact topological space X to a Hausdorff uniform space $(Y, \mathcal{V})$. Assume that $\mathcal{C}$ has the topology of uniform convergence on compact sets. Then a subfamily $\mathcal{E}$ of $\mathcal{C}$ is compact if and only if*

(a) *$\mathcal{E}$ is closed in $\mathcal{C}$;*

(b) *$\mathcal{E}[x]$ has compact closure for each $x \in X$;*

(c) *the family $\mathcal{E}$ is equicontinuous.*

Proof: This is immediate from what went before—especially 6.5.5 and 6.5.6. $\square$

For the record, we now record a version of Ascoli-Arzela that can be found in many textbooks. We leave the verification of this result for the reader.

THEOREM 6.5.8 *Let X be a compact metric space. Let $\mathcal{F} = \{f_\alpha\}_{\alpha \in A}$ be a family of real-valued functions on X that satisfies*

(a) *$\mathcal{F}$ is equicontinuous;*

(b) *$\mathcal{F}$ is uniformly bounded in the sense that there is an $M > 0$ such that $|f_\alpha(x)| \leq M$ for all $f_\alpha \in \mathcal{F}$ and all $x \in X$.*

Then there is a subsequence $\{f_{\alpha_j}\}$ that converges uniformly on X.

6.6   The Weierstrass Approximation Theorem

One of the startling results of nineteenth-century analysis was the celebrated approximation theorem of Weiestrass. A bit of context is in order so that we may appreciate the significance and meaning of the theorem.

Perhaps the simplest and easiest of all functions to understand is the polynomial. A polynomial is a function of the form

$$p(x) = a_0 + a_1 x + a_2 x^2 + \cdots + a_k x^k .$$

The nice thing about a polynomial is that you plug in a number x and you do some simple, straightforward arithmetic calculations and then you get an answer $p(x)$. Most functions are not like that. Even the familiar sine and cosine functions can only be calculated (by hand) at certain special values. The same for the logarithm and the exponential.

What Weierstrass tells us is that *any* continuous functions f on the interval $[0, 1]$ can be approximated by a polynomial. Even more striking is that this approximation is *uniform*: given an $\epsilon > 0$ there is a polynomial p such that

$$|f(x) - p(x)| < \epsilon$$

for all $x \in [0, 1]$.

Weierstrass's ideas grew out of his studies of trigonometric and Fourier series. His proof fits very naturally into that context. The argument that we present here is a modern and streamlined treatment that is more self-contained.

The name Weierstrass has occurred frequently in this chapter. In fact Karl Weierstrass (1815-1897) revolutionized analysis with his examples and theorems. This section is devoted to one of his most striking results. We introduce it with a motivating discussion.

It is natural to wonder whether the standard functions of calculus—$\sin x, \cos x$, and e^x, for instance—are actually polynomials of some very high degree. Since polynomials are so much easier to understand than these transcendental functions, an affirmative answer to this question would certainly simplify mathematics. Of course a moment's thought shows that this wish is impossible: a polynomial of degree k has at most k real roots. Since sine and cosine have infinitely many real roots, they cannot be polynomials. A polynomial of degree k has the property that if it is differentiated enough times (namely $k + 1$ times) then the derivative is zero. Since this is not the case for e^x, we conclude that e^x cannot be a polynomial. The exercises discuss other means for distinguishing the familiar transcendental functions of calculus from polynomial functions.

In calculus we learn of a formal procedure, called Taylor series, for associating polynomials with a given function f. In some instances these polynomials form a sequence that converges back to the original function. Of course the method of the Taylor expansion has no hope of working unless f is infinitely differentiable. Even then, it turns out that the Taylor series rarely converges back to the original function. Even when the Taylor series converges, there is no guarantee that it will converge to the original f. See [KRP1] for a detailed consideration of these matters.

Nevertheless, Taylor's theorem with remainder might cause us to hope that any reasonable function can be approximated in some fashion by polynomials. In fact the theorem of Weierstrass gives a spectacular affirmation of this speculation:

THEOREM 6.6.1 (The Weierstrass Approximation Theorem) *Let f be a continuous function on an interval $[a, b]$. Then there is a sequence of polynomials $p_j(x)$ with the property that the sequence p_j converges uniformly on $[a, b]$ to f.*

In a few moments we shall prove this theorem in detail. Let us first consider some of its consequences. A restatement of the theorem would be that, given a continuous function f on $[a, b]$ and an $\epsilon > 0$, there is a polynomial p such that
$$|f(x) - p(x)| < \epsilon$$
for every $x \in [a, b]$. If one were programming a computer to calculate values of a fairly wild function f, the theorem guarantees that, up to a given degree of accuracy, one could use a polynomial instead (which would in fact be much easier for the computer to handle). Advanced techniques can even tell what degree of polynomial is needed to achieve a given degree of accuracy. The proof that we shall present also suggests how this might be done.

Let f be the Weierstrass nowhere differentiable function. The theorem guarantees that, on any compact interval, f is the uniform limit of polynomials. Thus even the uniform limit of infinitely differentiable functions need not be differentiable—even at one point.

We shall break up the proof of the Weierstrass Approximation Theorem into a sequence of lemmas.

Lemma 6.6.2 *Let ψ_j be a sequence of continuous functions on the interval $[-1, 1]$ with the following properties:*

(i) $\psi_j(x) \geq 0$ *for all x;*

(ii) $\int_{-1}^{1} \psi_j(x)\, dx = 1$ *for each j;*

(iii) *For any $\delta > 0$ we have*

$$\lim_{j \to \infty} \int_{\delta \leq |x| \leq 1} \psi_j(x)\, dx = 0 \, .$$

If f is a continuous function on the real line which is identically zero off the interval $[0, 1]$, then the functions $f_j(x) = \int_{-1}^{1} \psi_j(t) f(x - t)\, dt$ converge uniformly on the interval $[0, 1]$ to $f(x)$.

Proof: By multiplying f by a constant we may assume that $\sup |f| = 1$. Let $\epsilon > 0$. Since f is uniformly continuous on the interval $[0, 1]$ we may choose a $\delta > 0$ such that if $|x - t| < \delta$ then $|f(x) - f(t)| < \epsilon/2$. By property **(iii)** above we may choose an N so large that $j > N$ implies that $\left| \int_{\delta \leq |t| \leq 1} \psi_j(t)\, dt \right| < \epsilon/4$. Then, for any $x \in [0, 1]$, we have

$$
\begin{aligned}
|f_j(x) - f(x)| &= \left| \int_{-1}^{1} \psi_j(t) f(x - t)\, dt - f(x) \right| \\
&= \left| \int_{-1}^{1} \psi_j(t) f(x - t)\, dt - \int_{-1}^{1} \psi_j(t) f(x)\, dt \right| \, .
\end{aligned}
$$

Notice that, in the last line, we have used fact **(ii)** about the functions ψ_j to multiply the term $f(x)$ by 1 in a clever way. Now we may combine the two integrals to find that the last line

$$
\begin{aligned}
&= \left| \int_{-1}^{1} (f(x - t) - f(x)) \psi_j(t)\, dt \right| \\
&\leq \int_{-\delta}^{\delta} |f(x - t) - f(x)| \psi_j(t)\, dt \\
&\quad + \int_{\delta \leq |t| \leq 1} |f(x - t) - f(x)| \psi_j(t)\, dt \\
&= A + B \, .
\end{aligned}
$$

To estimate term A, we recall that, for $|t| < \delta$, we have $|f(x-t)-f(x)| < \epsilon/2$; hence

$$A \leq \int_{-\delta}^{\delta} \frac{\epsilon}{2} \psi_j(t)\, dt \leq \frac{\epsilon}{2} \cdot \int_{-1}^{1} \psi_j(t)\, dt = \frac{\epsilon}{2}.$$

For B we write

$$
\begin{aligned}
B &\leq \int_{\delta \leq |t| \leq 1} 2 \cdot \sup |f| \cdot \psi_j(t)\, dt \\
&\leq 2 \cdot \int_{\delta \leq |t| \leq 1} \psi_j(t)\, dt \\
&< 2 \cdot \frac{\epsilon}{4} = \frac{\epsilon}{2},
\end{aligned}
$$

where in the penultimate line we have used the choice of j. Adding together our estimates for A and B, and noting that these estimates are independent of the choice of x, yields the result. $\qquad\square$

Lemma 6.6.3 *Define $\psi_j(t) = k_j \cdot (1 - t^2)^j$, where the positive constants k_j are chosen so that $\int_{-1}^{1} \psi_j(t)\, dt = 1$. Then the functions ψ_j satisfy the properties* **(i)**–**(iii)** *of the last lemma.*

Proof: Of course property **(ii)** is true by design. Property **(i)** is obvious. In order to verify property **(iii)**, we need to estimate the size of k_j.

Notice that

$$
\begin{aligned}
\int_{-1}^{1} (1 - t^2)^j\, dt &= 2 \cdot \int_{0}^{1} (1 - t^2)^j\, dt \\
&\geq 2 \cdot \int_{0}^{1/\sqrt{j}} (1 - t^2)^j\, dt \\
&\geq 2 \cdot \int_{0}^{1/\sqrt{j}} (1 - jt^2)\, dt,
\end{aligned}
$$

where we have used the binomial theorem. But this last integral is easily evaluated and equals $4/(3\sqrt{j})$. We conclude that

$$\int_{-1}^{1} (1 - t^2)^j\, dt > \frac{1}{\sqrt{j}}.$$

As a result, $k_j < \sqrt{j}$.

Now, to verify property (iii) of the lemma, we notice that, for $\delta > 0$ fixed and $\delta \leq |t| \leq 1$, it holds that

$$|\psi_j(t)| \leq k_j \cdot (1 - \delta^2)^j \leq \sqrt{j} \cdot (1 - \delta^2)^j$$

and this expression tends to 0 as $j \to \infty$. Thus $\psi_j \to 0$ uniformly on $\{t : \delta \leq |t| \leq 1\}$. It follows that the ψ_j satisfy property (iii) of the lemma.
$\square$

Proof of the Weierstrass Approximation Theorem: We may assume without loss of generality (just by changing coordinates) that f is a continuous function on the interval $[0,1]$. After adding a linear function (which is a polynomial) to f, we may assume that $f(0) = f(1) = 0$. Thus f may be continued to be a continuous function which is identically zero on the entire real line.

Let ψ_j be as in Lemma 9.1 and form f_j as in that lemma. Then we know that f_j converge uniformly on $[0,1]$ to f. Finally,

$$
\begin{aligned}
f_j(x) &= \int_{-1}^{1} \psi_j(t) f(x - t)\, dt \\
&= \int_{0}^{1} \psi_j(x - t) f(t)\, dt \\
&= k_j \int_{0}^{1} (1 + (x - t)^2)^j f(t)\, dt \, .
\end{aligned}
$$

But multiplying out the expression $(1+(x-t)^2)^j$ in the integrand then shows that f_j is a polynomial of degree at most $2j$ in x. Thus we have constructed a sequence of polynomials f_j that converges uniformly to f on the interval $[0,1]$.
$\square$

Exercises

1. Consider the set of all continuously differentiable functions on the unit interval $[0, 1]$. Let $\epsilon > 0$. Show that there is a polynomial p such that

$$\sup_{x \in [0,1]} |f(x) - p(x)| < \epsilon$$

$$\sup_{x \in [0,1]} |f'(x) - p'(x)| < \epsilon.$$

   [*Hint:* Apply the usual Weierstrass approximation theorem to f'.]

2. Show that an analogue of Exercise 1 is true in C^k for any k.

3. Let us say that a sequence $\{f_j\}$ of continuous functions on the reals *converges* to a limit f if, for each $\epsilon > 0$, and each compact $K \subseteq \mathbb{R}$, there is an $N > 0$ such that when $j > N$ then $|f_j(x) - f(x)| < \epsilon$ for every $x \in K$. We may call this topology "uniform convergence on compact sets." Show that it is just the same as the compact-open topology.

4. Let X be the space of all continuous functions on the real line, and $\mathcal{P}$ the subspace of polynomials. What is the closure of $\mathcal{P}$ in the compact-open topology?

5. Let us call a function $f : \mathbb{R} \to \mathbb{R}$ *piecewise constant* if we may write $\mathbb{R} = \cup_j I_j$, each I_j is an interval, and f is constant on each I_j. Let P be the space of all piecewise constant functions. Show that any continuous function on $\mathbb{R}$ is the limit of a sequence of elements of P in the compact-open topology.

6. Let f be any Riemann integrable function on the interval $[0, 1]$ and let $\epsilon > 0$. Show that there is a polynomial p such that

$$\int_0^1 |f(x) - p(x)| \, dx < \epsilon.$$

   [*Hint:* First show that there is a continuous function φ that satisfies a similar conclusion. Then use Weierstrass.]

7. Let

$$f(x) = \begin{cases} x & \text{if} \quad x \geq 0 \\ -x & \text{if} \quad x < 0. \end{cases}$$

Let p_j be a sequence of polynomials that converges uniformly to f on the interval $[0, 1]$. Show that the degrees of the p_j must tend to infinity as $j \to \infty$.

8. Let $f(x) = |x|$. Show that there is a sequence of polynomials that converges pointwise to f on the entire real line.

9. Consider the family $\mathcal{F}$ of functions of the form $\sum_{j=-N}^{N} a_j \sin jx$ on the interval $[0, 2\pi]$, for N any nonnegative integer. Show that $\mathcal{F}$ is *not* an equicontinuous family. However, if we restrict $N \leq 100$ and $|a_j| \leq 1$ for all j, then $\mathcal{F}$ *is* an equicontinuous family.

10. Let $\mathcal{G}$ be the family of continuously differentiable functions on the unit interval $[0, 1]$ such that **(i)** $|f(x)| \leq 1$ for all x and **(ii)** $|f'(x)| \leq 1$ for all x. Prove that $\mathcal{G}$ is an equicontinuous family.

11. Explain why the final (classical) statement of the Ascoli-Arzela theorem is a compactness statement for certain families of functions.

12. Let X be the continuous, real-valued functions on the real line, and let $\mathcal{P}$ be the subspace of polynomials. Show that the polynomials are dense in the topology of pointwise convergence.

13. On the unit interval $[0, 1]$, show that any polynomial is the uniform limit of a sequence of continuous functions *none of which* is a polynomial.

14. On the unit interval, show that any continuous function can be written as the *sum* of polynomials, where the series converges uniformly on the entire interval.

15. If we omit the monomial x^3 when forming our polynomials, then polynomials will *still* be dense in the continuous functions on the unit interval. Explain why that is so. What if we omit finitely many monomials?

16. If we form our polynomials only using monomials of the form x^{2^j}, then the polynomials will no longer be dense in the continuous functions on the unit interval. Explain why that is so.

17. Any polynomial will grow more slowly than e^x as $x \to +\infty$. Make this statement precise and prove it. Conclude that e^x cannot be a polynomial.

18. Any nonconstant polynomial will grow more rapidly than $\log x$ as $x \to +\infty$. Make this statement precise and prove it. Conclude that $\log x$ cannot be a polynomial.

19. Let f be a continuous function on the interval $[0, 1]$. Show that f is the uniform limit on the interval of polynomials p_j such that $p_j(0) = f(0)$ and $p_j(1) = f(1)$ for every j.

20. Let f be a continuously differentiable function on the interval $[0, 1]$. We assume here that $f'(0)$ is defined as a one-sided limit and likewise $f'(1)$ is defined as a one-sided limit. Show that f is the uniform limit on the interval of polynomials p_j such that $p'_j(0) = f'(0)$ and $p'_j(1) = f'(1)$ for every j. Can we strengthen the convergence so that $p_j \to f$ uniformly and also $p'_j \to f'$ uniformly?

21. Suppose that f_j are functions on the unit interval I satisfying a Lipschitz condition of the form

$$|f_j(x) - f_j(y)| \le |x - y|$$

for any $x, y \in I$. Assume that the sequence $\{f_j\}$ converges pointwise to a limit function f. Show that f satisfies the same Lipschitz condition.

22. Let f be a continuous function on the unit interval I and let $\epsilon > 0$. Show that there is a polynomial p of even degree such that

$$|f(x) - p(x)| < \epsilon$$

for every $x \in I$.

23. Consider the statement, "If $f_j \to f$ uniformly and $g_j \to g$ uniformly, then $f_j \cdot g_j \to f \cdot g$ uniformly." This assertion is true when the functions are continuous and defined on the closed unit interval. But it is false when the functions are defined on the real line. Explain.

24. If f_j are functions on the interval $[0, 1]$, then we may consider $\sum_j f_j$ and speak of uniform convergence of this series. Define uniform convergence in this context. Give a Cauchy condition for uniform convergence of a series of functions. Give a Cauchy condition for pointwise convergence. The Weierstrass M-test says that if $|f_j(x)| \le M_j$ for all x and if $\sum_j M_j$ is finite, then the series converges uniformly. Prove this result.

25. Dini's theorem states that if f_j are continuous functions on the unit interval $I = [0, 1] \subseteq \mathbb{R}$ and $f_1(x) \le f_2(x) \le \cdots$ for all x and if the pointwise limit function $\lim_j f_j(x)$ is a continuous function $f(x)$, then the convergence is uniform. Prove Dini's theorem.

26. Let f_j be continuous functions on the unit interval and suppose that $\lim_{j \to \infty} f_j(x) = f(x)$ exists at each x. Then it is not possible for f to be discontinuous at every point. Use the Baire category theorem to explore the continuity properties of f.

27. Let $\mathcal{C}$ be the continuous functions on the unit interval $I = [0, 1]$. Define a distance on $\mathcal{C}$ by

$$d(f, g) = \left[\int_0^1 |f(x) - g(x)|^2 dx \right]^{1/2}.$$

Then $\mathcal{C}$ is not complete when equipped with this metric. Give an explicit example of a function that is in the closure but not in $\mathcal{C}$.

28. A trigonometric polynomial is a function of the form

$$p(x) = \sum_{j=-N}^{N} a_j e^{2\pi i j x}.$$

Show that the trigonometric polynomials are uniformly dense in the continuous functions on the unit interval $[0, 1]$.

29. A continuous function on the unit interval I is called *piecewise linear* if it is the union of linear functions on compact subintervals I_j with the I_j having disjoint interiors and $\cup_j I_j = I$. Show that the piecewise linear functions are uniformly dense in the continuous functions on I.

30. Give an example of a sequence of differentiable functions $\{f_j\}$ on the unit interval so that $f_j(x)$ converges at every point but $f_j'(x)$ converges at no point.

Chapter 7

Knot Theory

7.1   What Is a Knot?

We all have an intuitive notion of what a knot is: it's what you get when you tangle up a piece of string. But how does a mathematician define a knot?

Definition 7.1.1 A *knot* is an embedding of the circle S^1 into Euclidean space $\mathbb{R}^3$.

Put in plain English, a knot is a curve in space, that does not cross itself, with its ends identified. The simplest knot of all is the *trivial knot* (or *unknot*), which is illustrated in Figure 7.1. Other knots, with their common names, are illustrated in Figures 7.2 and 7.3. Of course we cannot accurately depict 3-dimensional space on this page. What we are showing in each instance is a 2-dimensional *projection* of the knot. Observe that crossings are illustrated with a break in the curve. The broken curve lies *under*, while the unbroken curve lies *over* in each crossing.

The fundamental question in this subject is to decide when two knots are equivalent. When can one knot be deformed—by way of continuous movements (without cutting the knot!)—to another? While this question has held interest for pure mathematicians for many years, recent developments have tied knot theory to theoretical physics and engineering. Knot theory is intimately bound up with the Poincaré conjecture (recently resolved by Grigori Perelman), and tells us a good deal about the shape of space.

It is intuitively clear that the five knots exhibited in Figures 7.1–7.3 are distinct. They cannot be obtained one from the other by continuous deformation. But we need a language for making these assertions precise.

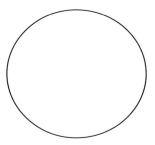

Figure 7.1: The trivial knot or unknot.

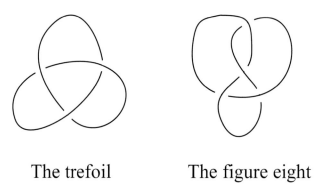

The trefoil   **The figure eight**

Figure 7.2: The trefoil and figure eight knots.

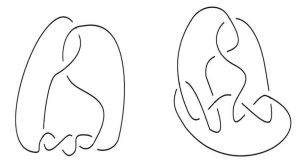

The Stevedore's knot   The true lover's knot

Figure 7.3: The Stevedore's and True Lover's knots.

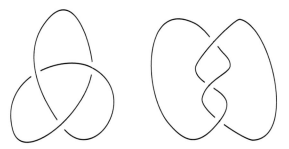

Figure 7.4: Two apparently inequivalent knots which are in fact equivalent.

Definition 7.1.2 We declare two knots $\mathbf{k}_1$ and $\mathbf{k}_2$ to be *equivalent* if there is a homeomorphism h of $\mathbb{R}^3$ (where the two knots live) such that $h(\mathbf{k}_1) = \mathbf{k}_2$.

What is important here is that the equivalence of two knots does not depend simply on the knots themselves, but also on how they sit in space. Any two knots themselves are homeomorphic, since they are both embeddings of the circle into $\mathbb{R}^3$. So the equivalence of knots must be determined in a more subtle fashion. An example helps to illustrate the ideas.

EXAMPLE 7.1.3 Let $\mathbf{k}$ be the trefoil knot, illustrated in Figure 7.2. Let $\mathbf{k}'$ be the reflection of $\mathbf{k}$ in a plane in 3-dimensional space. It is impossible to deform $\mathbf{k}$ into $\mathbf{k}'$. Yet the reflection is certainly a valid homeomorphism of space.

It is generally not at all clear whether two given knots are equivalent or not. For instance, the two knots in Figure 7.4 appear to be inequivalent, but they are in fact equivalent.

Ideally we would like to have some "knot invariants" which aid us in distinguishing knots. With that thought in mind, we give the following definition:

Definition 7.1.4 A *regular presentation* of a knot $\mathbf{k}$ is a depiction of the knot in the plane so that

- All crossings are transversal (i.e., nontangential);

- No two segments or arcs of the curve overlap;

- Three (or more) segments do not meet at a single point.

As Figure 7.4 illustrates, the regular presentation of a given knot is neither canonical nor unique. One possible invariant for a knot that we could consider is the *crossing number*: this is the minimal crossing number of all possible regular presentations of the knot. With this invariant, we can instantly tell that the trivial knot and the trefoil knot (see Figures 7.1 and 7.2) are distinct knots. For the first has crossing number 0 and the second has crossing number 3. End of story. The trouble is that, for many knots, it is quite difficult to calculate the crossing number. It turns out that there are just two knots with crossing number 3: the trefoil, and its reflection in a plane. Showing that these two are inequivalent is actually quite difficult, and was not achieved until 1930 by Max Dehn (1878–1952).

In recent years some more sophisticated invariants—the Alexander polynomial and the Jones polynomial—have been developed. We shall describe both of these in the ensuing sections.

7.2   The Alexander Polynomial

A fundamental and initially confusing fact about knots is this: all knots are the same because they are all homeomorphic to the circle. We cannot distinguish knots intrinsically; instead, we can tell two knots apart by how they sit in space. Put in other words, one approach to distinguishing two knots $\mathbf{k}_1$ and $\mathbf{k}_2$ is by comparing $\mathbb{R}^3 \setminus \mathbf{k}_1$ and $\mathbb{R}^3 \setminus \mathbf{k}_2$. This is how the Alexander polynomial comes about. In particular, we look at the homology generators of these two complementary spaces.

We present now J. Alexander's algorithm from 1928 for calculating the Alexander polynomial (James Alexander (1888–1971)). Afterward we shall give an indication of the mathematical justification behind the algorithm.

First we note that, if two curves cross transversally, then one of the curves crosses *under* the other. The picture is as in Figure 7.5. We take the curve to be oriented, so there is a sense of the curve that is crossing under proceeding from left to right (in this particular figure). Thus, in the figure, there are indicated four planar regions:

- The left-hand region *before* the crossing (labeled LB in the figure);

- The right-hand region *before* the crossing (labeled RB in the figure);

- The left-hand region *after* the crossing (labeled LA in the figure);

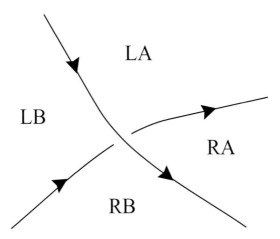

Figure 7.5: Regions delineated by two curves crossing.

- The right-hand region *after* the crossing (labeled RA in the figure).

Now examine a typical knot, such as the trefoil in Figure 7.6. In this figure the regions of the plane are labeled A, B, C, D, E and the crossings are labeled $1, 2, 3$. We have endowed this knot with a counterclockwise orientation. Now we shall "read" the knot and produce a matrix based on this knot. The Alexander polynomial will simply be the determinant of this Alexander matrix.

The *pro forma* for the Alexander matrix is as follows:

	A	B	C	D	E
1					
2					
3					

Now the rules for forming the Alexander matrix are these:

LB The matrix entry for this node and this face is $-t$;

RB The matrix entry for this node and this face is 1;

LA The matrix entry for this node and this face is t;

RA The matrix entry for this node and this face is -1.

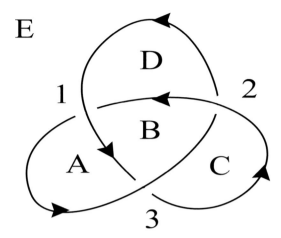

Figure 7.6: The oriented trefoil with nodes and faces labeled.

To illustrate the ideas, we now present the Alexander matrix for the trefoil knot:

	A	B	C	D	E
1	t	−t	0	1	−1
2	0	−t	1	t	−1
3	1	−t	t	0	−1

Just to make the point clear, the entry in the upper left-hand corner of this matrix is the **A** − **1** entry. So we look at crossing **1** and notice that region **A** is the *left, after* region for this crossing. That is why the entry in the matrix is t.

We see that this is a 5×3 matrix (ignoring the labels in bold and just concentrating on the entries). We now eliminate two columns corresponding to adjacent regions—we elect to remove the last two columns, corresponding to regions D and E (the mathematical justification for this step has to do with redundancy in the information—this will be explained below). Thus we obtain the matrix

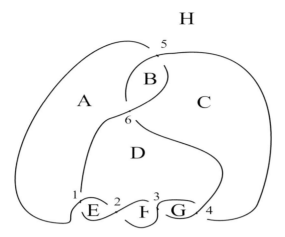

Figure 7.7: The Stevedore's knot.

$$\begin{matrix} t & -t & 0 \\ 0 & -t & 1 \\ 1 & -t & t \end{matrix}$$

The determinant of this matrix is $-t^3 + t^2 - t$. We typically divide out common powers of t, and normalize with a minus sign to make the leading coefficient 1. So we obtain Alexander polynomial $t^2 - t + 1$ for the trefoil.

The reader may find it of interest to eliminate other pairs of columns and calculate the determinant. You will not always get $t^2 - 2t + 1$, but you will instead sometimes get the product of a power of t times $t^2 - 2t + 1$. In practice we always divide out the highest power of t possible and declare the result to be the Alexander polynomial.

Next let us have another look at the Stevedore's knot. See Figure 7.7, which shows this knot with a counterclockwise orientation. The faces and the vertices are labeled as in the previous example (with faces running from A through H and vertices from 1 through 6). Following the rules already enunciated, the Alexander matrix is

	A	B	C	D	E	F	G	H
1	t	0	0	−1	1	0	0	−t
2	0	0	0	−t	1	t	0	−1
3	0	0	0	−1	0	t	1	−t
4	0	0	t	−t	0	0	1	−1
5	t	−t	1	0	0	0	0	−1
6	1	−t	t	−1	0	0	0	0

We strike out the last two columns because they correspond to two regions that are adjacent. The result is the matrix

$$
\begin{matrix}
t & 0 & 0 & -1 & 1 & 0 \\
0 & 0 & 0 & -t & 1 & t \\
0 & 0 & 0 & -1 & 0 & t \\
0 & 0 & t & -t & 0 & 0 \\
t & -t & 1 & 0 & 0 & 0 \\
1 & -t & t & -1 & 0 & 0
\end{matrix}
$$

Now a tedious calculation allows us to compute the determinant of the resulting square matrix. The result is $2t^2 - 5t + 2$ for the Alexander polynomial. In particular, we see that the trefoil and the Stevedore's knot have different Alexander polynomials. So we can be sure that they are inequivalent knots.

Why does the Alexander polynomial work? Why is it a knot invariant? Why is it the case that if two knots k_1 and k_2 have distinct Alexander polynomials, then they are inequivalent knots? Before we attempt to answer these questions, there are some points of logic that need to be addressed.

First of all, the Alexander polynomial is *not* a complete invariant. This means that if two knots k_1 and k_2 have the *same* Alexander polynomial, then that does *not* imply that the knots are equivalent. Only the converse is true: if the knots are equivalent then they have the same Alexander polynomial. Thus (by contrapositive), if the Alexander polynomials are different then the knots are inequivalent. The later, and more sophisticated, theory of the Jones polynomial has in fact led to complete invariants for knot theory. But the Alexander polynomial was historically the very first invariant. It is a wonderful tool, but it has its limitations.

Figure 7.8: A knot and its thickening.

In point of fact the rigorous mathematical contruction of the Alexander polynomial is rather technical and difficult. It involves a good deal of sophisticated topological and algebraic machinery. Thus we shall only give an informal description of the ideas here.

First, as already noted, we can only hope to characterize a knot by examining its complement in space. We shall do as professional knot theorists do and think of our knot as living not in $\mathbb{R}^3$ but rather in S^3, the sphere (of course the sphere is just the one-point compactification of space—see Section 2.5). So let **k** be a given knot in S^3. Thicken **k** slightly to form a knotted tube $\mathbf{k}^*$ (Figure 7.8).

Let $\mathbf{k}_o^*$ be the interior of this tube, and set $X = S^3 \setminus \mathbf{k}_o^*$. Of course X is compact. Notice that X must be connected and path-connected. Let G be the fundamental group of X, and let G' denote its commutator subgroup. It is known (see [ARM1, p. 222) that G/G' is infinite cyclic. Finally let $\widehat{X}$ be the regular covering space corresponding to the group G'—see Section 3.3. Then of course the fundamental group of $\widehat{X}$ is isomorphic to G', and the group of covering transformations is infinite cyclic. We call $\widehat{X}$ the *infinite cyclic covering space* of X.

Now we examine the first homology group $H_1(\widehat{X})$ of $\widehat{X}$ (of course this is nothing other than the abelianization of the fundamental group—see Theorem 3.2.20). Of course this is an abelian group. The covering transformation

$$h : \widehat{X} \to \widehat{X}$$

induces a group automorphism

$$h_* : H_1(\widehat{X}) \to H_1(\widehat{X}).$$

It is the automorphism h_* that will give rise to the Alexander polynomial.

Let G' denote the fundamental group of $\widehat{X}$ and G'' the commutator subgroup. Then we know that the monomorphism $\pi_* : \pi_1(\widehat{X}) \to \pi_1(X)$ induces an isomorphism $\pi_{**} : H_1(\widehat{X}) \to G'/G''$. The Alexander polynomial is then the determinant of a certain matrix that arises from the representation theory for G'. We refer the reader to [ARM1, pp. 236 ff.] for the details. The lesson that we may take from this discussion is that the Alexander polynomial, by the very nature of the construction described here, is clearly a topological invariant. It is a means of distinguishing knots.

7.3   The Jones Polynomial

After Alexander's insight of 1928, not much progress was made in the development of polynomial knot invariants until Vaughan Jones's breakthrough of 1985. Using ideas from the quite distant field of von Neumann algebras, Jones created a knot invariant that is much more refined than Alexander's—it can distinguish many more knots. While the so-called "Jones polynomial" is not a complete invariant (i.e., it is not the case that two knots are equivalent if and only if their Jones polynomials are equal), further research in the field has produced complete sets of invariants. Jones's ideas have had a profound impact on knot theory, on topology, and on mathematical physics. Here we shall give a brief description of the construction of the Jones polynomial.

The Jones polynomial is actually a "Laurent polynomial," meaning that it has both positive and negative powers of the variable. Thus a Jones polynomial has the form

$$p(z) = c_{-k}z^{-k}+c_{-(k-1)}z^{-(k-1)}+\cdots+c_{-1}z^{-1}+c_0z^0+c_1z^1+\cdots+c_{m-1}z^{m-1}+c_m z^m$$

for some positive integers k and m.

7.3.1   Knot Projections

In our examination of the Jones polynomial (and the associated bracket polynomial) we shall be considering the projection of a knot into the plane. This

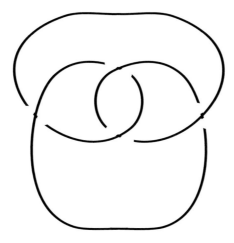

Figure 7.9: A knot projection.

projection will have certain crossings—see Figure 7.9. There are three rules for forming the bracket polynomial. We shall, for the moment, formulate these rules in terms of three indeterminate variables A, B, and C. Later on we shall use some algebraic tricks to transform the 3-variable polynomial into a 1-variable polynomial.

Rule 1: The trivial knot (or *unknot*) has bracket polynomial equal to the constant 1. See Figure 7.10.

Rule 2: Adding a trivial (unknotted) loop to a given knot multiplies the bracket polynomial for that given knot by C. See Figure 7.11.

Rule 3: If a given knot projection has a crossing (Figure 7.12) and we re-place that crossing by a non-crossing (in each of two different ways), then the polynomials P, P', P'' for the three knot projections are related by

$$P = A \cdot P' + B \cdot P''.$$

See Figure 7.13.

In a moment we shall use these three rules to calculate the bracket poly-nomial for the left-oriented trefoil knot. A nearly identical calculation gives

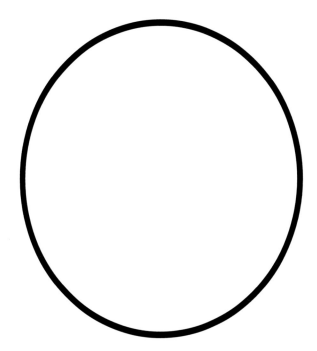

Figure 7.10: The trivial knot, or "unknot."

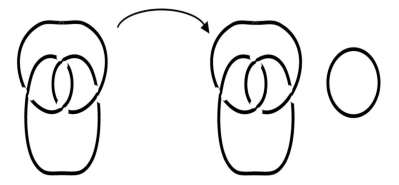

Figure 7.11: Adding a trivial (unknotted) loop to a knot.

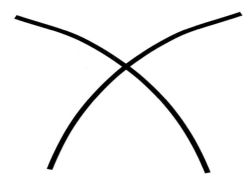

Figure 7.12: A knot with a crossing.

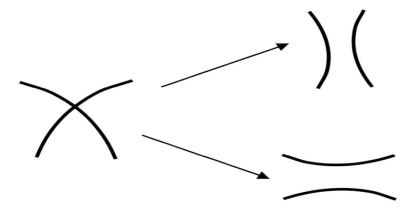

Figure 7.13: Replacing a crossing by two non-crossings.

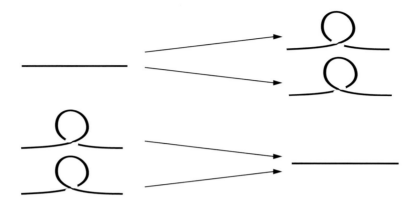

Figure 7.14: Adding or removing a loop from the knot.

the bracket polynomial for the right-oriented terfoil knot—and the result is different! This is a new result—unobtainable with the Alexander polynomial. First we need a little preliminary development.

7.3.2   Reidemeister Moves

One of the key ideas in the background of the Jones polynomial is the idea of Reidemeister move. These are three manipulations of knot projections under which we want the Jones construction to be invariant.

First Reidemeister Move: Add or remove a loop from the knot. See Figure 7.14. Notice that this move does not change the knot type. There are no new tangles or complexities in the knot. It is just a way to "rearrange" the knot projection in the plane.

Second Reidemeister Move: We either slide one of two adjacent strands under the other (Figure 7.15) or we slide one of two adjacent strands *out* from under the other (Figure 7.16).

Third Reidemeister Move: We slide a strand past a crossing, in one of two ways as illustrated in Figures 7.17 and 7.18.

It is a fact—we shall not treat the details here—that two knots are equivalent if and only if their respective planar projections can be transformed one to the other by a sequence of Reidemeister moves and planar isotopies (here a planar isotopy is a certain type of deformation of the plane—see the exercises below). We refer the reader to [ADF] for a more thorough discussion of

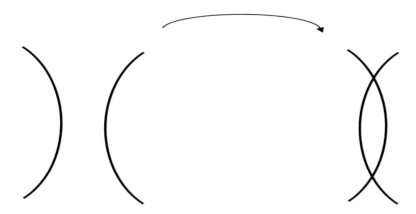

Figure 7.15: Sliding one of two adjacent strands under the other.

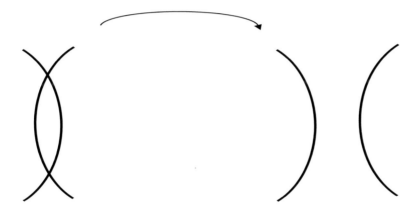

Figure 7.16: Sliding one of two adjacent strands out from under the other.

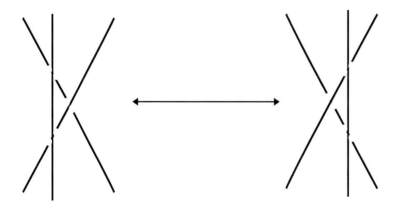

Figure 7.17: Sliding a strand past a crossing.

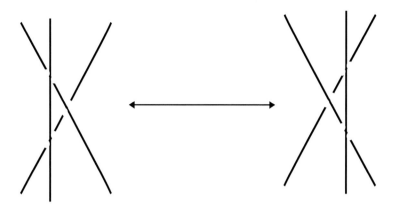

Figure 7.18: Sliding a strand past a crossing.

this concept.

Now we attempt to understand the Reidemeister moves from the point of view of the three rule enunciated above for the Jones polynomials. We first set up this table of notation (Figure 7.19).

We want to set up our polynomial theory so that it is invariant under the Reidemeister moves. So let us examine a double crossing as in the Second Reidemeister move. In what follows, we denote the bracket polynomial associated to a knot projection P by $\langle P \rangle$. We calculate, using the notation from the table in Figure 7.19, as follows (be sure to read the explanation afterwards):

$$
\begin{aligned}
\langle L \rangle &= A\langle Q \rangle + B\langle X \rangle \\
&= A\left[A\langle T \rangle + B\langle O \rangle\right] + B\left[A\langle R \rangle + B\langle T \rangle\right] \\
&= AA\langle T \rangle + ABC\langle T \rangle + BA\langle R \rangle + BB\langle T \rangle \\
&= \left[A^2 + ABC + B^2\right]\langle T \rangle + BA\langle R \rangle .
\end{aligned}
$$

Observe here that, in the first equality, we have separated the lower crossing of the knot L from the table according to Rule 3 of the Jones polynomial. In the second equality, we have again applied Rule 3 to each of the knots from the last step. In the third equality, we have applied Rule 2 to the knot O from the last step. In the last equality we have simply performed some elementary algebra.

In order to be able to say that the polynomial $\langle L \rangle$ is unaffected by Reidemeister moves of type 2, we need to know that $\langle L \rangle = \langle R \rangle$. From the calculation just performed, it is thus clear that we must have $A^2 + ABC +$

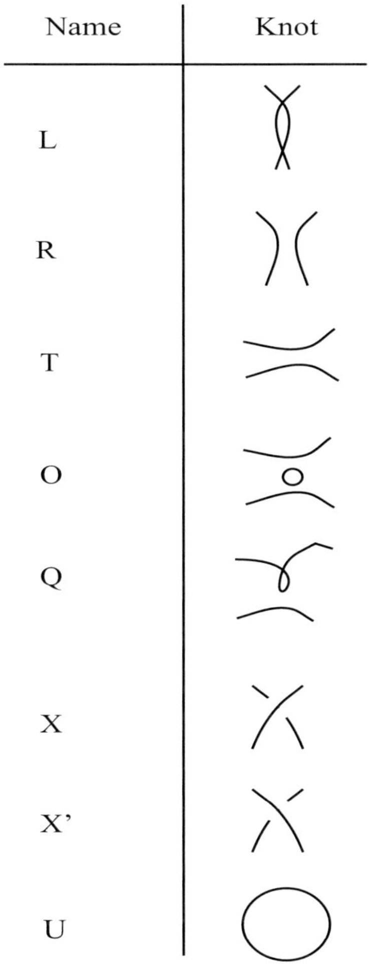

Name	Knot
L	
R	
T	
O	
Q	
X	
X'	
U	

Figure 7.19: Lexicon of certain basic knots.

$B^2 = 0$ and $BA = 1$. So it is propitious for us to set $B = A^{-1}$ and $C = -A^2 - A^{-2}$. Now our original three rules for the Jones polynomial (with U denoting the unknot) can be expressed as

Rule 1: $\langle U \rangle = 1$.

Rule 2: $\langle P \cup U \rangle = (-A^2 - A^{-2})\langle P \rangle$.

Rule 3: $\langle X \rangle = A\langle R \rangle + A^{-1}\langle T \rangle$
     and
$$\langle X' \rangle = A\langle T \rangle + A^{-1}\langle R \rangle.$$

Now it is clear that Reidemeister moves leave our polynomials invariant.

7.3.3   Bracket Polynomials

We now prepare to take advantage of our new machinery for calculating bracket polynomials by calculating some preliminary polynomials for some fundamental knots.

Lemma 7.3.1 *Define knots as in the next table.*
     Then the polynomial entries in this table are valid.

Proof: Our job is to confirm each line of this table: that each given knot has the specified bracket polynomial. Let us begin with line 1.

Of course a single unknot has, according to Rule 1, bracket polynomial 1. Then, according to Rule 2, adjoining a separate loop or unknot to that will multiply the given polynomial by C, which we now know is equal to $-A^2 - A^{-2}$. The result, trivially, is $-A^2 - A^{-2}$. This is line 1.

For line 2, notice that we may decompose the given knot into an unknot plus the union of two disjoint unknots (following the paradigm for Rule 3). The bracket polynomial for the first of these is 1 and that for the second of these is $-A^2 - A^{-2}$. Then, according to Rule 3, the result of this decomposition is the polynomial

$$A \cdot 1 + B \cdot (-A^2 - A^{-2}) = A \cdot 1 + A^{-1} \cdot (-A^2 - A^{-2}) = -A^{-3}.$$

That is line 3.

Line 3 is similar and we omit the details.

Knot	Name	Bracket Polynomial
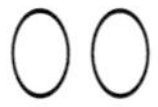	W	$-A^2-A^{-2}$
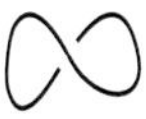	G	$-A^{-3}$
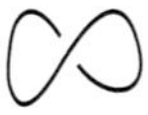	H	$-A^{3}$
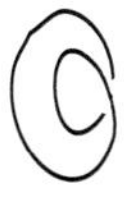	J	$-A^{-3}$
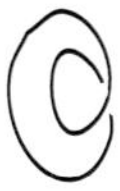	K	$-A^{3}$

Figure 7.20: Some fundamental knots.

For line 4, we observe that the given knot may be decomposed into the union of two unknots together with a single unknot. The first of these has bracket polynomial $-A^2 - A^{-2}$ and the second has bracket polynomial 1. Thus the aggregate of the two has bracket polynomial

$$A \cdot (1) + B \cdot (-A^2 - A^{-2}) = A + A^{-1} \cdot (-A^2 - A^{-2}) = A - A - A^{-3} = -A^{-3} .$$

That is line 4 of the table. Line 5 is verified similarly. □

Let us now calculate the bracket polynomial of the trefoil knot shown in Figure 7.21.

EXAMPLE 7.3.2 We need to begin again with a concordance of certain basic knots. Refer to Figure 7.22. Now we calculate that

$$\begin{aligned}
\langle B\rangle &= A\langle C\rangle + A^{-1}\langle D\rangle \\
&= A\left[A\langle E\rangle + A^{-1}\langle F\rangle\right] + A^{-1}\left[A\langle M\rangle + A^{-1}\langle S\rangle\right] \\
&= A^2\left[-A^2 - A^{-2}\right]\langle H\rangle + 2\langle H\rangle + A^{-2}\langle J\rangle .
\end{aligned}$$

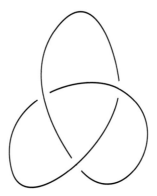

Figure 7.21: The trefoil.

The natural thing to do now is to substitute in the bracket polynomials from Figure 7.20. Thus we obtain

$$
\begin{aligned}
\langle B \rangle &= A^2 \left[-A^2 - A^{-2} \right] (-A^3) + 2(-A^3) + (A^{-2})(-A^{-3}) \\
&= A^7 - A^3 - A^{-5} .
\end{aligned}
$$

We conclude that the bracket polynomial of the projection B of the trefoil is

$$
A^7 - A^3 - A^{-5} .
$$

EXERCISE FOR THE READER 7.3.3 Let Z be the knot exhibited on the left in Figure 7.23 and let Z' be the unknotted curve on the right of that figure. Calculate that

$$
\langle Z \rangle = -A^{-3} \langle Z' \rangle .
$$

7.3.4   Creation of a New Polynomial Invariant

Now it is a notable fact that the bracket polynomial is *not* a knot invariant. We see that it is a knot invariant precisely if and only if it is invariant under the Reidemeister moves and under isotopies. It turns out that bracket polynomials are not invariant under Reidemeister moves of the first type. In fact Rule 1 for the bracket polynomial tells us right away that adding a loop will change the bracket polynomial.

So we need a new idea to create an invariant.

Name	Knot

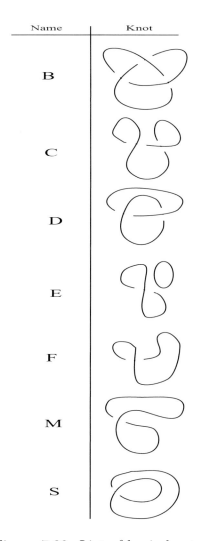

B

C

D

E

F

M

S

Figure 7.22: List of basic knots.

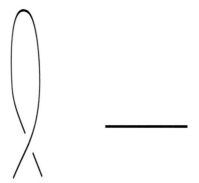

Figure 7.23: A link.

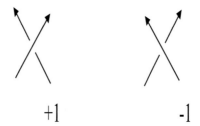

Figure 7.24: Assignment of $+1$ or -1 to a crossing.

Definition 7.3.4 Let P be a knot projection. We assign to a crossing of this projection the number $+1$ or -1 according to the rule given in Figure 7.24.

The *writhe* $w(P)$ of the knot is the sum of all these labels.

It is clear that with a type I Reidemeister move the writhe will either increase or decrease by 1 (depending on the orientation of the move). Now we will present Louis Kauffman's version of Vaughan Jones's construction of a polynomial knot invariant.

Definition 7.3.5 Consider the projection P of an oriented knot. The *Kauffman polynomial* of P is defined to be

$$K(P) = (-A^3)^{-w(P)} \cdot < P > .$$

THEOREM 7.3.6 *The Kauffman polynomial $K(P)$ is a knot invariant.*

This result is of course the culmination of our work. It creates a new knot invariant, considerably more powerful than the original Alexander polynomial.

Proof: Our job is to show that the Kauffman polynomial is unaffected by Reidemeister moves. We shall sketch what happens for the three types.

Reidemeister moves of type I: If we perform a Reidemeister move of this type, then the writhe goes down by 1. So the polynomial K is multiplied by $-A^3$. But also the bracket polynomial is multiplied by $-A^{-3}$, as the preceding Exercise for the Reader shows. The net effect of these two changes is multiplication by 1, so the polynomial is invariant.

Reidemeister moves of type II: This type of move either adds two crossings of opposite sign or removes two such crossings. The net effect on K is not to change it.

Reidemeister moves of type III: We already know that the bracket polynomial is unchanged by this type of move. Also this move just rearranges crossings, without changing the ± 1 labels. So this move will not alter K.

Thus we see that all three types of Reidemeister moves leave the Kauffman polynomial unchanged. You can check for yourself that planar isotopies do not affect the polynomial. Thus K is a knot invariant.   □

Proposition 7.3.7 *The trefoil knot and the unknot are not equivalent. In other words, the trefoil cannot be "untied."*

Proof: We have calculated the Kauffman bracket polynomial of the trefoil to be $A^7 - A^3 - A^{-5}$. The writhe of the trefoil, as depicted in Figure 7.21, is $+1$. Thus the Kauffman polynomial is

$$K = (-A^3)^{-1} \cdot (A^7 - A^3 - A^{-5}) = -A^4 + 1 + A^{-8} \,.$$

On the other hand, the bracket polynomial of the unknot is 1 and the writhe of the unknot is 0. Thus its Kauffman polynomial is

$$K = (-A^3)^0 \cdot 1 = 1 \,.$$

Since the Kauffman polynomials are unequal, the knots must be inequivalent.
□

EXERCISE FOR THE READER 7.3.8 We have calculated the Kauffman polynomial of the trefoil exhibited in Figure 7.21 in the proof of the last proposition. Now calculate the Kauffman polynomial of the "reflected" trefoil shown in Figure 7.25. Are these knots equivalent or not?

Exercises

1. A homotopy $F : X \times I \to Y$ of two homeomorphisms f and g on X is called an *ambient isotopy* if, for each fixed $t \in I$, $F(\cdot, t)$ is a homeomorphism (here $F(x, 0) = f(x)$ and $F(x, 1) = g(x)$) and also $F(x, 0)$ is the identity mapping on X. Construct an ambient isotopy of the identity map on $\mathbb{R}^2$ (taking, in particular, the unit square) to the unit square to the homeomorphism of $\mathbb{R}^2$ that takes the unit square to the unit circle.

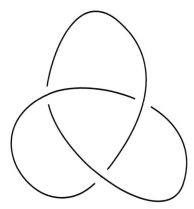

Figure 7.25: The reflected trefoil.

2. The unit circle and the figure eight—both in the plane—are *not* isotopic (refer to Exercise 1 for terminology). Explain why.

3. Refer to Exercise 1 for terminology. Explain why two knots in $\mathbb{R}^3$ are equivalent—as we have defined the concept in the text—if and only if their respective embedding maps into $\mathbb{R}^3$ are ambient isotopic.

4. Knot theorists find it useful to think of a knot as living in the sphere S^3 rather than in Euclidean space $\mathbb{R}^3$—just because S^3 is compact. And of course the former is simply the one-point compactification of the latter. Show that two knots are equivalent in $\mathbb{R}^3$ if and only if they are equivalent in S^3.

5. Prove that the notion of ambient isotopy (see Exercise 1) gives an equivalence relation on the collection of embeddings of a topological space X into a topological space Y.

6. You already examined the two knots in Figure 7.26 using the Kauffman polynomial in the Exercise for the Reader at the end of the last section. See now whether you can give a heuristic reason why these two knots are equivalent or not equivalent.

7. Calculate the Alexander polynomials of the two knots shown in Figure 7.3.

8. A *link* is an embedding of a finite collection of circles into $\mathbb{R}^3$. Two links are considered to be equivalent if there is an ambient isotopy from one

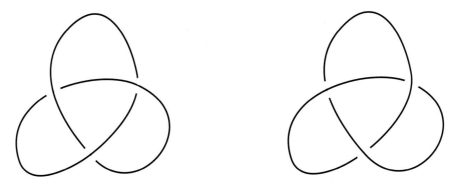

Figure 7.26: Two versions of the trefoil.

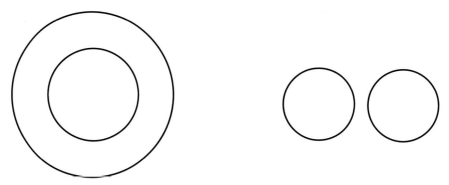

Figure 7.27: Two possibly equivalent links.

to the other. Show that two disjoint circles in $\mathbb{R}^3$ is a configuration that is *not* equivalent to two linked circles in $\mathbb{R}^3$.

9. Figure 7.27 shows two different links. Determine whether they are equivalent (in the language of Exercise 8).

10. A knot can be studied by looking at its projection into a plane. The projection should be selected so that the crossings are distinct, and they are all transversal. Show that a knot that has a projection with just one crossing is in fact equivalent to the unknot (or trivial knot).

11. Refer to the last exercise for terminology. Are there any nontrivial knots with two crossings?

12. Refer to Exercise 10 for terminology. Exhibit a knot with 4 crossings that is equivalent to the unknot (or trivial knot).

13. Calculate the bracket polynomial of the right-oriented trefoil knot. Confirm that this is distinct from the bracket polynomial calculated in Example 7.3.2 for the left-oriented trefoil knot. It follows that the Kauffman polynomials are unequal, so these two knots are inequivalent.

14. Calculate the Kauffman polynomial of the figure eight knot.

15. Calculate the Kauffman polynomial of a knot with two crossings. Verify that it is in fact the same as the Kauffman polynomial for the unknot.

16. Look up—on the Internet or elsewhere—the square knot. What is the Kauffman polynomial of the square knot?

17. Consider the Olympian rings (three rings linked together in the celebrated fashion). What is the Kauffman polynomial of this configuration?

18. Calculate the Alexander polynomial for the trefoil and the reflected trefoil, and confirm that they are the same. So the Alexander polynomial does not distinguish these two trefoils. But the Jones polynomial does.

19. Are there any knots with Jones polynomial A? How about A^{-1}?

20. Classify all the knots whose projections have just three crossings.

21. Explain in the language of Reidemeister moves why the unknot and the trefoil are inequivalent.

22. What is the fundamental group of the complement in space of the unknot?

23. What is the fundamental group of the complement in space of the trefoil?

Chapter 8

Graph Theory

8.1 Introduction

We learn even in high school about graphs of functions. The graph of a function is usually a curve drawn in the x-y plane. See Figure 8.1. But the word "graph" has other meanings. In finite or discrete mathematics, a graph is a collection of points and edges or arcs in the plane. Figure 8.2 illustrates a graph as we are now discussing the concept.

Leonhard Euler (1707–1783) is considered to have been the father of graph theory. His paper in 1736 on the seven bridges of Königsberg is considered to have been the foundational paper in the subject. It is worthwhile now to review that topic.

Königsberg is a town, founded in 1256, that was originally in Prussia. After a stormy history, the town became part of the Soviet Union and was renamed Kaliningrad in 1946. In any event, during Euler's time the town

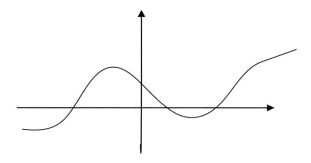

Figure 8.1: The graph of a function in the plane.

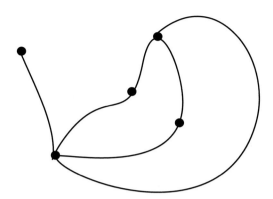

Figure 8.2: A graph as a combinatorial object.

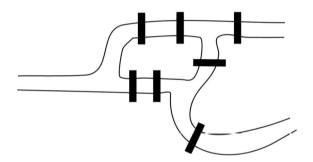

Figure 8.3: The seven bridges of Königsberg.

had seven bridges (named Krämer, Schmiede, Holz, Hohe, Honig, Köttel, and Grünespanning) spanning the Pregel River. Figure 8.3 gives a simplified picture of how the bridges were originally configured (two of the bridges were later destroyed during World War II, and two others demolished by the Russians). The question that fascinated people in the eighteenth century was whether it was possible to walk a route that never repeats any part of the path and that crosses each bridge exactly once.

Euler in effect invented graph theory and used his ideas to show that it is impossible to devise such a route. We shall, in the subsequent sections, devise a broader version of Euler's ideas and explain his solution of the Königsberg bridge problem in the process.

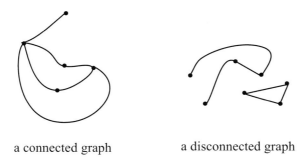

a connected graph   a disconnected graph

Figure 8.4: Connected and disconnected graphs.

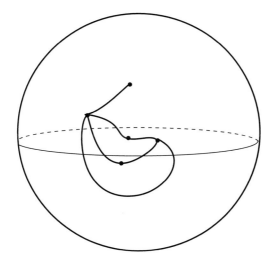

Figure 8.5: A graph on the sphere.

8.2 Fundamental Ideas of Graph Theory

A graph consists of vertices and edges. The edges are not allowed to intersect or cross. A graph may be *connected*, i.e., consist of one continuous piece or *disconnected*, i.e., consist of more than one contiguous piece. See Figure 8.4. Notice in the figure that the edges of the graph determine certain two-dimensional regions, or faces, in the graph. The graph on the left defines 3 faces, and the graph on the right defines 2 faces (we count the large region outside the graph as a face). It is customary in this subject to think of the graph as living on a *sphere* rather than in the plane, so that the exterior region (see Figure 8.5) is more plainly a face.

Euler's first fundamental insight about graphs is the following theorem:

Figure 8.6: Beginning of the proof of Euler's formula.

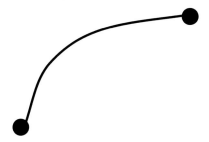

Figure 8.7: Euler's formula for a more complicated graph.

THEOREM 8.2.1 *Let $\mathcal{G}$ be any connected planar graph. Let V be the number of vertices, E the number of edges, and F the number of two-dimensional regions (or faces) defined by the edges. Then*

$$V - E + F = 2.$$

We should like to spend some time explaining why this important theorem is true. Begin with the simplest possible graph—see Figure 8.6. It has just one vertex. There are no edges. And there is one face—which is the entire region of the sphere complementary to the single vertex. Thus $V = 1$, $E = 0$, and $F = 1$. As a result, we see that

$$V - E + F = 1 - 0 + 1 = 2.$$

So Euler's theorem is true in this very simple case.

Now imagine making the graph more complex. We add a single edge, as shown in Figure 8.7. How have the numbers changed? Well, now $E = 1$. But there is an additional vertex (i.e., there is a vertex at each end of the edge), so $V = 2$. And there is still a single face, so $F = 1$. Now

$$V - E + F = 2 - 1 + 1 = 2.$$

Thus Euler's theorem remains true.

Now the fundamental insight is that we can build up *any* graph by adding one edge at a time. And there are only three ways that we may add an edge. Let us discuss them one at a time:

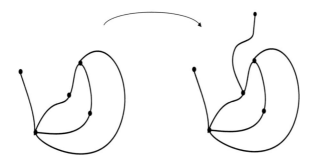

Figure 8.8: The inductive step in the proof of Euler's formula.

- We can add a new edge so that it has one vertex on the existing graph and one vertex free—see Figure 8.8. In doing so, we add one vertices, add one edge, and do not change the number of faces. Thus, in the formula $V - E + F$, we have increased E by 1 and increased V by 1. These two increments cancel out, so the sum of 2 remains unchanged.

- We can add the new edge so that both ends (the vertices) are at the same point on the existing graph—see Figure 8.9. Thus we have added one edge, no vertices, and one face. As a result, in the formula $V - E + F$, we have increased E by 1 and increased F by 1. These two increments cancel out, so the sum of 2 remains unchanged.

- We can add the new edge so that two ends are at two different vertices of the existing graph—see Figure 8.10. so we have added one edge and one face, but no vertices. As a consequence, in the formula $V - E + F$, we have increased E by 1 and increased F by 1. But there are no new vertices. Therefore the two increments cancel out, and the sum of 2 remains.

This exhausts all the cases, and shows that, as we build up any graph, the Euler sum $V - E + F$ will always be 2.

We call 2 the *Euler characteristic* of the sphere. It is a fundamental geometric invariant of this surface. It turns out that the Euler characteristic of a torus—see Figure 8.11—is *not* 2. It is in fact 0, as the figure indicates.

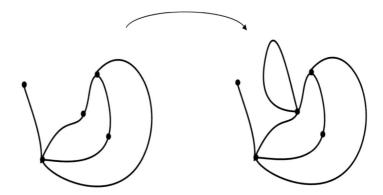

Figure 8.9: The inductive step in the proof of Euler's formula.

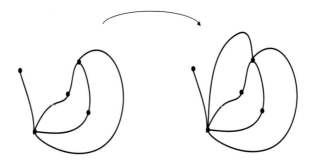

Figure 8.10: The next step in the proof of Euler's formula.

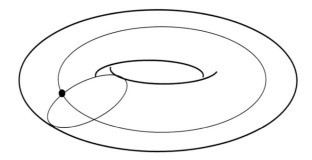

Figure 8.11: The Euler characteristic of a torus.

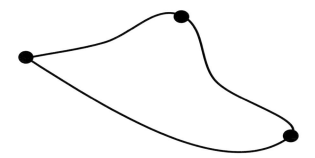

Figure 8.12: The complete graph on three vertices.

8.3 Application to the Königsberg Bridge Problem

Before returning to Euler's original problem, let us look at an even more fundamental question. Let $\{v_1, v_2, \ldots, v_k\}$ be a collection of vertices in the plane (or on the sphere). The *complete graph* on these vertices is the graph that has an edge connecting any two of the vertices. As an instance, the complete graph on three vertices is shown in Figure 8.12. The complete graph on four vertices is shown in Figure 8.13. We may ask whether the complete graph on five vertices can be drawn in the plane—or on the sphere—without any edges crossing any others. If that were possible, then the resulting graph would have 5 vertices (so $V = 5$) and $\binom{5}{2} = 10$ edges (so $E = 10$) and $\binom{5}{3} = 10$ faces (because every face would have to be a triangle). But then $V - E + F = 5 - 10 + 10 = 5$, and that is impossible. The answer is supposed to be 2! We say that the complete graph on five vertices *cannot be embedded in the plane* (or in the sphere). This simple example already illustrates the power of Euler's formula.

Now let us examine the seven bridges of Königsberg. In Figure 8.14 we convert the original Königsberg bridge configuration into a planar graph. We do so by planting a flag in each land mass defined by the river (there are four land masses in Figure 8.3) and connecting two flags if there is a bridge between the two corresponding land masses. If the *order* of a vertex in a graph is the number of edges that meets at that vertex, then we see that the graph in Figure 8.14 has three vertices of order 3. This fact has the following significance.

Imagine a traveler endeavoring to satisfy the stipulations of the Königsberg

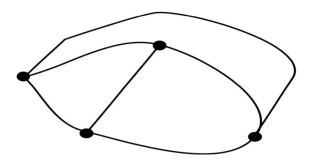

Figure 8.13: The complete graph on four vertices.

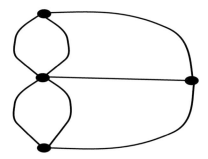

Figure 8.14: The graph corresponding to the Königsberg bridge configuration.

bridge problem. If this traveler enters a vertex of order 3, then that traveler will leave that vertex on a different edge (since the traveler is not allowed to traverse the same edge twice in this journey). But then, if the traveler ever enters that vertex again, he/she cannot leave. There is no edge left on which the traveler can leave (since there are only 3 edges total at that vertex). So the journey would have to end at that vertex. Similar reasoning shows that the journey must begin at a vertex of order 3. That is OK, but there are three vertices of order 3. The journey can only begin at one of those vertices and it can only end at one of those vertices. There is one left over. That is a contradiction.

We see therefore, by Euler's original analysis, that it is impossible to find a journey that traverses all 7 bridges while not repeating any part of the path.

EXERCISE FOR THE READER 8.3.1 Remove one of the seven bridges from the Pregel River. How does this affect the Königsberg bridge problem? Is it now possible to chart a path, never repeating any part of the route and crossing each bridge precisely once? Does it matter which one of the seven bridges you remove?

Let us say that a graph has an *Euler path* if the path traces each edge once and only once. Such a graph is termed an *Eulerian graph*. We have shown that the graph corresponding to the original seven Königsberg bridges does *not* have an Euler path.

In a given graph, call a vertex *odd* (or of *odd degree*, or *odd valence*) if an odd number of edges meet at that vertex. Call the vertex *even* (or of *even degree*, or *even valence*) if an even number of edges meet at that vertex.

EXERCISE FOR THE READER 8.3.2 Explain why, if a graph has more than two odd vertices, then it does not have an Euler path.

EXERCISE FOR THE READER 8.3.3 Examine each of the graphs in Figure 8.15. Which of these has an Euler path? If it does, then find this path.

EXERCISE FOR THE READER 8.3.4 Examine each of the graphs in Figure 8.16. Which of these has an Euler path? If it does, then find this path.

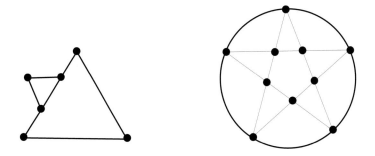

Figure 8.15: Two graphs for Euler analysis.

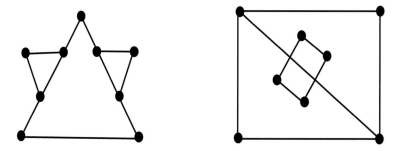

Figure 8.16: Two more graphs for Euler analysis.

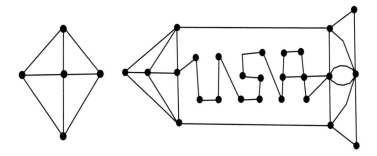

Figure 8.17: Yet two more graphs for Euler analysis.

EXERCISE FOR THE READER 8.3.5 Examine each of the graphs in Figure 8.17. Which of these has an Euler path? If it does, then find this path.

8.4 Coloring Problems

Many mathematic problems originate among professional mathematicians at universities. After all, they are the folks who spend all day every day thinking about mathematics. They are well qualified to identify and develop interesting directions to investigate. But it also happens that some fascinating and long-standing mathematics problems will originate with laymen. The celebrated 4-color problem is an example of such.

In 1852 Francis W. Guthrie, a graduate of University College London, posed the following question to his brother Frederick:

> Imagine a geographic map on the earth (i.e., a sphere) consisting of countries only—no oceans, lakes, rivers, or other bodies of water. The only rule is that a country must be a single contiguous mass—in one piece, and with no holes—see Figure 8.18. As cartographers, we wish to *color* the map so that no two adjacent countries will be of the same color (Figure 8.19—note that R, G, B, Y stand for red, green, blue, and yellow). How many colors should the map-maker keep in stock so that he can be sure he can color any map?

Frederick Guthrie was a student of Augustus De Morgan (1806–1871), and ultimately communicated the problem to his mentor. The problem was

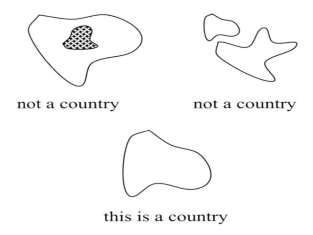

not a country   not a country

this is a country

Figure 8.18: Map coloring.

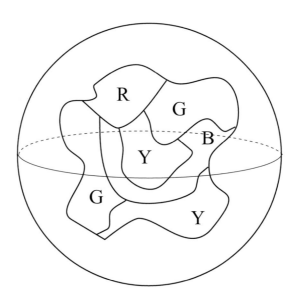

Figure 8.19: The 4-color problem.

passed around among academic mathematicians for a number of years (in fact De Morgan communicated the problem to William Rowan Hamilton (1805–1865)). The first allusion in print to the problem was by Arthur Cayley (1821–1895) in 1878.

The eminent mathematician Felix Klein (1849–1925) in Göttingen heard of the problem and declared that the only reason the problem had never been solved is that no capable mathematician had ever worked on it. *He*, Felix Klein, would offer a class, the culmination of which would be a solution of the problem. He failed.

In 1879, Alfred Kempe (1845–1922) published a solution of the 4-color problem. That is to say, he showed that any map whatever could be colored with four colors. Kempe's proof stood for eleven years. Then a mistake was discovered by Percy Heawood (1861–1955). Heawood studied the problem further and came to a number of fascinating conclusions:

- Kempe's proof, particularly his device of "Kempe chains," *does* suffice to show that any map whatever can be colored with five colors.

- Heawood showed that if the number of edges around each region in the map is divisible by 3, then the map is 4-colorable.

- Heawood found a formula that gives an estimate for the "chromatic number" of any surface. Here the chromatic number $\chi(g)$ of a surface is the least number of colors it will take to color *any* map on that surface. We write the chromatic number as $\chi(g)$. In fact the formula is

$$\chi(g) \leq \left\lfloor \frac{1}{2} \left(7 + \sqrt{48g + 1} \right) \right\rfloor$$

so long as $g \geq 1$.

Here is how to read this formula. It is known, thanks to the work of Camille Jordan (1838–1922) and August Möbius (1790–1868), that any surface in space is a sphere with handles attached. See Figure 8.20. The number of handles is called the *genus*, and we denote it by g. The Greek letter chi (χ) is the chromatic number of the surface—the least number of colors that it will take to color any map on the surface. Thus $\chi(g)$ is the number of colors that it will take to color any map on a surface that consists of the sphere with g handles. Next, the symbols $\lfloor \ \rfloor$ stand for the "greatest integer function." For example $\lfloor \frac{9}{2} \rfloor = 4$

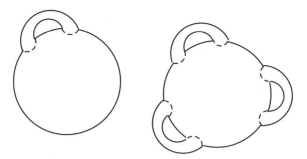

Figure 8.20: The structure of a closed surface in space.

just because the greatest integer in the number "four and a half" is 4. Also $\lfloor \pi \rfloor = 3$ because $\pi = 3.14159\ldots$ and the greatest integer in the number pi is 3.

Now a sphere is a sphere with no handles, so $g = 0$. We may calculate that

$$\chi(g) \leq \left\lfloor \frac{1}{2} \left(7 + \sqrt{48 \cdot 0 + 1} \right) \right\rfloor = \left\lfloor \frac{1}{2} (8) \right\rfloor = 4 \,.$$

This is the 4-color theorem! Unfortunately, Heawood's proof was only valid when the genus is at least 1. It gives no information about the sphere.

The torus (see Figure 8.21) is topologically equivalent to a sphere with one handle. Thus the torus has genus $g = 1$. Then Heawood's formula gives the estimate 7 for the chromatic number. And in fact we can give an example—see Figure 8.22—of a map on the torus that requires 7 colors. Here is what Figure 8.21 shows. It is convenient to take a pair of scissors and cut the torus apart. With one cut, the torus becomes a cylinder; with the second cut it becomes a rectangle. The arrows on the edges indicate that the left and right edges are to be identified (with the same orientation), and the upper and lower edges are to be identified (with the same orientation). We call our colors "1," "2," "3," "4," "5," "6," and "7". The reader may verify that there are 7 countries shown in our Figure 8.22, and every country is adjacent to (i.e., touches) every other. Thus they all must have different colors! This is a map on the torus that requires 7 colors; it shows that Heawood's estimate is sharp for this surface. See Figure 8.23.

Heawood was unable to decide whether the chromatic number of the sphere is 4 or 5. He was also unable to determine whether any of his estimates

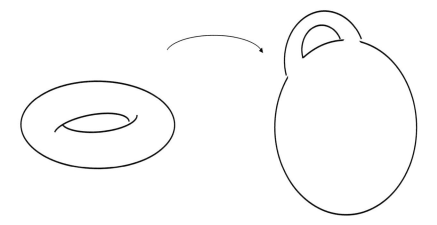

Figure 8.21: The torus is a sphere with one handle.

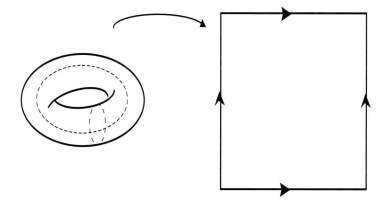

Figure 8.22: The torus as a rectangle with identifications.

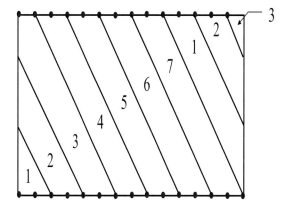

Figure 8.23: A map on the torus that requires 7 colors.

for the chromatic numbers of various surfaces of genus $g \geq 1$ were sharp or accurate. That is to say, for the torus (the closed surface of genus 1), Heawood's formula says that the chromatic number does not exceed 7. Is that in fact the best number? Is there a map on the torus that really requires 7 colors? And for the torus with two handles (genus 2), Heawood's estimate gives an estimate of 8. Is that the best number? Is there a map on the double torus that actually *requires* 8 colors? And so forth: we can ask the same question for every surface of every genus. Heawood could not answer these questions.

8.4.1   Modern Developments

The late nineteenth century saw more alleged solutions of the 4-color problems, many of which stood for as long as eleven years. Eventually errors were found, and the problem remained open on into the twentieth century.

What is particularly striking is that Gerhard Ringel (1919–2008) and J. W. T. Youngs (1910–1970) were able to prove in 1968 that all of Heawood's estimates, for the chromatic number of any surface of genus at least 1, are sharp. So the chromatic number of a torus is indeed 7. The chromatic number of a "double-torus" with two holes is 8. And so forth. But the Ringel/Youngs proof, just like the Heawood formula, does not apply to the sphere. They could not improve on Heawood's result that 5 colors will always suffice. The 4-color problem remained unsolved.

Then in 1974 there was blockbuster news. Using 1200 hours of computer time on the University of Illinois supercomputer, Kenneth Appel and Wolfgang Haken showed that in fact 4 colors will always work to color any map on the sphere. Their technique is to identify 633 fundamental configurations of maps (to which all others can be reduced) and to prove that each of them is reducible to a simpler configuration. But the number of "fundamental configurations" was very large, and the number of reductions required was beyond the ability of any human to count. And the reasoning is extremely intricate and complicated. Enter the computer.

In those days computing time was expensive and not readily available, and Appel and Haken certainly could not get a 1200-hour contiguous time slice for their work. So the calculations were done late at night, "off the record," during various down times. In fact, Appel and Haken did not know for certain whether the calculation would ever cease. Their point of view was this:

(1) If the computer finally stopped, then it will have checked all the cases and the 4-color problem was solved.

(2) If the computer never stopped, then they could draw no conclusion.

Well, the computer stopped. But the level of discussion and gossip and disagreement in the mathematical community did not. Was this really a proof? The computer had performed tens of millions of calculations. Nobody could ever check them all.

But now the plot thickens. Because in 1975 a mistake was found in the proof. Specifically, there was something amiss with the algorithm that Appel and Haken fed into the computer. It was later repaired. The paper was published in 1976. The 4-color problem was declared to be solved.

In a 1986 article, Appel and Haken point out that the reader of their seminal 1976 article must face

(1) 50 pages containing text and diagrams;

(2) 85 pages filled with almost 2500 additional diagrams;

(3) 400 microfiche pages that contain further diagrams and thousands of individual verifications of claims made in the 24 statements in the main section of the text.

But it seems as though there is always trouble in paradise. Errors continued to be discovered in the Appel/Haken proof. Invariably the errors were fixed. But the stream of errors never seemed to cease. So is the Appel/Haken work really a proof?

Well, there is hardly anything more reassuring than another, independent proof. Paul Seymour and his group at Princeton University found another way to attack the problem. In fact they found a new algorithm that seems to be more stable. They also needed to rely on computer assistance. But by the time they did their work computers were *much*, much faster. So they required much less computer time. In any event, this paper appeared in 1994. See [SEY] for some of the details.

8.4.2   Denouement

It is still the case that mathematicians are most familiar with, and most comfortable with, a traditional, self-contained proof that consists of a sequence of logical steps recorded on a piece of paper. We still hope that some day there will be such a proof of the 4-color theorem. After all, it is only a traditional, Euclidean-style proof that offers the understanding, the insight, and the sense of completion that all scholars seek.

And there are new societal needs: theoretical computer science and engineering and even modern applied mathematics require certain pieces of information and certain techniques. The need for a workable device often far exceeds the need to be *certain* that the technique can stand up to the rigorous rules of logic. The result may be that we shall re-evaluate the foundations of our subject. The way that mathematics is practiced in the year 2100 may be quite different from the way that it is practiced today.

8.5   The Traveling Salesman Problem

It is a charming fact of life that some of the most fascinating mathematical problems have utterly simple statements that can be understood by most anyone. Fermat's last theorem is such a problem (see [SIN]). The 4-color problem (see Section 8.4 above) is another. A problem that is fairly old, and is still of pre-eminent importance both for mathematics and logic and also for theoretical computer science, is the celebrated Traveling Salesman Problem (TSP). We shall discuss this problem in the present section.

First studied in the mid-nineteenth century by William Rowan Hamilton (1805–1865) and Thomas Kirkman (1806–1895), the question concerns traveling a circuit in the most efficient fashion. The question is often formulated in terms of a traveling salesman who must visit cities $C_1, C_2, \ldots, C_k$. There is a path connecting any city to any other, and a cost assigned to each path. The goal of the salesman is to begin at some city—say C_1—and to visit every city precisely once. The trip is to end again at C_1. And obviously the salesman wants to minimize his cost. See Figure 8.24.

We may use our knowledge of counting techniques to quickly get an estimate of the number of possible paths that the salesman can take. For the first leg of the trip, the salesman (beginning at C_1) may choose to go to C_2 or C_3 or ...or C_k. Thus there are $(k-1)$ choices. For the next leg, the

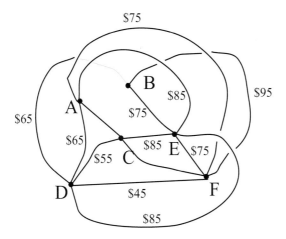

Figure 8.24: The traveling salesman problem.

salesman may choose any of the remaining $(k-2)$ cities. So there are $(k-2)$ choices. And so forth. In summary, there are

$$(k-1) \cdot (k-2) \cdots 3 \cdot 2 \cdot 1 = (k-1)!$$

possible paths that the traveling salesman might take. According to a formula of Stirling,

$$(k-1)! \approx \sqrt{2\pi(k-1)} \cdot \frac{(k-1)^{k-1}}{e^{k-1}} . \tag{8.5.1}$$

In particular, the number is exponential in k. This means that the problem is difficult, and takes a great many steps.

A variety of techniques are known for *estimating* the correct solution to the traveling salesman problem. In particular, if one is willing to settle for a path that costs *not more than twice* the optimal path, then one may find a solution rather efficiently. But the truly optimal solution can be quite complex to find. As an instance, in 2001 the optimal tour of 15,112 German cities was found—and it was shown that the given solution was indeed optimal. The calculation was performed on a 110-CPU parallel-processing computer and required 22.6 years of computer time on a single 500 MHz Alpha processor. In 2006 the optimal circuit for 85,900 cities was found. It took over 136 computing years on a CONCORDE chip.

A current and timely instance of the traveling salesman problem occurs in the manufacture of electronic circuits. These days such circuits are solid state, and all the components are mounted on a board. There are often thousands of units, and much of the manufacture is automated. Certainly part of the process is that a drill head must travel around the circuit board making holes for the various electronic components. Since many tens of thousands of circuit boards will be manufactured, one wants the tour of the drill head to be as efficient as possible.

It has been argued that the single most important problem today in the mathematical sciences is the **P** vs. **NP** problem. Roughly speaking, this is a problem of considering when a problem can be solved in polynomial time and when it will take exponential time. As a simple instance, the problem of taking N randomly shuffled cards and putting them in order is of polynomial time because one first goes through all N cards to find the first card, then one looks through the remaining $(N - 1)$ cards to find the second card, and so forth. In short, it takes at most

$$N + (N - 1) + (N - 2) + \cdots 3 + 2 + 1 = \frac{N(N + 1)}{2} < N^2$$

steps to sort the cards. This is a polynomial estimate on the number of steps. By contrast, the traveling salesman problem with N cities takes about $(N - 1)! \approx (N/e)^{N-1}$ steps, and hence is of exponential complexity. It is known that a complete solution to the traveling salesman problem is logically equivalent to a solution of the **P/NP** problem. This is one of the Clay Mathematics Institute's $1 million dollar Millenium Prize Problems!

We conclude with a few words about how one might find an efficient (though not necessarily optimal) circuit for the traveling salesman problem. There is a commonly used algorithm, both in tree theory and in graph theory, for finding optimal paths and circuits. It is called the *greedy algorithm*. The idea is simple and intuitive. Suppose we are given a layout of cities and paths connecting them, with a cost connected with each path—see Figure 8.24. Notice in the figure that every city (node) is connected to every other city. We begin our circuit with the cheapest path between two cities. In the figure that is path from F to D that costs $45. So, for our first step, we pass from F to D. The next arc must begin at D. We choose the cheapest remaining arc emanating from D (but of course *not* the one we have already traversed). In the figure, that takes us to C over the arc that costs $55. And we continue in this fashion, at every step choosing the cheapest arc possible. This is the greedy algorithm.

There is no guarantee that the greedy algorithm produces the truly *optimal* circuit for the traveling salesman. But it is a theorem that it produces a circuit that costs no more than twice as much as the optimal amount. For many applications this is adequate.

Exercises

1. Give an example of a graph on five vertices without an Euler path.

2. Give an example of a graph on five vertices with two distinct Euler paths.

3. Imagine a torus with two handles. What would be the correct Euler formula for this surface? It should have the form

$$V - E + F = (\text{some number}).$$

   What is that number? The number χ on the right-hand side is called the *Euler characteristic* of the surface.

4. Consider the complete graph on 5 vertices. How many edges does it have? How many faces?

5. How many edges does the complete graph on k vertices have? How many faces?

6. Consider a graph built on two rows of three vertices for a total of 6 vertices. Construct a graph by connecting every vertex in the first row to every vertex of the second row and vice versa. How many edges does this graph have? Can this graph be realized as a planar graph?

7. Consider the standard picture of a 5-pointed star. This can be thought of as a graph. How many vertices does it have? How many edges?

8. Give an example of a graph with more vertices than edges. Give an example of a graph with more edges than vertices.

9. Suppose that the 2-dimensional surface M has Euler characteristic 0, is non-orientable, and has no boundary components. Can you describe M?

10. Suppose that the 2-dimensional surface M has Euler characteristic -2, is orientable, and has empty boundary. Can you describe M?

11. If a surface can be described as a "sphere with g handles," then we say it has genus g. Thus a lone sphere has genus 0, a torus has genus 1, and so forth. Based on your experience with Exercise 3 above, posit a formula that relates the Euler characteristic χ with the genus g.

12. Consider two rows of 4 vertices, and connect every vertex in the first row to every vertex in the second row (and vice versa). Is the resulting graph a planar graph? What happens when you calculate the Euler number?

13. Let G be a graph. Show that G has an even number of vertices having odd degree.

14. Suppose that G is a connected graph. Show that G has an Euler circuit (i.e., a *closed* Euler path in which no edge repeats) if and only if all of the vertices of G have even degree.

15. A *circuit* in a graph is a closed path in which no edge repeats. A circuit in which no vertex repeats is called a *cycle*. Show that if every vertex in a graph G has degree at least 2, then G has a cycle.

16. Let m, n be positive integers. The *complete bipartite graph* $K_{m,n}$ is constructed as follows. Consider a group of m vertices and a group of n vertices. Join each vertex of the first group with an edge to each vertex of the second group (and vice versa). Show that $K_{3,3}$ cannot be embedded in the plane.

17. How many people does it take to form a group so that either 3 of the people are mutually acquainted or 3 of the people are mutually unacquainted? The answer is 6, and you should construe the question as a graph theory problem (the people are vertices, and two people are connected by an edge if they are acquainted, etc.). Then prove the result. When the number 3 is replaced by 5, the problem is unsolved and considered to be impossibly difficult.

18. Show that if a map on the sphere is formed just by drawing circles (see Figure 8.25), then the map can always be colored with three colors.

19. *Every* graph can be embedded in $\mathbb{R}^3$. Explain why this is so.

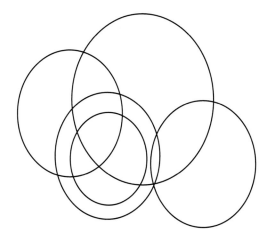

Figure 8.25: A map formed with circles.

20. Refer to Exercise 13 for terminology. Let G be a graph. An edge that lies in no cycle will be the boundary edge of just one face of G. An edge that lies in a cycle will be the boundary edge of two faces of G. Prove these statements.

21. Use the Euler characteristic to show that the sphere and the torus cannot be homeomorphic.

22. Use the Euler characteristic to show that the sphere with k handles attached and the sphere with ℓ handles attached, $k \neq \ell$, are not homeomorphic.

23. A graph is called *simple* if no pair of vertices is joined by two edges, and if the graph also contains no loop. For a simple graph, show that

$$E \leq \binom{V}{2}.$$

24. Two graphs are called *isomorphic* if there is a one-to-one and onto mapping of vertices to vertices and another one-to-one and onto mapping of edges to edges so that if two vertices belong to an edge in the first graph then their images belong to the image edge in the second graph. Show that the 5-pointed star and the complete graph on 5 vertices are *not* isomorphic. Exhibit two graphs on 4 vertices that are not isomorphic and prove your assertion.

25. Refer to the last two exercises for terminology. Show that there are 11 nonisomorphic simple graphs on 4 vertices.

26. Let G be a simple graph. Show that $E = \binom{V}{2}$ if and only if G is the complete graph on k vertices for some k.

27. Refer to Exercise 24 for terminology. An *automorphism* of a graph is an isomorphism from the graph to itself. How many distinct automorphisms does the 5-pointed star have? How many distinct automorphisms does the complete graph on 5 vertices have?

28. Show that, in any group of two or more people, there are always two with exactly the same number of friends in the group.

29. Let $\{P_1, P_2, \ldots, P_k\}$ be distinct points in the plane such that any pair has distance at least 1. Show that the number of pairs with distance exactly 1 is at most $3k$.

30. Draw, if possible, an Eulerian graph with an even number of vertices and an odd number of edges. Or else explain why such a thing is impossible.

31. A mouse eats its way through a $3 \times 3 \times 3$ cube of cheese by tunneling through all 27 of the $1 \times 1 \times 1$ subcubes. If it starts at a corner cube, and always moves from a finished cube to an uneaten subcube, then can it finish in the center cube?

Chapter 9

Dynamical Systems

Dynamical systems were first conceived by Henri Poincaré in order to study certain questions of celestial mechanics. He was able to solve some very important problems of long standing by using this powerful new technique. Today, dynamical systems form one of the hottest and most central areas of mathematical research. Dynamical systems are used to study meteorology, cosmology, particle physics, signal processing, and many other aspects of science and technology.

One of the fun parts of studying dynamical systems is that they can be used to generate attractive pictures. These are sometimes called *fractals*. We shall not say much about fractals in the present chapter, but we give a few sample graphics of the Mandelbrot set (Benoit Mandelbrot (1924–)).

The main purpose of this brief and elementary chapter is to give the flavor of some of the basic ideas of dynamical systems theory. We shall learn how iterating a simple function can generate complex behavior, and how that behavior can model a variety of systems.

9.1   Flows

Consider the system of ordinary differential equations given by

$$\begin{cases} \dfrac{dx}{dt} &= f(x, y, t) \\[2mm] \dfrac{dy}{dt} &= g(x, y, t) \,. \end{cases} \qquad (9.1.1)$$

Notice that we allow the data functions f and g to depend on a parameter t. So the system (9.1.1) is nonautonomous. The role of t will become clearer as our discussion develops. Let us assume that the functions f and g are continuously differentiable.

If a point (x_0, y_0) in $\mathbb{R}^2$ is fixed and $t_0 \in \mathbb{R}$ is fixed, then we may seek a solution $(x(t), y(t))$ to the system (9.1.1) that satisfies $x(t_0) = x_0$ and $y(t_0) = y_0$. Of course this is a standard initial value problem for a first-order system like (9.1.1). We denote such a solution by

$$\begin{aligned} x &= \phi(t; t_0, x_0, y_0) \\ y &= \psi(t; t_0, x_0, y_0). \end{aligned}$$

Do not be confused by the notation here. We wish to explicitly emphasize the dependence on the initial condition.

If t_1 is another point in the domain of x and y, then we may consider the value of our solution pair when evaluated at the point t_1. This is $\phi(t_1; t_0, x_0, y_0)$, $\psi(t_1; t_0, x_0, y_0)$. Thus we think of

$$F_{t_0, t_1}(x_0, y_0) = \big(\phi(t_1; t_0, x_0, y_0), \psi(t_1; t_0, x_0, y_0) \big).$$

It must be clearly understood that this is just a new way of writing something that we already know: we are considering the solution of an initial value problem for a system that is based at the point (x_0, y_0) and seeing where it goes as t passes from t_0 to t_1. Thus we think of the process of solving our system as moving the point (x_0, y_0) to a new location. For simplicity, in the ensuing discussion, we shall drop the subscript 0 on x and y.

If in fact $t_0 < t_1 < t_2$, then it can be checked that

$$F_{t_0, t_2}(x, y) = F_{t_1, t_2} \circ F_{t_0, t_1}(x, y). \tag{9.1.2}$$

This is an immediate consequence of the uniqueness part of the fundamental theorem for systems of ordinary differential equations (see [SKR]). We leave the details to the reader. A family of mappings with the "semi-group" property described by Equation (9.1.2) is called a *flow*.

EXAMPLE 9.1.3 Let us describe the flow associated with the system

$$\begin{cases} \dfrac{dx}{dt} = -y \\[2mm] \dfrac{dy}{dt} = x. \end{cases} \tag{9.1.3.1}$$

Solution: It is convenient to write this system of ordinary differential equations in matrix form as

$$\frac{dx}{dt} = A \cdot \begin{pmatrix} x \\ y \end{pmatrix}$$

or

$$\frac{dx}{dt} = A \cdot \mathbf{v},$$

where

$$\mathbf{v} = \begin{pmatrix} x \\ y \end{pmatrix} = (x, y)$$

and

$$A = \begin{pmatrix} 0 & -1 \\ 1 & 0 \end{pmatrix}.$$

Of course we know how to solve such a system, but now we shall learn to write the solution in a new and useful way.

If $\mathbf{c} = (c_1, c_2)$ is a vector of arbitrary constants, then we write the solution as

$$\mathbf{v} = e^{At}\mathbf{c}.$$

We interpret this equation as follows. The exponential of the matrix At is defined using the power series expansion of exp:

$$
\begin{aligned}
\exp(At) &= \sum_{j=0}^{\infty} \frac{(At)^j}{j!} \\
&= I + \frac{At}{1!} + (At)^2 2! + \frac{(At)^3}{3!} + \cdots \\
&= \begin{pmatrix} 1 & 0 \\ 0 & 1 \end{pmatrix} + \begin{pmatrix} 0 & -t \\ t & 0 \end{pmatrix} + \begin{pmatrix} -t^2/2! & 0 \\ 0 & -t^2/2! \end{pmatrix} \\
&\quad \begin{pmatrix} 0 & t^3/3! \\ -t^3/3! & 0 \end{pmatrix} + \begin{pmatrix} t^4/4! & 0 \\ 0 & t^4/4! \end{pmatrix} + \begin{pmatrix} 0 & -t^5/5! \\ t^5/5! & 0 \end{pmatrix} + \cdots \\
&= \begin{pmatrix} \cos t & -\sin t \\ \sin t & \cos t \end{pmatrix}
\end{aligned}
$$

As a result, our solution is

$$\mathbf{v} = e^{At}\mathbf{c} = (c_1 \cos t - c_2 \sin t, c_1 \sin t + c_2 \cos t).$$

We invite the reader to check that this really solves the system (9.1.3.1).

We shall not take the time here to develop the full machinery of exponentials of matrices. But the student should at least see that this makes solving a system like (9.1.3.1) just the same, and just as simple, as solving the single ordinary differential equation $y' = a \cdot y$. The exponential notation will simplify what we now have to say about dynamics.

To come to the point of this example, we find that the transition map from time t_0 to time t_1 is

$$
\begin{aligned}
F_{t_0,t_1}(x,y) &= \begin{pmatrix} \cos(t_1 - t_0) & -\sin(t_1 - t_0) \\ \sin(t_1 - t_0) & \cos(t_1 - t_0) \end{pmatrix} \cdot \begin{pmatrix} x \\ y \end{pmatrix} \\
&= \big([\cos(t_1 - t_0)]x - [\sin(t_1 - t_0)]y, \\
&\qquad [\sin(t_1 - t_0)]x + [\cos(t_1 - t_0)]y \big) \,.
\end{aligned}
$$

Of course this flow is merely a counterclockwise rotation of the plane through an angle of $t_1 - t_0$ radians.

9.1.1   Dynamical Systems

We have used the phrase "dynamical system" earlier in this book without ever actually saying what one was. Let us now repair that omission. In the subject of dynamical systems we study *iteration of mappings*. In Example 9.1.3 this would be iterations of the rotations induced by the system of differential equations.

To be precise, a *dynamical system* is a one-to-one mapping M from a set of points A to itself. We can form iterations, or compositions, of such a mapping: $M^2 = M \circ M$, $M^3 = M \circ M \circ M$, etc. We can also define M^{-1}, $M^{-2} = M^{-1} \circ M^{-1}$, etc.

We continue to restrict attention to $\mathbb{R}^2$. For any dynamical system M, if $p = (x_0, y_0)$ is a point in the domain of M then the *orbit* of p is the bi-infinite sequence of points

$$
x_j = M^j p \,.
$$

It is worth noting that if the dynamical system happens to come from the flow of a system of ordinary differential equations, as in the first part of this section, then any orbit is a subset of a path of that system.

The focus of our discussion here will be dynamical systems, and these are always one-to-one mappings. Occasionally, however, we shall find it useful to illustrate an idea using an M that is *not one-to-one*. We shall simply refer to such an M as a *mapping*. When M is only a mapping, and hence not one-to-one, then we cannot calculate its inverse. So an orbit of a point p will then only consist of $M^j p$ for $j \geq 0$.

Sometimes the orbit of a dynamical system or a mapping M will consist of just one point $\mathbf{p}^*$. This obviously means that the point $\mathbf{p}^*$ satisfies $M\mathbf{p}^* = \mathbf{p}^*$. We call such a $\mathbf{p}^*$ a *fixed point* of the mapping. This is a point that is never moved; its orbit is just $\{\mathbf{p}^*\}$.

EXAMPLE 9.1.4 Find the fixed points of the mapping $M(x, y) = (2x^2 - y^2, xy)$.

Solution: First notice that M is not a dynamical system; it is only a mapping. Even so, the question makes sense and we can answer it.

We want to solve the equation $M(x, y) = (x, y)$. The problem reduces to solving the system

$$2x^2 - y^2 = x$$
$$xy = y.$$

If $y \neq 0$ then the second equation yields $x = 1$. Putting this value into the first equation gives $2 - y^2 = 1$ hence $y = \pm 1$.

If $y = 0$ then, substituting into the first equation, we find that $x = 0$ or $1/2$.

Altogether, the fixed points we have found are $(1, 1)$, $(1, -1)$, $(0, 0)$, and $(1/2, 0)$.

Now we pass to the slightly more general idea of a periodic point. A point $p = (x, y)$ is called a *k-periodic point* of a mapping M if

$$\underbrace{M \circ M \circ \cdots \circ M}_{k \text{ times}}(p) = p.$$

EXAMPLE 9.1.5 Consider the mapping $M(x, y) = (2xy, y^2 - x^2)$. Then

$$M\left(-\frac{\sqrt{3}}{2}, -\frac{1}{2}\right) = \left(\frac{\sqrt{3}}{2}, -\frac{1}{2}\right)$$

and

$$M\left(\frac{\sqrt{3}}{2}, -\frac{1}{2}\right) = \left(-\frac{\sqrt{3}}{2}, -\frac{1}{2}\right).$$

As a result, the point $(-\sqrt{3}/2, -1/2)$ is 2-periodic and the point $(\sqrt{3}/2, -1/2)$ is also 2-periodic.

Of course fixed points and k-periodic points are closely related. If $\mathbf{p}^*$ is a k-periodic point for the mapping M, then $\mathbf{p}^*$ is a fixed point for the mapping M^k. Also notice that if $\mathbf{p}^*$ is a 2-periodic point, then it is also a 4-periodic point, a 6-periodic point, and so forth. We shall always be interested in knowing the *least* k for which a point is k-periodic.

Much in the spirit of flows in the plane and Poincaré-Bendixson theory, we wish now to consider when an orbit converges, when it exhibits stable behavior, and when it exhibits unstable behavior.

Let M be a mapping and $\mathbf{p}_0 = (x_0, y_0)$ a point of its domain. If $\mathbf{p}_j = (x_j, y_j) = M^j(x_0, y_0) = M^j(\mathbf{p}_0)$ satisfies $\mathbf{p}_j \to \mathbf{p}^* \equiv (\mathbf{x}^*, \mathbf{y}^*)$ as $j \to +\infty$, then we say that $\mathbf{p}_j$ *converges* to $\mathbf{p}^*$. In case M is one-to-one (and hence a dynamical system) then we may also speak of $\mathbf{p}_j \to \mathbf{p}^*$ as $j \to -\infty$.

Proposition 9.1.6 *If M is a mapping (which, as usual, is supposed to be continuous) and if $\mathbf{p}_j = (x_j, y_j) = M^j(x_0, y_0) \to (x^*, y^*) \equiv \mathbf{p}^*$, then $\mathbf{p}^*$ is a fixed point of M.*

Proof: We write

$$
\begin{aligned}
M(\mathbf{p}^*) &= \lim_{j \to \infty} M[\mathbf{p}_j] \\
&= \lim_{j \to \infty} M[M^j(\mathbf{p}_0)] \\
&= \lim_{j \to \infty} M^{j+1}(\mathbf{p}_0) \\
&= \mathbf{p}^*.
\end{aligned}
$$

Thus $\mathbf{p}^*$ is a fixed point.   □

Definition 9.1.7 We say that a fixed point $\mathbf{p}^*$ is *asymptotically stable* if there exists a circle $\mathcal{C}$ centered at $\mathbf{p}^*$ such that if $(x_0, y_0) = \mathbf{p}_0$ lies inside $\mathcal{C}$ then the orbit of (x_0, y_0) converges to $\mathbf{p}^*$ as $j \to +\infty$.

If instead $\mathbf{p}^* = (x^*, y^*)$ is k-periodic, then it is a fixed point of M^k and is thus said to be *asymptotically stable* if it is an asymptotically stable fixed point of M^k.

Definition 9.1.8 We say that a fixed point $\mathbf{p}^*$ of the mapping M is *(asymptotically) unstable* if there exists a circle $\mathcal{C}$ centered at $\mathbf{p}^*$ such that, given any $r > 0$, there is a point $(x_0, y_0) = \mathbf{p}_0$ inside $\mathcal{C}$ such that the orbit of $\mathbf{p}_0$ eventually escapes $\mathcal{C}$. [Here we mean *not* that the orbit leaves $\mathcal{C}$ and never returns. Rather, we mean that at least one iterate of the orbit lands outside $\mathcal{C}$.]

9.1.2 Stable and Unstable Fixed Points

It turns out that we can test for stability or instability of an orbit by using the derivative. This is a fundamental property of orbits, and it is useful to have a simple and direct test. We first formulate a result of this kind in a model case in one real variable.

Proposition 9.1.9 *Let $I \subseteq \mathbb{R}$ be an open interval and assume that $\mathbf{p}^* \in I$. Assume that $\mathbf{p}^*$ is a fixed point of a continuously differentiable mapping $f : I \to \mathbb{R}$.*

(a) *If $|f'(\mathbf{p}^*)| < 1$, then $\mathbf{p}^*$ is asymptotically stable.*

(b) *If $|f'(\mathbf{p}^*)| > 1$, then $\mathbf{p}^*$ is asymptotically unstable.*

Proof: We begin with the proof of **(a)**. Assume that $|f'(\mathbf{p}^*)| < 1$. Select a number $0 < m < 1$ such that $|f'(\mathbf{p}^*)| < m < 1$. By the continuity of the derivative we may choose $\delta > 0$ such that if $|p - \mathbf{p}^*| < \delta$ then $|f'(p)| < m$.

Now select $\mathbf{p}_0$ (the starting place of an orbit) so that $|\mathbf{p}_0 - \mathbf{p}^*| < \delta$. Since $f(\mathbf{p}^*) = \mathbf{p}^*$, we have (by the mean value theorem) that

$$
\begin{aligned}
|\mathbf{p}_1 - \mathbf{p}^*| &= |f(\mathbf{p}_0) - f(\mathbf{p}^*)| \\
&= |f'(\xi_1)| \cdot |\mathbf{p}_0 - \mathbf{p}^*|
\end{aligned}
$$

for some ξ_1 between $\mathbf{p}_0$ and $\mathbf{p}^*$. Note that $|\xi_1 - \mathbf{p}^*| < \delta$ so that $|f'(\xi_1)| < m$. It follows that

$$
|\mathbf{p}_1 - \mathbf{p}^*| < m \cdot |\mathbf{p}_0 - \mathbf{p}^*| < m \cdot \delta .
$$

But now we may estimate again:

$$
\begin{aligned}
|\mathbf{p}_2 - \mathbf{p}^*| \;&=\; |f(\mathbf{p}_1) - f(\mathbf{p}^*)| \\
&=\; |f'(\xi_2)| \cdot |\mathbf{p}_1 - \mathbf{p}^*| \\
&<\; m \cdot (m\delta) = m^2 \cdot \delta \,.
\end{aligned}
$$

Inductively, we find that

$$
|\mathbf{p}_j - \mathbf{p}^*| < m^j \cdot \delta \,.
$$

Thus, since $0 < m < 1$, we see that $\mathbf{p}_j \to \mathbf{p}^*$.

A similar argument shows that, if $|f'(\mathbf{p}^*)| > 1$, then $\mathbf{p}^*$ is unstable. □

EXAMPLE 9.1.10 Let $f : \mathbb{R} \to \mathbb{R}$ be given by

$$
f(x) = 2.2x(1 + x) \,.
$$

Find the fixed points of f and test them for stability.

Solution: The fixed points are those values of x for which

$$
x = f(x) = 2.2x(1 + x)
$$

or

$$
1.2x + 2.2x^2 = 0 \,.
$$

Factoring, we find that

$$
x(1.2 + 2.2x) = 0 \,,
$$

hence the fixed points are $x = 0$ and $x = -6/11$.

Now $f'(x) = 2.2 + 4.4x$. Thus

$$
f'(0) = 2.2 \qquad \text{and} \qquad f'(-6/11) = -1/5 \,.
$$

We conclude that 0 is an unstable fixed point and $-6/11$ is a stable fixed point.

In order to find a version of Proposition 9.1.9 that is valid in two dimensions, we are going to have to proceed in stages. We first consider *linear* dynamical systems in $\mathbb{R}^2$.

9.1.3 Linear Dynamics in the Plane

We shall now let the mapping M be an invertible linear transformation of the plane. Such a mapping is given by a 2×2 matrix A with nonvanishing determinant (we sometimes call such mappings *nondegenerate*). Thus we may write a typical orbit as

$$\begin{pmatrix} x_j \\ y_j \end{pmatrix} = A^j \begin{pmatrix} x_0 \\ y_0 \end{pmatrix}.$$

Of course $\mathbf{0} = (0,0)$ is a fixed point of any linear mapping. We shall be able to characterize $\mathbf{0}$ as an asymptotically stable fixed point of the dynamical system in terms of the *eigenvalues*, or *characteristic values*, of the matrix. We shall see that the stability condition is just that all the eigenvalues lie inside the unit circle of the complex plane.

We begin with some terminology. A matrix Q is called a *logarithm* of the matrix A if $A = e^Q$. In case all the entries of Q are real, then we call Q a *real matrix*, and we say further that Q is a *real logarithm* of A. We want to keep this chapter as concrete as possible, and to avoid as much theory as we can. So we shall nail down the idea of logarithm by way of some examples:

EXAMPLE 9.1.11 Verify that the matrix

$$Q = \begin{pmatrix} 0 & 0 \\ 1 & 0 \end{pmatrix}$$

is a logarithm of the matrix

$$A = \begin{pmatrix} 1 & 0 \\ 1 & 1 \end{pmatrix}.$$

Solution: We calculate that

$$\begin{aligned} e^Q &= I + \frac{Q}{1!} + \frac{Q^2}{2!} + \cdots \\ &= \begin{pmatrix} 1 & 0 \\ 0 & 1 \end{pmatrix} + \begin{pmatrix} 0 & 0 \\ 1 & 0 \end{pmatrix} + \begin{pmatrix} 0 & 0 \\ 0 & 0 \end{pmatrix} + \cdots, \end{aligned}$$

and all higher powers of Q are $\mathbf{0}$. Thus

$$e^Q = \begin{pmatrix} 1 & 0 \\ 1 & 1 \end{pmatrix} = A.$$

EXAMPLE 9.1.12 Verify that the matrix

$$P = \begin{pmatrix} 0 & \pi \\ -\pi & 0 \end{pmatrix}$$

is a logarithm of

$$A = \begin{pmatrix} -1 & 0 \\ 0 & -1 \end{pmatrix}.$$

Solution: We calculate that

$$
\begin{aligned}
e^P &= \begin{pmatrix} 1 & 0 \\ 0 & 1 \end{pmatrix} + \begin{pmatrix} 0 & \pi \\ -\pi & 0 \end{pmatrix} + \begin{pmatrix} -\pi^2/2! & 0 \\ 0 & -\pi^2/2! \end{pmatrix} \\
&\quad + \begin{pmatrix} 0 & -\pi^3/3! \\ \pi^3/3! & 0 \end{pmatrix} + \begin{pmatrix} \pi^4/4! & 0 \\ 0 & \pi^4/4! \end{pmatrix} + \cdots \\
&= \begin{pmatrix} \cos\pi & \sin\pi \\ -\sin\pi & \cos\pi \end{pmatrix} \\
&= \begin{pmatrix} -1 & 0 \\ 0 & -1 \end{pmatrix} \\
&= A.
\end{aligned}
$$

Consider the dynamical system

$$\mathbf{p}_j = M^j \mathbf{p}_0, \, -\infty < j < \infty,$$

where M is given by a real, 2×2 matrix. In case M has a logarithm L then we can view this dynamical system in the following way:

$$\mathbf{p}_j = M^j \mathbf{p}_0 = [e^L]^j \mathbf{p}_0 = e^{Lj} \mathbf{p}_0. \tag{9.1.13}$$

Now compare this equation with the flow arising from the linear system

$$\frac{d\mathbf{x}}{dt} = L\mathbf{x}. \tag{9.1.14}$$

That flow is

$$F_{0,\tau}(\mathbf{p}_0) = e^{\tau L}\mathbf{p}_0 \,, \tag{9.1.15}$$

as we learned in Example 9.1.1.

Comparing Equations (9.1.13) and (9.1.15) now yields that

$$\mathbf{p}_j = F_{0,j}(\mathbf{p}_0) \,. \tag{9.1.16}$$

As noted previously, the orbit $\{\mathbf{p}_j\}$ of the linear dynamical system is a subset of the corresponding path of the system (9.1.14) of ordinary differential equations.

The upshot of these calculations is that one can embed a linear dynamical system into a flow provided one can find a logarithm of the matrix M. Thus we may ask which matrices have logarithms. We may certainly notice that if M has a logarithm L, then $M = e^L$, and e^L has the inverse e^{-L}. Thus M must be invertible, i.e., have nonzero determinant. We see that

If M has a logarithm then M is invertible.

The converse is not obviously true: the matrix

$$M = \begin{pmatrix} -1 & -1 \\ 0 & 1 \end{pmatrix}$$

is certainly invertible, but it has no real logarithm. But it *does* have a complex logarithm.

It turns out that the question of whether a real matrix has a real logarithm is closely related to the question of whether the matrix has a real square root. Here the matrix M is said to have a square root if there is a matrix N such that $N^2 = M$. If M has a logarithm L then of course $e^L = M$ and hence $[e^{L/2}]^2 = M$, so M has a square root. In fact we have the following useful, and rather complete, result:

Proposition 9.1.13 *A square, real matrix M has a (real or complex) logarithm if and only if M is nonsingular (i.e., invertible). If A has a nonsingular, real square root, then A has a real logarithm.*

This result is too complicated to prove here, but we can get good use from it. For it is much easier to check for square roots of matrices than it is to check for logarithms.

THEOREM 9.1.14 *Let s_1, s_2 be the characteristic roots (i.e., the eigen-values) of a 2×2 matrix M. If $|s_1| < 1$ and $|s_2| < 1$, then the origin is an asymptotically stable fixed point of the dynamical system defined by M. If instead either $|s_1| > 1$ or $|s_2| > 1$, then the origin is an unstable fixed point.*

Proof: For simplicity, we shall treat the case that $s_1 \neq s_2$, hence these eigenvalues have linearly independent characteristic vectors (eigenvectors) $\mathbf{v}_1$, $\mathbf{v}_2$. Thus we may write

$$\mathbf{p}_0 = c_1 \mathbf{v}_1 + c_2 \mathbf{v}_2 \,.$$

Thus, exploiting linearity, we see that

$$\mathbf{p}_j = M^j \mathbf{p}_0 = c_1 s_1^j \mathbf{v}1 + c_1 s_2^j \mathbf{v}_2 \,. \tag{9.1.14.1}$$

Since $|s_1| < 1$, we see that $s_1^j \to 0$ and likewise $s_2^j \to 0$. Thus line (9.1.14.1) shows that $\mathbf{p}_j \to 0$. That proves the result.

We omit the argument for $|s_1| > 1$ or $|s_2| > 1$. $\square$

Now let us pass to the situation that really interests us. Suppose that $M(x, y) = (f(x, y), g(x, y))$ is a mapping that sends an open subset of the plane to itself. We suppose that f and g, and therefore M itself, are continuously differentiable. Assume that $\mathbf{p}^* = (x^*, y^*)$ is a fixed point of M. Then we can approximated P near $\mathbf{p}^*$ by the derivative matrix (this is just Taylor's theorem). Namely, the derivative matrix is

$$A = \left(\begin{array}{cc} f_x(x^*, y^*) & f_y(x^*, y^*) \\ g_x(x^*, y^*) & g_y(x^*, y^*) \end{array} \right) \,.$$

[This derivative matrix is also known as the *Jacobian*.] The approximation is then

$$M(p) = M(\mathbf{p}^*) + A \cdot (p - \mathbf{p}^*) + \mathcal{O}(|p - \mathbf{p}^*|^2) \,.$$

Now we can relate the dynamics of M to the dynamics of its approximating matrix A.

THEOREM 9.1.15 *Suppose that $\mathbf{p}^*$ is a fixed point of a continuously differentiable mapping M, and let A be the matrix of first derivatives (the Jacobian) of M at $\mathbf{p}^*$. Then*

(a) *If all of the eigenvalues of A lie inside the unit circle of the complex plane, then then $\mathbf{p}^*$ is asymptotically stable.*

(b) *If either one (or both) of the eigenvalues of A lie outside the unit circle of the complex plane, then $\mathbf{p}^*$ is asymptotically unstable.*

We shall not prove this theorem, as this would take us far afield. We instead illustrate the ideas with some examples.

EXAMPLE 9.1.16 For the mapping $M(x, y) = (2xy, y^2 - x^2)$, test the stability of the fixed points $(0, 0)$ and $(0, 1)$. Also test the stability of the periodic point $(-\sqrt{3}/2, -1/2)$.

Solution: The Jacobian of M at the point (x, y) is

$$A(x, y) = \begin{pmatrix} 2y & 2x \\ -2x & 2y \end{pmatrix}.$$

Since $A(0, 0)$ is the 0-matrix (with eigenvalues 0 and 0), we see immediately that $(0, 0)$ is a stable point. On the other hand, the point $(0, 1)$ has

$$A(0, 1) = \begin{pmatrix} 2 & 0 \\ 0 & 2 \end{pmatrix}.$$

The eigenvalues, both equal to 2, are greater than 1. So the point is unstable.
    For the periodic point, we calculate that

$$A\left(\frac{-\sqrt{3}}{2}, -\frac{1}{2}\right) = \begin{pmatrix} -1 & -\sqrt{3} \\ -\sqrt{3} & -1 \end{pmatrix}.$$

Since the point in question is a 2-periodic point, it is therefore a fixed point of M^2. So we should study the eigenvalues of the Jacobian matrix of M^2. That matrix is just A^2. Thus the required eigenvalues are the squares of the eigenvalues of A itself. Now

$$\text{trace}(A) = [\text{sum of eigenvalues of } A] = -2$$

and

$$\text{determinant}(A) = [\text{product of eigenvalues of } A] = 4.$$

It follows that the eigenvalues of A must be $-1 \pm \sqrt{3}$. Each has absolute value 2. Thus the eigenvalues of A^2 have absolute value 4. We conclude that the periodic point is unstable.

EXAMPLE 9.1.17 Find the stability properties of the fixed points $(0,0)$, $(0,1)$, $(1, 1/2+\sqrt{5}/2)$, $(1, 1/2-\sqrt{5}/2)$ for the mapping $M(x,y) = (x^2, y^2-x)$.

Solution: The Jacobian matrix of M is

$$A = \begin{pmatrix} 2x & 0 \\ -1 & 2y \end{pmatrix}.$$

Thus we see that

$$A(0,0) = \begin{pmatrix} 0 & 0 \\ -1 & 0 \end{pmatrix}.$$

The eigenvalues are 0 and 0, so the point $(0,0)$ is stable.

Next,

$$A(0,1) = \begin{pmatrix} 0 & 0 \\ -1 & 2 \end{pmatrix}.$$

The eigenvalues of this matrix are 0 and 2. We see that the point is unstable.

For the point $(1, 1/2 + \sqrt{5}/2)$, we see that

$$A(1, 1/2 + \sqrt{5}/2) = \begin{pmatrix} 2 & 0 \\ -1 & 1 + \sqrt{5} \end{pmatrix}.$$

It is not difficult to see that at least one of the eigenvalues of this matrix exceeds 1. Thus the point is unstable.

The analysis of the fixed point $(1, 1/2 - \sqrt{5}/2)$ is similar to this last. The point is unstable.

9.2   Planar Autonomous Systems

In the present section we develop some ideas about autonomous systems in the plane using ideas from dynamical systems. One result will be that we shall actually be able to *prove* the Poincaré-Bendixson theorem. This proof will involve not only dynamical systems, but also a number of other quite interesting ideas from geometric analysis. We shall present each of those in a self-contained way. Then we shall cap off the section with the desired proof.

The dominating idea of the Poincaré-Bendixson theory is that the topology of the plane controls the dynamics of the plane. In particular, autonomous systems of ordinary differential equations cannot have chaotic solutions.

9.2.1   Ingredients of the Proof of Poincaré-Bendixson

Consider the autonomous system

$$\begin{cases} \dfrac{dx}{dt} & = \quad f(x,y) \\[2mm] \dfrac{dy}{dt} & = \quad g(x,y) \,. \end{cases} \qquad (9.2.1)$$

As usual, we assume that f and g are continuously differentiable. Our analysis throughout this section will be of the system (9.2.1).

Recall that we have declared a point $\mathbf{p}$ to be a *forward limit point* of an orbit (or path) $\mathbf{v}(t)$ for (10) if $\mathbf{p} = \lim_{j \to \infty} v(t_j)$ for some $v_1 < v_2 < \ldots \to \infty$. We let $\lim_+ \mathbf{v}$ denote the set of forward limit points, and we call $\lim_+ \mathbf{v}$ the *forward limit set* of (9.2.1).

One of the basic ideas in modern mathematics is the notion of invariance. We say that a set $G \subseteq \mathbb{R}^2$ is *invariant* if every orbit that intersects G is actually a subset of G.

Proposition 9.2.2 *The set $\lim_+ \mathbf{v}$ is a closed, invariant set.*

Proof: We cannot provide all the details. The idea is that if some $M^j(\mathbf{p}_0)$ is actually a forward limit point, then it is of course a limit of other elements of the flow. But then, by applying the (inverse of the) flow action to this limit point, we see that all the elements of the orbit are forward limit points. So they are all in $\lim_+ \mathbf{v}$,

For the closedness, suppose that $D(\mathbf{p}, \epsilon) \cap \lim_+ \mathbf{v} \neq \emptyset$ for each $\epsilon > 0$. For each j, choose $\mathbf{p}_j$ to be a point in $D(\mathbf{p}, 1/j) \cap \lim_+ \mathbf{v}$. Using the definition of $\lim_+ \mathbf{v}$, we now choose t_1 with $|\mathbf{p}_1 - \mathbf{v}(t_1)| < 1/1$, $t_2 > t_1$ with $|\mathbf{p}_2 - \mathbf{v}(t_2)| < 1/2$, and so forth. Then, by the triangle inequality,

$$|\mathbf{p} - \mathbf{v}(t_j)| \leq |\mathbf{p} - \mathbf{p}_j| + |\mathbf{p}_j - \mathbf{v}(t_j)| < \frac{2}{j} \,.$$

This shows explicitly that $\mathbf{p} \in \lim_+ \mathbf{v}$. □

Proposition 9.2.3 *Assume that there is a number K such that $\mathbf{v}(t) \in D(0, K)$ for all $t \geq 0$. Then $\lim_+ \mathbf{v}$ is a nonempty, connected set.*

Proof: Obviously the sequence $\{\mathbf{v}(j) : j = 0, 1, 2, \dots\}$ is a bounded sequence, since $|\mathbf{v}(j)| \leq K$ for all j. The Bolzano-Weierstrass theorem (see [KRA1]) then tells us that there is a subsequence $\{\mathbf{v}_{j_k}\}$ that converges to a forward limit point $\mathbf{p}$.

Seeking a contradiction, we suppose that there is a separation $U_1 \cup U_2$ of the forward limits set $\lim_+ \mathbf{v}$. Let $\mathbf{p}_1 \in U_1 \cap \lim_+ \mathbf{v}$ and $\mathbf{p}_2 \in U_2 \cap \lim_+ \mathbf{v}$. Choose $\epsilon > 0$ such that $D(\mathbf{p}_j, \epsilon) \subseteq U_j$, $j = 1, 2$. Then there exist $t_1 < t_2 < t_3 < \cdots \to \infty$ such that, when j is odd, $|\mathbf{p}(t_j) - \mathbf{p}_1| < \epsilon$ and, when j is even, $|\mathbf{v}(t_j) - \mathbf{p}_2| < \epsilon$.

We may think of $\mathbf{v} : [t_{2j-1}, t_{2j}] \to \mathbb{R}^2$. Because the sets U_1, U_2 are disjoint and open, and since $\mathbf{v}(t_{2j-1}) \in U_1$ and $\mathbf{v}(t_{2j}) \in U_2$, we may conclude that there exists a $\widetilde{t}_j \in (t_{2j-1}, t_{2j})$ such that $\widetilde{t}_j \notin U_1 \cup U_2$. [Here we have used the connectedness of $[t_{2j-1}, t_{2j}]$.] The Bolzano-Weierstrass theorem may again be applied to see that there is an increasing sequence $\widetilde{t}_{j_k}$ such that $\mathbf{v}(\widetilde{t}_{j_k})$ converges.

Now the complement of $U_1 \cup U_2$ is closed and contains the sequence $\{\mathbf{v}(\widetilde{t}_{j_k})\}$; so it also must contain the limit point of that sequence. That limit point also belongs to $\lim_+ \mathbf{v}$. This contradiction to the hypothesis that U_1, U_2 is a separation establishes our result. $\qquad\square$

We need one last piece of terminology before we launch our attack on proving the Poincaré-Bendixson theorem. We call a line segment $T \subseteq \mathbb{R}^2$ a *transversal* for the system (9.2.1) provided it contains no critical points of the system and that, for each point $(x, y) \in T$, the vector $f(x, y)\mathbf{i} + g(x, y)\mathbf{j}$ (called the *phase vector*) is not tangent to T. In other words, T *crosses* the flow. See Figure 9.1.

Usually our transversals will in fact be directed line segments. This point will be clear from context.

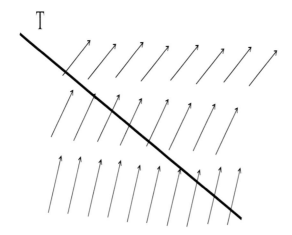

Figure 9.1: The phase vector is transversal.

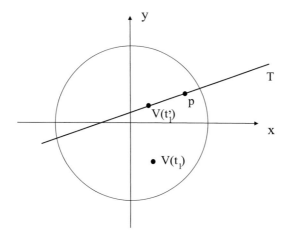

Figure 9.2: The geometry of transversals.

Proposition 9.2.4 *The following statements are true about transversals:*

(a) *All paths cross a transversal in the same direction.*

(b) *If* **p** *is a point in the plane that is not a critical point of the system (9.2.1), then there is a transversal T such that* **p** $\in T$.

(c) *If* **p** *is a point that lies on a transversal T, then there is an $\epsilon > 0$ such that the disc $D(\mathbf{p}, \epsilon)$ has this property: For every solution* **v** *of the system (9.2.1) and each point t_1 such that $\mathbf{v}(t_1) \in D(\mathbf{p}, \epsilon)$, there is a unique point t_1' such that $\mathbf{v}(t_1') \in T$ and $\mathbf{v}(t) \in D(\mathbf{p}, \epsilon)$ for all values of t lying between t_1 and t_1'. See Figure 9.2.*

Proof:
(a)   If T is transversal and $\mathbf{p} \in T$, then the counterclockwise angle $\theta(\mathbf{p})$ made by T and the vector $f(\mathbf{v})\mathbf{i} + g(\mathbf{v})\mathbf{j}$ is not a multiple of π. We may assume that $0 < \theta(\mathbf{p}) < \pi$. The function θ is of course continuous on T. If there is another point $\mathbf{q}$ on T at which $\theta(\mathbf{q}) < 0$, then the intermediate value property of continuous functions ([KRA1] or Appendix 3) implies that θ vanishes at some point between $\mathbf{p}$ and $\mathbf{q}$ on T. But that is impossible, since T is transversal. Thus θ is never negative, and so $0 < \theta < \pi$ at all points of T. It follows that all paths cross T in the same direction.

(b)   Let T' be a line through $\mathbf{p}$ that is not tangent to $f(\mathbf{p})\mathbf{i}+g(\mathbf{p})\mathbf{j}$. As in the proof of part **(a)**, we may assume that the angle measure of $\theta(\mathbf{p})$—between T' and $\mathbf{v}(\mathbf{p})$—lies in the interval $(0, \pi)$. Since θ is a continuous function on T', we may find a closed, bounded subinterval $T \subseteq T'$ on which θ takes values that are bounded away from 0 and π. This T will be the transversal that we seek.

(c)   After a rotation of coordinates, we may as well assume that T is an interval lying in the x-axis. Let $F_{t_1, t_2}(x, y)$ be the flow associated with the system (9.2.1). Define

$$(h(u, t), k(u, t)) = F_{0,t}(u, 0) \,.$$

If $\mathbf{w}(t)$ is the unique solution of (9.2.1) having initial value $\mathbf{w}(0) = (u, 0)$, then $(h(u, t), k(u, t)) = \mathbf{w}(t)$. It follows that $h(u, 0) \equiv u$ and $k(u, 0) \equiv 0$, so that

$$\frac{\partial h}{\partial u}(u, 0) = 1 \quad \text{and} \quad \frac{\partial k}{\partial u}(u, 0) = 0 \,.$$

In addition, when u is held constant and t is the variable, then $(h(u,t), k(u,t))$ traces a path of (9.2.1). In conclusion,

$$\frac{\partial h}{\partial t}(u,0) = f(u,0) \quad \text{and} \quad \frac{\partial k}{\partial t}(u,0) = g(u,0).$$

Now $g(u,0) \neq 0$ when $(u,0) \in T$, just because the vector $\mathbf{i}$ (which gives the direction of the transversal) cannot be parallel to $f(u,0)\mathbf{i} + g(u,0)\mathbf{j}$. Thus, when $(u,0) \in T$, the matrix

$$D(u,0) = \begin{pmatrix} \dfrac{\partial h}{\partial u}(u,0) & \dfrac{\partial k}{\partial u}(u,0) \\ \dfrac{\partial h}{\partial t}(u,0) & \dfrac{\partial k}{\partial t}(u,0) \end{pmatrix} = \begin{pmatrix} 1 & 0 \\ f(u,0) & g(u,0) \end{pmatrix}$$

has nonzero determinant (i.e., is nonsingular). By the very important *inverse function theorem* (see [KRA1] and [KRP2]), there must now be a number $\epsilon > 0$ such that for each point $\mathbf{q} \in D(\mathbf{p}, \epsilon)$ there is a unique (u,t) such that $\mathbf{q} = (h(u,t), g(u,t))$.

Finally suppose that $\mathbf{v}$ is a solution of (9.2.1) with $\mathbf{v}(t_1) \in D(\mathbf{p}, \epsilon)$. Then there is a unique (u_2, t_2) such that

$$\mathbf{v}(t_1) = (h(u_2, t_2), g(u_2, t_2)).$$

Since $\mathbf{z}(t) = \mathbf{v}(t + t_1 - t_2)$ and $(h(u_2, t), g(u_2, t))$ are solutions which agree at the point t_2, it follows from the uniqueness theorem that they are in fact the same solution. As a result,

$$\mathbf{z}(0) = (h(u_2, 0), g(u_2, 0)) = (u_2, 0) \in T.$$

In summary, $\mathbf{v}(t_1 - t_2) \in T$. $\qquad\qquad\qquad\qquad\qquad\qquad\qquad\qquad\square$

Now we need another topological idea. Let $\mathbf{q}$ be a point on the transversal T. Then the solution $\mathbf{v}(t)$ of (9.2.1) having $\mathbf{v}(0) = \mathbf{q}$ may in fact return later to the transversal T. That is to say, for some $t_1 > 0$ it may happen that $\mathbf{v}(t_1) \in T$. For any point $\mathbf{q} \in T$ with this property, let $P(\mathbf{q})$ denote the point $\mathbf{v}(t_1)$, where t_1 is the first value of the parameter for which $\mathbf{v}$ returns to T. Thus P is a mapping that takes subsets of T to subsets of T. We call T the *first return mapping*.

Now we shall make use of the Jordan curve theorem.

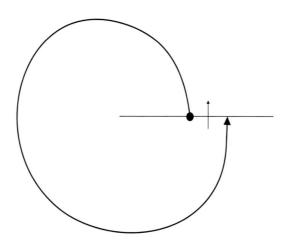

Figure 9.3: Poincaré-Bendixson theory.

Lemma 9.2.5 *The path of a point* $\mathbf{q}$ *on a transversal* T *is a cycle if and only if* $\mathbf{q}$ *is a fixed point of the first return mapping, that is* $P(\mathbf{q}) = \mathbf{q}$.

In case $P(\mathbf{q}) > \mathbf{q}$ *(we are still thinking of* T *as a subset of the* x-*axis) then* $P(\mathbf{q}_1) > P(\mathbf{q})$ *for each point* $\mathbf{q}_1$ *satisfying* $\mathbf{q} \le \mathbf{q}_1$ *and such that* $P(\mathbf{q}_1)$ *is well defined.*

In case $P(\mathbf{q}) < \mathbf{q}$ *(we are still thinking of* T *as a subset of the* x-*axis) then* $P(\mathbf{q}_1) < P(\mathbf{q})$ *for each point* $\mathbf{q}_1$ *satisfying* $\mathbf{q} \ge \mathbf{q}_1$ *and such that* $P(\mathbf{q}_1)$ *is well defined.*

Proof: If $P(\mathbf{q}) = \mathbf{q}$ then the solution $\mathbf{v}$ to the system (9.2.1) is periodic. Thus it returns to its starting point. We conclude that the path is a cycle.

We continue to assume that T lies in the x-axis. Suppose now that $P(\mathbf{q}) > \mathbf{q}$ and that $\mathbf{q}_1 > \mathbf{q}$ and $P(\mathbf{q}_1)$ exists. We are assuming that the picture is as in Figure 9.3. Notice that, as we have arranged before, T lies in the x-axis and the paths cross T in the upward (positive y-) direction. The pieces of the path that connects $\mathbf{q}$ with $P(\mathbf{q})$ and the segment $T_1 \subseteq T$ with the endpoints $\mathbf{q}$ and $P(\mathbf{q})$ form a Jordan curve $\mathcal{C}$. Observe that no path can cross $\mathcal{C}$ except in a point of T', just because paths of the system (9.2.1) cannot cross each other. Furthermore, paths that cross T' must do so in the upward, outward direction. The path through $\mathbf{q}_1$

(1) Points out of $\mathcal{C}$ if $\mathbf{q}_1 < P(\mathbf{q})$;

(2) Originates outside of $\mathcal{C}$ if $\mathbf{q}_1 > P(\mathbf{q})$.

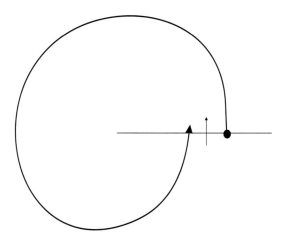

Figure 9.4: Proof when $P(\mathbf{q}) < \mathbf{q}$.

In conclusion, the path can never enter the bounded (or interior) component of $\mathbb{R}^2 \setminus \mathcal{C}$ in order to reach a point of T that lies to the left of $P(\mathbf{q})$.

The proof of the statement when $P(\mathbf{q}) < \mathbf{q}$ follows similar lines, and is left as an exercise. Refer to Figure 9.4. □

Corollary 9.2.6 *Let $\mathbf{q}$ be a point on a transversal T. Then the path*

$$\mathbf{q}, P(\mathbf{q}), P^2(\mathbf{q}), P^3(\mathbf{q}), \ldots$$

of $\mathbf{q}$ under the first return mapping P is either a constant sequence or else it is a monotone sequence (which may be increasing or decreasing on T).

Proof: Immediate from the preceding result. □

The following idea is decisive towards the proof of Poincaré-Bendixson.

Proposition 9.2.7 *Let $\mathbf{v}(t)$ be a path of the system (9.2.1) and let T be a transversal. Then $\lim_+ \mathbf{v} \cap T$ can contain at most one point.*

Proof: Let $\mathbf{p} \in \lim_+ \mathbf{v} \cap \mathbf{T}$. Then Proposition 9.2.4 tells us that there is an $\epsilon > 0$ such that each path that enters $D(\mathbf{p}, \epsilon)$ crosses T at precisely one point before it leaves $D(\mathbf{p}, \epsilon)$. Of course there is an increasing sequence t_j such that $\mathbf{v}(t_j) \to \mathbf{p}$. Hence there exists a $K > 0$ such that $\mathbf{v}(t_j) \in D(\mathbf{p}, \epsilon)$ whenever $j > K$. Thus we may assume that $\mathbf{v}(t_j) \in T$ for each $j > K$.

We may not necessarily suppose that $\mathbf{v}(t_{j+1}) = P(\mathbf{v}(t_j))$ because the path traced by $\mathbf{v}(t)$ may cross T many times between these points. But if there are k such intermediate crossings, then $\mathbf{v}(t_{j+1}) = P^{k+1}(\mathbf{v}(t_j))$ so that the sequence

$$\mathbf{v}(t_{K+1}), \mathbf{v}(t_{K+2}), \mathbf{v}(t_{K+3}), \ldots$$

is a subsequence of the full path

$$\mathbf{v}(t_{N+1}), P(\mathbf{v}(t_{N+1})), P^2(\mathbf{v}(t_{N+1})), \ldots . \qquad (9.2.7.1)$$

The full path is monotone of course by Corollary 9.2.5. Thus the sequence (9.2.7.1) converges to $\mathbf{p}$ just because the limit of a monotone sequence (or, indeed, of any convergent sequence) must coincide with the limit of any of its sequences.

If $\mathbf{q}$ were some other point in $T \cap \lim_+(\mathbf{v})$, then a sequence of points at which the path of $\mathbf{v}$ crosses T would have to converge to $\mathbf{q}$. This sequence would be a subsequence of (9.2.7.1), which certainly converges to $\mathbf{p}$. A convergent sequence can have only one limit, so it must follow that $\mathbf{q} = \mathbf{p}$. □

Corollary 9.2.8 *If $\mathbf{v}(t)$ is a path of the system (9.2.1) such that $L = \lim_+ \mathbf{v}$ contains a cycle $\mathcal{C}$, then $L = \mathcal{C}$.*

Proof: Certainly the complement of a cycle is an open set. If $\mathcal{C} \neq L$ then $\mathcal{C}$ is a proper subset of L. So every open set that contains $\mathcal{C}$ also contains points of $L \backslash \mathcal{C}$ (otherwise L would be disconnected, contradicting Proposition 9.2.3). Thus there is a sequence of point $\mathbf{p}_j \in L \backslash \mathcal{C}$ that converges to a point $\mathbf{p} \in \mathcal{C}$.

If the segment T is transversal at $\mathbf{p} \in T$, then there is a number K such that, for $j > K$, the path of $\{\mathbf{p}_j\}$ crosses T. Since $L \backslash \mathcal{C}$ is invariant, these crossings also belong to $L \backslash \mathcal{C}$. But this means that T will meet L at an infinite set of points, contradicting Proposition 9.2.7. □

Let us now provide the formal statement of the Poincaré-Bendixson theorem.

THEOREM 9.2.9 (Poincaré-Bendixson) *Let R be a bounded region of the phase plane together with its boundary. Assume that R does not contain any critical points of the system (9.2.1). If $C = [x(t), y(t)]$ is a path of (9.2.1) that lies in R for some t_0 and remains in R for all $t \geq t_0$, then C is either itself a closed path or it spirals toward a closed path as $t \to \infty$. Thus, in either case, the system (9.2.1) has a closed path in R.*

Proof of the Poincaré-Bendixson Theorem:

We are given a solution $\mathbf{v}(t)$ such that $L = \lim_+ \mathbf{v}$ is nonempty and contains no stationary points. Let $\mathbf{p} \in L$, and let $\widetilde{\mathbf{v}}$ be that solution satisfying $\widetilde{\mathbf{v}}(0) = \mathbf{p}$. Since L is invariant, the path of $\widetilde{\mathbf{v}}$ is a subset of L. Thus $N \equiv \lim_+ \widetilde{\mathbf{p}} \subseteq L$. But N is nonempty because the path of $\widetilde{\mathbf{p}}$ is bounded.

Let $\mathbf{p}' \in N$. By Proposition 9.2.3, there is a transversal T containing p'. The intersection of the path of $\widetilde{\mathbf{p}}$ with T can only be at $\mathbf{p}'$ because $\mathbf{v}$ has only one limit point on T (by Proposition 9.2.6) and because every point of the path is a limit point of $\mathbf{v}$. In conclusion, N contains a cycle, and Corollary 9.2.8 tells us that N is equal to this cycle. Since $N \subseteq L$, we see that Corollary 9.2.8 once again tells us that $L = N$.

If $P(\mathbf{p}) = \mathbf{p}$, then the path of $\mathbf{v}$ is a cycle. If $P(\mathbf{p}) \neq \mathbf{p}$ then $\mathbf{v}$ converges to a limit cycle. That completes the proof. $\square$

Our purpose here has been to show how to use the theory of dynamical systems to provide a proof of this beautiful theorem. This methodology is similar to the original approach of Poincaré and Bendixson.

We conclude this discussion with some pictures of graphics that arise from dynamical systems (see Figures 9.5, 9.6, 9.7). Specifically, these are pictures of the set of points which, under the action of a specific dynamical system, have a bounded orbit. Such sets are sometimes called *Mandelbrot sets*, although they were discovered somewhat earlier by Brooks and Matelski [BRM].

Figure 9.5: A dynamical graphic.

Figure 9.6: A second dynamical graphic.

Figure 9.7: A third dynamical graphic.

9.3 Lagrange's Equations

In classical mechanics, we may apply Hamilton's principle to derive Lagrange's equations of motion. We present this important circle of ideas in the present section. The first order of business is to come up with a palpable understanding of "degrees of freedom" and "generalized coordinates."

Imagine a single particle moving freely in 3-dimensional space. It is said to have *three degrees of freedom* because its position is specified by the three independent coordinates x, y, and z. By constraining the particle to move on a surface $\{(x, y, z) : G(x, y, z) = 0\}$, we reduce the degrees of freedom to two—since one of its coordinates can be expressed in terms of the other two.

Similarly, an unconstrained system of n particles has $3n$ degrees of freedom; the effect of introducing constraints is to reduce the number of independent coordinates needed to describe the configuration of the system, and hence to reduce the degrees of freedom. In particular, if the rectangular coordinates of the particles are given by x_j, y_j, z_j for $j = 1, \ldots, n$, and if the constraints are described by k consistent and independent equations of the form

$$G_\ell(x_1, y_1, z_1, \ldots, x_n, y_n, z_n) = 0 \ , \ell = 1, 2, \ldots, k \,, \qquad (9.3.1)$$

then the number of degrees of freedom is reduced from $3n$ to $m = 3n - k$. In principle, Equation (9.3.1) can be used to reduce the number of coordinates needed to specify the positions of the n points from $3n$ to m by expressing the $3n$ numbers x_j, y_j, z_j in terms of m parameters.

It is in fact more convenient, in the preceding discussion, to introduce *Lagrange's generalized coordinates* $q_1, \ldots, q_m$. These are any independent coordinates whatsoever whose values determine the configurations of the system. This idea gives us the freedom to choose any coordinate system adapted to the problem at hand—rectangular, cylindrical, spherical, or some other—and renders our analysis independent of any particular coordinate system. We shall now express the rectangular coordinates in terms of the generalized coordinates and we shall see that the resulting formulas have the constraints built in:

$$
\begin{aligned}
x_j &= x_j(q_1, \ldots, q_m) \\
y_j &= y_j(q_1, \ldots, q_m) \\
z_j &= z_j(q_1, \ldots, q_m)
\end{aligned}
$$

for $j = 1, \ldots n$.

If m_j is the mass of the jth particle, then the kinetic energy of the system is

$$
T = \frac{1}{2} \sum_{j=1}^{n} m_j \left\{ \left(\frac{dx_j}{dt} \right)^2 + \left(\frac{dy_j}{dt} \right)^2 + \left(\frac{dz_j}{dt} \right)^2 \right\}.
$$

In terms of the generalized coordinates, this last can be written as

$$
T = \frac{1}{2} \sum_{j=1}^{n} m_j \left\{ \left(\sum_{k=1}^{m} \frac{\partial x_j}{\partial q_k} \dot{q}_k \right)^2 + \left(\sum_{k=1}^{m} \frac{\partial y_j}{\partial q_k} \dot{q}_k \right)^2 + \left(\sum_{k=1}^{m} \frac{\partial z_j}{\partial q_k} \dot{q}_k \right)^2 \right\},
$$

$$(9.3.2)$$

where $\dot{q}_k = dq_k/dt$. Note that T is a homogeneous function of degree 2 in the variables $\dot{q}_k$. The potential energy V of the system is hypothesized to be a function of the q_j alone, hence the Lagrangian $L = T - V$ is a function of the form

$$
L = l(q_1, q_2, \ldots, q_m, \dot{q}_1, \dot{q}_2, \ldots, \dot{q}_m).
$$

Hamilton's principle tells us that the motion proceeds in such a way that the action $\int_{t_1}^{t_2} L \, dt$ is stationary over any interval of time $t_1 \le t \le t_2$, thus Euler's equations must be satisfied. In the present instance these are

$$
\frac{d}{dt} \left(\frac{\partial L}{\partial \dot{q}_k} \right) - \frac{\partial L}{\partial q_k} = 0, \quad k = 1, 2, \ldots, m. \tag{9.3.3}
$$

These are called *Lagrange's equations*. They constitute a system of m second-order differential equations whose solution yields the q_k as functions of t.

We shall now use Lagrange's equations to derive the *law of conservation of energy*. We first note the following identity, which holds (just by the chain rule of calculus) for any function L of the variables t, q_1, q_2, ..., q_m, $\dot{q}_1$, $\dot{q}_2$, ..., $\dot{q}_m$:

$$\frac{d}{dt}\left\{\sum_{k=1}^{m}\dot{q}_k\frac{\partial L}{\partial \dot{q}_k} - L\right\} = \sum_{k=1}^{m}\dot{q}_,\left\{\frac{d}{dt}\left(\frac{\partial L}{\partial \dot{q}_k}\right) - \frac{\partial L}{\partial q_k}\right\} - \frac{\partial L}{\partial t}. \qquad (9.3.4)$$

Since the Lagrangian L of our system satisfies Equation (9.3.3) and does not explicitly depend on t, the right side of (9.3.4) vanishes for this L and we have

$$\sum_{k=1}^{m}\dot{q}_k\frac{\partial L}{\partial \dot{q}_k} - L = E \qquad (9.3.5)$$

for some constant E.

We next observe that $\partial V/\partial \dot{q}_k = 0$, so

$$\frac{\partial L}{\partial \dot{q}_k} = \frac{\partial T}{\partial \dot{q}_k}$$

for every $k = 1, \ldots, m$. We have already noted that formula (9.3.2) shows that T is a homogeneous function of degree two in the $\dot{q}_k$, so we have

$$\sum_{k=1}^{m}\dot{q}_k\frac{\partial L}{\partial \dot{q}_k} = \sum_{k=1}^{m}\dot{q}_k\frac{\partial T}{\partial \dot{q}_k} = 2T$$

by a result of Euler on homogeneous functions.[1]

Thus Equation (9.3.5) becomes $2T - l = E$ or $2T - (T - V) = E$, hence

$$T + V = E.$$

This equation states that, during motion, the sum of the kinetic and potential energies is constant. That is the law of conservation of energy.

Now we look at an example in which Lagrange's equations are applied to dynamics.

---

[1]Euler's theorem says this: Let $g(x, y)$ be a function that is homogeneous of degree r in x and y. So we have $g(\lambda x, \lambda y) = \lambda^r g(x, y)$. If both sides of this equation are differentiated in λ and then λ is set equal to 1, then we obtain

$$x\frac{\partial f}{\partial x} + y\frac{\partial f}{\partial y} = r \cdot f(x, y).$$

This result holds for homogeneous functions of any number of variables.

EXAMPLE 9.3.6 If a particle of mass m moves in a plane under the influence of a gravitational force of magnitude km/r^2 directed toward the origin in the plane, then we choose polar coordinates to analyze the situation. These will be our generalized coordinates. So let $q_1 = r$, $q_2 = \theta$. It is easy to see that

$$T = \frac{m}{2}\left(\dot{r}^2 + r^2\dot{\theta}^2\right)$$

and

$$V = -\frac{km}{r}.$$

Thus the Lagrangian is

$$L = T - V = \frac{m}{2}\left(\dot{r}^2 + r^2\dot{\theta}^2\right) + \frac{km}{r}.$$

Lagrange's equations are therefore

$$\frac{d}{dt}\left(\frac{dL}{d\dot{r}}\right) - \frac{\partial L}{\partial r} = 0, \tag{9.3.6.1}$$

$$\frac{d}{dt}\left(\frac{dL}{d\dot{\theta}}\right) - \frac{\partial L}{\partial \theta} = 0. \tag{9.3.6.2}$$

Since L does not depend explicitly on θ, Equation (9.3.6.2) shows that $\partial L/\partial\dot{\theta} = mr^2\dot{\theta}$ is constant, hence

$$r^2\frac{d\theta}{dt} = h$$

for some positive constant h. We next notice that (9.3.6.1) can be written in the form

$$\frac{d^2r}{dt^2} - r\left(\frac{d\theta}{dt}\right)^2 = -\frac{k}{r^2}. \tag{9.3.6.3}$$

With some ingenuity, this differential equation can be solved to produce the path of the motion of the particle. We shall outline the method in the next paragraphs.

First notice that Newton's second law of motion, applied in the presence of a central force (and realizing that there will be no component of force in the angular direction) tells us that

$$r\frac{d^2\theta}{dt^2} + 2\frac{dr}{dt}\frac{d\theta}{dt} = 0.$$

Multiplying through by r and integrating, we obtain

$$r^2 \frac{d\theta}{dt} = 0. \tag{9.3.6.4}$$

Now the radial component of Newton's second law gives

$$m \left\{ \frac{d^2 r}{dt^2} - r \left(\frac{d\theta}{dt} \right)^2 \right\} = \mathbf{F}_r, \tag{9.3.6.5}$$

where $\mathbf{F}_r$ is the radial force. Newton's law of gravitation allows us to rewrite the last equation as

$$\frac{d^2 r}{dt^2} - r \left(\frac{d\theta}{dt} \right)^2 = -\frac{k}{r^2}. \tag{9.3.6.6}$$

But now we may use (9.3.6.4) to put (9.3.6.6) in the form

$$\frac{d^2 r}{dt^2} - \frac{h^2}{r^3} = -\frac{k}{r^2}. \tag{9.3.6.7}$$

We introduce a new dependent variable $z = 1/r$. Then we can express d^2/dt^2 in terms of $dz^2/d\theta^2$:

$$\frac{dr}{dt} = -h \frac{dz}{d\theta}$$

and

$$\frac{d^2 r}{dt^2} = -h^2 z^2 \frac{d^2 z}{d\theta^2}.$$

When the last expression is substituted into (9.3.6.7) and $1/r$ is replaced by z, we find that

$$-h^2 z^2 \frac{d^2 z}{d\theta^2} - h^2 z^3 = -k z^2$$

or

$$\frac{d^2 z}{d\theta^2} + z = \frac{k}{h^2}.$$

Of course this differential equation may be solved immediately. We find that

$$z = A \sin \theta + B \cos \theta + \frac{k}{h^2}.$$

Some shifting of the coordinate system and some algebraic manipulation finally give us that

$$r = \frac{h^2/k}{1 + e \cos \theta}.$$

This is the polar equation of a conic section.

Studying planetary motion—which of course is the paradigm for a system in which a mass moves under the influence of a central force—was the wellspring of dynamical systems theory. Henri Poincaré used dynamics to solve some important old problems in celestial mechanics. In this section, the celestial mechanics play a prominent role and the theory of dynamical systems is tacit.

Exercises

In Exercises 1–2, find a logarithm of the matrix A and then determine a flow $P_\tau(x, y)$ such that P_1 is the linear dynamical system defined by A.

1. $A = \begin{pmatrix} 3 & 1 \\ 0 & 3 \end{pmatrix}$

2. $A = \begin{pmatrix} 0 & 2 \\ -2 & 0 \end{pmatrix}$

In each of Exercises 3–4, find the fixed points of the given mapping and then identify those that are stable or unstable.

3. $G(x, y) = (x - y, x - y - 4y^2)$

4. $G(x, y) = (4xy, x^2 - y^2)$

5. One of the basic ideas in one-dimensional dynamical systems theory is that of the "staircase representation" for orbits. Let $f : I \to I$ be a mapping of an interval I to itself and fix a point $\mathbf{p_0} \in I$. Draw the graphs of $y = f(x)$ and $y = x$ on the same set of axes in the x-y plane. See Figure 9.8. The staircase representation of the orbit $\mathbf{p}_j = f^j(\mathbf{p_0})$ is generated as follows. Draw a vertical line from the point $Q_0 = (\mathbf{p_0}, \mathbf{p_0})$ on the line $y = x$ to the graph of $y = f(x)$; the intersection will be at the point $P_1 = (\mathbf{p_0}, \mathbf{p_1})$. Next, draw a horizontal line from P_1 to the graph of $y = x$; the intersection will be at the point $Q_1 = (\mathbf{p_1}, \mathbf{p_1})$. We repeat this process. Draw a vertical line from Q_1 to the graph

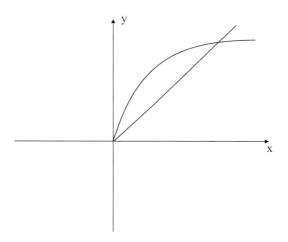

Figure 9.8: The staircase representation.

of $y = f(x)$; the intersection will be at $P_2 = (\mathbf{p}_1, \mathbf{p}_2)$. Next, draw a horizontal line from P_2 to $Q_2 = (\mathbf{p}_2, \mathbf{p}_2)$ on the line $y = x$. And so forth. Refer to Figure 9.9. Explain why this process may be continued indefinitely, and that the points generated are simply the orbit of $\mathbf{p}_0$.

6. Refer to Exercise 5 for terminology. Show that, if f is a positive, decreasing map, $f : \{x \in \mathbb{R} : x > 0\} \to \{x \in \mathbb{R} : x > 0\}$, then the staircase representation of any orbit is in fact a spiral.

Refer to Exercise 5 for terminology. In each of Exercises 7 and 8, sketch the staircase representation of the orbit of the given mapping beginning at the given point. If you have suitable computing equipment available, you may wish to make use of it. Otherwise just make the drawings by hand.

7. $f(x) = \cos x + \dfrac{1}{3}\cos \pi x \quad , \quad \mathbf{p}_0 = 1/8$

8. $g(x) = \dfrac{1}{2}(x^2 + 2x + 1) \quad , \quad \mathbf{p}_0 = 1$

9. The mapping $f : \mathbb{R} \to \mathbb{R}$ given by $f(x) = x^2$ has two fixed points. Which of these is stable and which is not?

10. Suppose that A is a matrix that has at least one eigenvalue of modulus greater than 1. Show that the dynamical system $\mathbf{p}_j = A^j \mathbf{p}_0$ has the origin as an unstable fixed point.

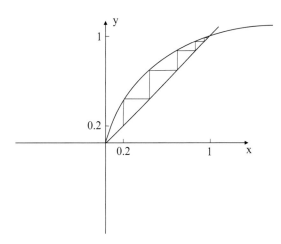

Figure 9.9: Representation of the orbit.

11. In each exercise, find a logarithm of the matrix A and then determine a flow $P_\tau(x, y)$ such that P_1 is the linear dynamical system defined by A.

(a) $A = \begin{pmatrix} 2 & 1 \\ 0 & 2 \end{pmatrix}$

(b) $A = \begin{pmatrix} 0 & 1 \\ -1 & 0 \end{pmatrix}$

12. In each exercise, find the fixed points of the given mapping and then identify those that are stable or unstable.

(a) $G(x, y) = (x - y - 4x^2, x - y)$

(b) $G(x, y) = (x^2 - y^2, 3xy)$

13. Apply the staircase representation technique of Exercise 5 to the mapping $f : [0, 1] \to [0, 1]$ given by $f(x) = x - x^2$. Describe what your diagram depicts. Now repeat this process for the map $g(x) = [x + x^2]/2$.

14. Describe the orbits for the van der Pol equation

$$\frac{d^2x}{dt^2} + \mu(x^2 - 1)\frac{dx}{dt} + x = 0$$

when $\mu < 0$.

15. Explain why the van der Pol equation has a cycle in the phase plane when $\mu > 0$.

16. Give an example of a mapping with two stable fixed points and two unstable fixed points.

17. The equation

$$w'' - \mu(1 - \frac{1}{3}w'^2)w' + u = 0$$

is called the *Rayleigh equation*. Write this ordinary differential equation as a system of two first-order differential equations. Next, show that the origin is the only critical point of this system. Determine the type of this critical point, and state whether it is stable or unstable.

Now take the parameter $\mu = 1$. Choose some initial conditions and calculate the solution of the resulting system on the interval $0 \leq t \leq 40$. Plot μ vs. t; also plot the trajectory in the phase plane. Argue that the trajectory approaches a limit cycle (i.e., a closed curve). Estimate the amplitude and the period of the limit cycle.

18. The dynamics of a continuously differentiable mapping from $\mathbb{R}^2$ to $\mathbb{R}^2$ is modeled on the dynamics of a linear map from $\mathbb{R}^2$ to $\mathbb{R}^2$. Explain what this assertion means, and give examples to illustrate when the contention is correct and examples to illustrate when it is not.

19. Explore, and see whether you can prove, this assertion: Let F and G be continuously differentiable functions in a region $U \subseteq \mathbb{R}^2$ that has no holes. If $\partial F/\partial x + \partial G/\partial y$ has just one sign throughout U, then there is no closed trajectory of the system

$$\frac{\partial x}{\partial t} = F(x, y)$$
$$\frac{\partial y}{\partial t} = G(x, y)$$

that lies entirely in U.

20. Give an example of a dynamical system with two fixed points.

21. In the complex plane, suppose that a constant c is given. Begin with a complex number z_0. Define

$$z_1 = z_0^2 + c,$$

$$z_2 = z_1^2 + c,$$

and, in general,

$$z_{j+1} = z_j^2 + c.$$

If c is a value such that the orbit $\{z_j\}$ is a bounded set, then we say that c lies in the *Mandelbrot set*. The Mandelbrot set is one of the most complicated objects in mathematics. It is a bounded, self-similar set. Give three values of c that lie in the Mandelbrot set.

22. Let

$$f(z_1, z_2) = (z_1, z_2 + z_1^2).$$

This type of mapping is known as a *shear*. What can you say about the dynamics of f? Does it have any fixed points? Is the fixed point stable? Is it unstable?

23. Give an example of the system (9.2.1) for which there is a path C which is closed.

24. Give an example of the system (9.2.1) for which there is a path C which spirals towards a closed path.

25. Give explicit transversals for the two paths in Exercises 23 and 24.

Appendices

Appendix 1: Principles of Logic

Strictly speaking, our approach to logic is "intuitive" or "naïve." Whereas in ordinary conversation these emotion-charged words may be used to downgrade the value of that which is being described, our use of these words is more technical. What is meant is that we shall prescribe in this chapter certain rules of logic which are to be followed in the rest of the book. They will be presented to you in such a way that their validity should be intuitively appealing and self-evident. We cannot *prove* these rules. The rules of logic are the point where our learning begins. A more advanced course in logic will explore other logical methods. The ones that we present here are universally accepted in mathematics and in most of science.

We shall begin with sentential logic and elementary connectives. This material is called the *propositional calculus* (to distinguish it from the predicate calculus, which will be treated later). In other words, we shall be discussing *propositions*—which are built up from atomic statements and connectives. The elementary connectives include "and," "or," "not," "if-then," and "if and only if." Each of these will have a precise meaning and will have exact relationships with the other connectives. The proof of the completeness of this system of elementary sentential logic is considered elsewhere (see [STO, p. 147 ff.] for a discussion of the work of Frege, Whitehead and Russell, Bernays, and Gödel in this regard).

An *elementary statement* (or *atomic statement*) is a sentence with a subject and a verb (and sometimes an object) but no connectives (and, or, not, if-then, if-and-only-if). For example,

John is good.

Mary has bread.

Ethel reads books.

are all atomic statements. We build up sentences, or propositions, from atomic statements using connectives.

Next we shall consider the quantifiers "for all" and "there exists" and their relationships with the connectives from the last paragraph. The quantifiers will give rise to the so-called *predicate calculus*. Connectives and quantifiers will prove to be the building blocks of all future statements in this book, indeed in all of mathematics.

A1.1   Truth

In everyday conversation, people sometimes argue about whether a statement is true or not. In mathematics there is nothing to argue about. In practice a sensible statement in mathematics is either true or false, and there is no room for opinion about this attribute. How do we determine which statements are true and which are false?

The modern methodology in mathematics works as follows.

- We *define* certain terms.

- We *assume* that these terms have certain properties or truth attributes (these assumptions are called axioms).

- We specify certain rules of logic.

Any statement that can be derived from the axioms, using the rules of logic, is understood to be true. It is not necessarily the case that every true statement can be derived in this fashion. However, in practice this is our method for verifying that a statement is true.

On the other hand, a statement is false if it is inconsistent with the axioms and the rules of logic. That is to say, a statement is false if the assumption that it is true leads to a contradiction. Alternatively, a statement **P** is false if the negation of **P** can be established or proved. While it is possible for a statement to be false without our being able to derive a contradiction in this fashion, in practice we establish falsity by the method of contradiction or by giving a counterexample (which is another aspect of the method of contradiction).

The point of view being described here is special to mathematics. While it is indeed true that mathematics is used to model the world around us—in physics, engineering, and in other sciences—the subject of mathematics itself is a man-made system. Its internal coherence is guaranteed by the axiomatic method that we have just described.

It is reasonable to ask whether mathematical truth is a construct of the human mind or an immutable part of nature. For instance, is the assertion that "the area of a circle is π times the radius squared" actually a fact of nature just like Newton's inverse square law of gravitation? Our point of view is that mathematical truth is relative. The formula for the area of a circle is a logical consequence of the axioms of mathematics, nothing more. The fact that the formula seems to describe what is going on in nature is convenient, and is part of what makes mathematics useful. But that aspect is something over which we as mathematicians have no control. Our concern is with the internal coherence of our logical system.

It can be asserted that a "proof" (a concept to be discussed later in the book) is a psychological device for convincing the reader that an assertion is true. However our view in this book is more rigid: a proof is a sequence of applications of the rules of logic to derive the assertion from the axioms. There is no room for opinion here. The axioms are plain. The rules are rigid. A proof is like a sequence of moves in a game of chess. If the rules are followed, then the proof is correct, otherwise it is not.

A1.2 "And" and "Or"

Let **A** and **B** be statements such as "Chelsea is smart" or "The earth is flat." The statement

$$\text{"A and B"}$$

means that both **A** is true *and* **B** is true. For instance,

Arvid is old and Arvid is fat.

This means both that Arvid is old *and* Arvid is fat. If we meet Arvid and he turns out to be young and fat, then the statement is false. If he is old and thin, then the statement is false. Finally, if Arvid is *both* young and thin, then the statement is false. The statement is *true* precisely when both properties—oldness and fatness—hold. We may summarize these assertions with a *truth table*. We let

A = Arvid is old.

and

B = Arvid is fat.

The expression

$$\mathbf{A} \wedge \mathbf{B}$$

will denote the phrase "**A and B**." We call this statement the *conjunction* of **A** and **B**. The letters "T" and "F" denote "True" and "False," respectively. Then we have

A	B	A ∧ B
T	T	T
T	F	F
F	T	F
F	F	F

Notice that we have listed all possible truth values of **A** and **B** and the corresponding values of the *conjunction* **A** ∧ **B**.

In a restaurant the menu often contains phrases such as

soup or salad

This means that we may select soup *or* select salad, but we may not select both. This use of "or" is called the *exclusive* "or"; it is not the meaning of "or" that we use in mathematics and logic. In mathematics we instead say that "**A or B**" is true provided that **A** is true or **B** is true or *both* are true. This is the *inclusive* "or." If we let **A** ∨ **B** denote "**A or B**," then the truth table is

A	B	A ∨ B
T	T	T
T	F	T
F	T	T
F	F	F

We call the statement **A** ∨ **B** the *disjunction* of **A** and **B**.

We see from the truth table that the only way that "**A or B**" can be false is if *both* **A** is false and **B** is false. For instance, the statement

Hillary is beautiful or Hillary is poor.

means that Hillary is either beautiful or poor or both. In particular, she will not be both ugly and rich. Another way of saying this is that if she is ugly she will compensate by being poor; if she is rich she will compensate by being beautiful. *But she could be both beautiful and poor.*

EXAMPLE A1.1 The statement

$$x > 2 \quad \text{and} \quad x < 5$$

is true for the number $x = 3$ because this value of x is both greater than 2 *and* less than 5. It is false for $x = 6$ because this x value is greater than 2 but not less than 5. It is false for $x = 1$ because this x is less than 5 but not greater than 2.

EXAMPLE A1.2 The statement

$$x \quad \text{is odd and} \quad x \quad \text{is a perfect cube}$$

is true for $x = 27$ because both assertions hold. It is false for $x = 7$ because this x, while odd, is not a cube. It is false for $x = 8$ because this x, while a cube, is not odd. It is false for $x = 10$ because this x is neither odd nor is it a cube.

EXAMPLE A1.3 The statement

$$x < 3 \quad \text{or} \quad x > 6$$

is true for $x = 2$ since this x is < 3 (even though it is not > 6). It holds (that is, it is true) for $x = 9$ because this x is > 6 (even though it is not < 3). The statement fails (that is, it is false) for $x = 4$ since this x is neither < 3 nor > 6.

EXAMPLE A1.4 The statement

$$x > 1 \quad \text{or} \quad x < 4$$

is true for every real x.

EXAMPLE A1.5   The statement $(\mathbf{A} \vee \mathbf{B}) \wedge \mathbf{B}$ has the following truth table:

$\mathbf{A}$	$\mathbf{B}$	$\mathbf{A} \vee \mathbf{B}$	$(\mathbf{A} \vee \mathbf{B}) \wedge \mathbf{B}$
T	T	T	T
T	F	T	F
F	T	T	T
F	F	F	F

Notice in Example A1.5 that the statement $(\mathbf{A} \vee \mathbf{B}) \wedge \mathbf{B}$ has the same truth values as the simpler statement $\mathbf{B}$. In what follows, we shall call such pairs of statements (having the same truth values) *logically equivalent*.

The words "and" and "or" are called *connectives*: their role in sentential logic is to enable us to build up (or to connect together) pairs of statements. The idea is to use very simple statements, like "**Jennifer is swift**" as building blocks; then we compose more complex statements from these building blocks by using connectives.

In the next two sections we will become acquainted with the other two basic connectives "not" and "if–then."

A1.3   "Not"

The statement "not $\mathbf{A}$", written $\sim \mathbf{A}$, is true whenever $\mathbf{A}$ is false. For example, the statement

Charles is not happily married.

is true provided the statement "Charles is happily married" is false. The truth table for $\sim \mathbf{A}$ is as follows:

$\mathbf{A}$	$\sim \mathbf{A}$
T	F
F	T

Greater understanding is obtained by combining connectives:

EXAMPLE A1.6   Here is the truth table for $\sim (\mathbf{A} \wedge \mathbf{B})$:

A	B	A $\wedge$ B	$\sim (A \wedge B)$
T	T	T	F
T	F	F	T
F	T	F	T
F	F	F	T

EXAMPLE A1.7 Now we look at the truth table for $(\sim A) \vee (\sim B)$:

A	B	$\sim A$	$\sim B$	$(\sim A) \vee (\sim B)$
T	T	F	F	F
T	F	F	T	T
F	T	T	F	T
F	F	T	T	T

Notice that the statements $\sim (A \wedge B)$ and $(\sim A) \vee (\sim B)$ have the *same truth table.* As previously noted, such pairs of statements are called *logically equivalent.*

The logical equivalence of $\sim (A \wedge B)$ with $(\sim A) \vee (\sim B)$ makes good intuitive sense: the statement $A \wedge B$ fails precisely when either A is false *or* B is false. Since in mathematics we cannot rely on our intuition to establish facts, it is important to have the truth table technique for establishing logical equivalence. The exercise set will give you further practice with this notion.

One of the main reasons that we use the *inclusive* definition of "or" rather than the exclusive one is so that the connectives "and" and "or" have the nice relationship just discussed. It is also the case that $\sim (A \vee B)$ and $(\sim A) \wedge (\sim B)$ are logically equivalent. These logical equivalences are sometimes referred to as *de Morgan's laws.*

A1.4 "If-Then"

A statement of the form "If A then B" asserts that whenever A is true then B is also true. This assertion (or "promise") is tested when A is true, because it is then claimed that something else (namely B) is true as well. *However* when A is false then the statement "If A then B" *claims nothing.* Using the symbols $A \Rightarrow B$ to denote "If A then B," we obtain the following truth table:

A	B	A $\Rightarrow$ B
T	T	T
T	F	F
F	T	T
F	F	T

Notice that we use here an important principle of Aristotelian logic: every sensible statement is either true or false. There is no "in between" status. When **A** is false we can hardly assert that **A** $\Rightarrow$ **B** is false. For **A** $\Rightarrow$ **B** asserts that "whenever **A** is true then **B** is true," and **A** is not true!

Put in other words, when **A** is false then the statement **A** $\Rightarrow$ **B** is not tested. It therefore cannot be false. So it must be true.

EXAMPLE A1.8   The statement "If $2 = 4$ then Calvin Coolidge was our greatest president" is true. This is the case no matter what you think of Calvin Coolidge.

The statement "If fish have hair, then chickens have lips" is true.

The statement "If $9 > 5$, then dogs don't fly" is true.

[Notice that the "if" part of the sentence and the "then" part of the sentence need not be related in any intuitive sense. The truth or falsity of an "if-then" statement is simply a fact about the logical values of its hypothesis and of its conclusion.]

EXAMPLE A1.9   The statement **A** $\Rightarrow$ **B** is logically equivalent with ($\sim$ **A**) $\vee$ **B**. The truth table for the latter is

A	B	$\sim$ A	($\sim$ A) $\vee$ B
T	T	F	T
T	F	F	F
F	T	T	T
F	F	T	T

which is the same as the truth table for **A** $\Rightarrow$ **B**.

You should think for a bit to see that ($\sim$ **A**) $\vee$ **B** *says the same thing* as **A** $\Rightarrow$ **B**. To wit, assume that the statement ($\sim$ **A**) $\vee$ **B** is true. Now suppose that **A** is true. Then, according to the disjunction, **B** must be true. But that says that $\mathbf{A} \Rightarrow \mathbf{B}$. For the converse, assume that $\mathbf{A} \Rightarrow \mathbf{B}$ is true. This means that if $\mathbf{A}$ holds, then $\mathbf{B}$ must follow. But that just says $(\sim \mathbf{A}) \vee \mathbf{B}$. So the two statements are equivalent, i.e., they say the same thing.

Once you believe that assertion, then the truth table for $(\sim \mathbf{A}) \vee \mathbf{B}$ gives us another way to understand the truth table for $\mathbf{A} \Rightarrow \mathbf{B}$.

There are in fact infinitely many pairs of logically equivalent statements. But just a few of these equivalences are really important in practice—most others are built up from these few basic ones. Some of the other basic pairs of logically equivalent statements are explored in the exercises.

EXAMPLE A1.10 The statement

$$\textbf{If } x \textbf{ is negative then } -5 \cdot x \textbf{ is positive}$$

is true. For if $x < 0$ then $-5 \cdot x$ is indeed > 0; if $x \geq 0$ then the statement is unchallenged.

EXAMPLE A1.11 The statement

$$\textbf{If } \{x > 0 \textbf{ and } x^2 < 0\} \textbf{ then } x \geq 10$$

is true since the hypothesis "$x > 0$ and $x^2 < 0$" is never true.

EXAMPLE A1.12 The statement

$$\textbf{If } x > 0 \textbf{ then } \{x^2 < 0 \textbf{ or } 2x < 0\}$$

is false since the conclusion "$x^2 < 0$ or $2x < 0$" is false whenever the hypothesis $x > 0$ is true.

EXAMPLE A1.13 Let us construct a truth table for the statement $\left(\mathbf{A} \vee (\sim \mathbf{B})\right) \Rightarrow \left((\sim \mathbf{A}) \wedge \mathbf{B}\right)$.

$\mathbf{A}$	$\mathbf{B}$	$\sim \mathbf{A}$	$\sim \mathbf{B}$	$\left(\mathbf{A} \vee (\sim \mathbf{B})\right)$	$\left((\sim \mathbf{A}) \wedge \mathbf{B}\right)$
T	T	F	F	T	F
T	F	F	T	T	F
F	T	T	F	F	T
F	F	T	T	T	F

$\big(\mathbf{A} \vee (\sim \mathbf{B})\big) \Rightarrow \big((\sim \mathbf{A}) \wedge \mathbf{B}\big)$
F
F
T
F

Notice that the statement $\big(\mathbf{A} \vee (\sim \mathbf{B})\big) \Rightarrow \big((\sim \mathbf{A}) \wedge \mathbf{B}\big)$ has the same truth table as $\sim \big(\mathbf{B} \Rightarrow \mathbf{A}\big)$. Can you comment on the logical equivalence of these two statements?

Perhaps the most commonly used logical syllogism is the following. Suppose that we know the truth of $\mathbf{A}$ and of $\mathbf{A} \Rightarrow \mathbf{B}$. We wish to conclude $\mathbf{B}$. Examine the truth table for $\mathbf{A} \Rightarrow \mathbf{B}$. The only line in which both $\mathbf{A}$ is true and $\mathbf{A} \Rightarrow \mathbf{B}$ is true is the line in which $\mathbf{B}$ is true. That justifies our reasoning. In logic texts, the syllogism we are discussing is known as *modus ponendo ponens*.

A1.5   Contrapositive, Converse, and "Iff"

The statement

$$\textbf{If A then B}$$

is the same as

$$\mathbf{A} \Rightarrow \mathbf{B}$$

or

$$\textbf{A suffices for B}$$

or as saying

$$\textbf{A only if B}$$

All these forms are encountered in practice, and you should think about them long enough to realize that they say the same thing.

On the other hand,

$$\textbf{If B then A}$$

is the same as saying

$$B \Rightarrow A$$

or

A is necessary for B

or as saying

A if B

We call the statement **B ⇒ A** the *converse* of **A ⇒ B**.

EXAMPLE A1.14   The converse of the statement

If x is a healthy horse, then x has four legs.

is the statement

If x has four legs, then x is a healthy horse.

Notice that these statements have very different meanings: the first statement is true while the second (its converse) is false. For instance, a chair has four legs but it is not a healthy horse.

The statement

A if and only if B

is a brief way of saying

If A then B   *and*   **If B then A**

We abbreviate **A if and only if B** as **A ⇔ B** or as **A iff B**. Here is a truth table for **A ⇔ B**.

A	B	A ⇒ B	B ⇒ A	A ⇔ B
T	T	T	T	T
T	F	F	T	F
F	T	T	F	F
F	F	T	T	T

Notice that we can say that $\mathbf{A} \Leftrightarrow \mathbf{B}$ is true only when both $\mathbf{A} \Rightarrow \mathbf{B}$ and $\mathbf{B} \Rightarrow \mathbf{A}$ are true. An examination of the truth table reveals that $\mathbf{A} \Leftrightarrow \mathbf{B}$ is true precisely when $\mathbf{A}$ and $\mathbf{B}$ are either both true or both false. Thus $\mathbf{A} \Leftrightarrow \mathbf{B}$ means precisely that $\mathbf{A}$ and $\mathbf{B}$ are logically equivalent. One is true when and *only when* the other is true.

EXAMPLE A1.15 The statement

$$x > 0 \Leftrightarrow 2x > 0$$

is true. For if $x > 0$ then $2x > 0$; and if $2x > 0$ then $x > 0$.

EXAMPLE A1.16 The statement

$$x > 0 \Leftrightarrow x^2 > 0$$

is false. For $x > 0 \Rightarrow x^2 > 0$ is certainly true while $x^2 > 0 \Rightarrow x > 0$ is false ($(-3)^2 > 0$ but $-3 \not> 0$).

EXAMPLE A1.17 The statement

$$\{\sim (\mathbf{A} \vee \mathbf{B})\} \Leftrightarrow \{(\sim \mathbf{A}) \wedge (\sim \mathbf{B})\} \qquad (*)$$

is true because the truth table for $\sim(\mathbf{A} \vee \mathbf{B})$ and that for $(\sim \mathbf{A}) \wedge (\sim \mathbf{B})$ are the same. Thus they are logically equivalent: one statement is true precisely when the other is. Another way to see the truth of $(*)$ is to examine the truth table:

$\mathbf{A}$	$\mathbf{B}$	$\sim (\mathbf{A} \vee \mathbf{B})$	$(\sim \mathbf{A}) \wedge (\sim \mathbf{B})$	$\sim (\mathbf{A} \vee \mathbf{B}) \Leftrightarrow \{(\sim \mathbf{A}) \wedge (\sim \mathbf{B})\}$
T	T	F	F	T
T	F	F	F	T
F	T	F	F	T
F	F	T	T	T

Given an implication

$$\mathbf{A} \Rightarrow \mathbf{B},$$

the *contrapositive* statement is defined to be the implication

$$\sim \mathbf{B} \Rightarrow \sim \mathbf{A}.$$

The contrapositive is logically equivalent to the original implication, as we see by examining their truth tables:

A	**B**	**A ⇒ B**
T	T	T
T	F	F
F	T	T
F	F	T

and

A	**B**	**∼ A**	**∼ B**	**(∼ B) ⇒ (∼ A)**
T	T	F	F	T
T	F	F	T	F
F	T	T	F	T
F	F	T	T	T

EXAMPLE A1.18   The statement

If it is raining, then it is cloudy.

has, as its contrapositive, the statement

If there are no clouds, then it is not raining.

A moment's thought convinces us that these two statements say the same thing: if there are no clouds, then it could not be raining; for the presence of rain implies the presence of clouds.

The main point to keep in mind is that, given an implication $\mathbf{A} \Rightarrow \mathbf{B}$, its *converse* $\mathbf{B} \Rightarrow \mathbf{A}$ and its *contrapositive* $(\sim \mathbf{B}) \Rightarrow (\sim \mathbf{A})$ are entirely different statements. The converse is distinct from, and *logically independent from*, the original statement. The contrapositive is distinct from, but *logically equivalent to*, the original statement.

Some classical treatments augment the concept of *modus ponens* with the idea of *modus tollens*. It is in fact logically equivalent to *modus ponens*. *Modus tollens* says

If $\sim B$ and $A \Rightarrow B$ then $\sim A$.

Modus tollens actualizes the fact that $\sim B \Rightarrow \sim A$ is logically equivalent to $A \Rightarrow B$. The first of these implications is of course the *contrapositive* of the second.

A1.6   Quantifiers

The mathematical statements that we will encounter in practice will use the *connectives* "and," "or," "not," "if–then," and "iff." They will also use *quantifiers*. The two basic quantifiers are "for all" and "there exists."

EXAMPLE A1.19   Consider the statement

All automobiles have wheels.

This statement makes an assertion about *all* automobiles. It is true, because every automobile does have wheels.

Compare this statement with the next one:

There exists a woman who is blonde.

This statement is of a different nature. It does not claim that all women have blonde hair—merely that there exists *at least one* woman who does. Since that is true, the statement is true.

EXAMPLE A1.20   Consider the statement

All positive real numbers are integers.

This sentence asserts that something is true for all positive real numbers. It is indeed true for *some* positive numbers, such as 1 and 2 and 193. However it is false for at least one positive number (such as $1/10$ or π), so the entire statement is false.

Here is a more extreme example:

The square of any real number is positive.

This assertion is *almost* true—the only exception is the real number 0: $0^2 = 0$ is not positive. But it only takes one exception to falsify a "for all" statement. So the assertion is false.

This example illustrates the principle that the negation of a "for all" statement is a "there exists" statement.

EXAMPLE A1.21   Look at the statement

There exists a real number which is greater than 5.

In fact there are lots of numbers which are greater than 5; some examples are $7, 42, 2\pi$, and $97/3$. Other numbers, such as 1, 2, and $\pi/6$, are not greater than 5. Since there is *at least one* number satisfying the assertion, the assertion is true.

EXAMPLE A1.22 Consider the statement

There is a man who is at least ten feet tall.

This statement is false. To *verify* that it is false, we must demonstrate that *there does not exist a man who is at least ten feet tall*. In other words, we must show that all men are shorter than ten feet.

The negation of a "there exists" statement is a "for all" statement.

A somewhat different example is the sentence

There exists a real number which satisfies the equation
$$x^3 - 2x^2 + 3x - 6 = 0.$$

There is in fact only one real number which satisfies the equation, and that is $x = 2$. Yet that information is sufficient to show that the statement is true.

We often use the symbol $\forall$ to denote "for all" and the symbol $\exists$ to denote "there exists." The assertion

$$\forall x, \ x + 1 < x$$

claims that for every x, the number $x + 1$ is less than x. If we take our universe to be the standard real number system, then this statement is false. The assertion

$$\exists x, \ x^2 = x$$

claims that there is a number whose square equals itself. If we take our universe to be the real numbers, then the assertion is satisfied by $x = 0$ and by $x = 1$. Therefore the assertion is true.

In all the examples of quantifiers that we have discussed thus far, we were careful to specify our *universe*. That is, "There is a woman such that ..." or "All positive real numbers are ..." or "All automobiles have ...". The quantified statement makes no sense unless we specify the universe of objects from which we are making our specification. In the discussion that follows, we will always interpret quantified statements in terms of a universe.

Sometimes the universe will be explicitly specified, while other times it will be understood from context.

Quite often we will encounter $\forall$ and $\exists$ used together. The following examples are typical:

EXAMPLE A1.23   The statement

$$\forall x \; \exists y, \; y > x$$

claims that for any number x there is a number y which is greater than it. In the realm of the real numbers this is true. In fact $y = x + 1$ will always do the trick.

The statement

$$\exists x \; \forall y, \; y > x$$

has quite a different meaning from the first one. It claims that there is an x which is less than *every* y. This is absurd. For instance, x is *not* less than $y = x - 1$.

EXAMPLE A1.24   The statement

$$\forall x \; \forall y, \; x^2 + y^2 \geq 0$$

is true in the realm of the real numbers: it claims that the sum of two squares is always greater than or equal to zero. [This statement happens to be *false* in the realm of the complex numbers. We shall learn about that number system later. When we interpret a logical statement, it will always be important to understand the context, or universe, in which we are working.]

The statement

$$\exists x \; \exists y, \; x + 2y = 7$$

is true in the realm of the real numbers: it claims that there exist x and y such that $x + 2y = 7$. Certainly the numbers $x = 3, y = 2$ will do the job (although there are many other choices that work as well).

We conclude by noting that $\forall$ and $\exists$ are closely related. The statements

$$\forall x, \; B(x) \qquad \text{and} \qquad \sim \exists x, \; \sim B(x)$$

are logically equivalent. The first asserts that the statement $B(x)$ is true for all values of x. The second asserts that there exists no value of x for which $B(x)$ fails, which is the same thing.

Likewise, the statements

$$\exists x, B(x) \qquad \text{and} \qquad \sim \forall x, \; \sim B(x)$$

are logically equivalent. The first asserts that there is some x for which $B(x)$ is true. The second claims that it is not the case that $B(x)$ fails for every x, which is the same thing. The books [HAL] and [GIH] explore the algebraic structures inspired by these quantifiers.

A "for all" statement is something like the conjunction of a very large number of simpler statements. For example, the statement

For every non-zero integer n, n² > 0.

is actually an efficient way of saying that **1² > 0 and (−1)² > 0 and 2² > 0,** etc. It is not feasible to apply truth tables to "for all" statements, and we usually do not do so.

A "there exists" statement is something like the disjunction of a very large number of statements (the word "disjunction" in the present context means an "or" statement). For example, the statement

There exists an integer n such that P(n) = 2n² − 5n + 2 = 0.

is actually an efficient way of saying that "$P(1) = 0$ **or** $P(-1) = 0$ **or** $P(2) = 0$, etc.". It is not feasible to apply truth tables to "there exist" statements, and we usually do not do so.

It is common to say that *first-order logic* consists of the connectives $\wedge$, $\vee$, $\sim$, $\Rightarrow$, $\Longleftrightarrow$, the equality symbol $=$, and the quantifiers $\forall$ and $\exists$, together with an infinite string of variables $x, y, z, \ldots, x', y', z', \ldots$ and, finally, parentheses (, ,) to keep things readable (see [BAR, p. 7]). The word "first" here is used to distinguish the discussion from second-order and higher-order logics. In first-order logic the quantifiers $\forall$ and $\exists$ always range over elements of the domain M of discourse. Second-order logic, by contrast, allows us to quantify over subsets of M and functions F mapping $M \times M$ into M. Third-order logic treats sets of function and more abstract constructs. The distinction among these different orders is often moot.

A1.7   Truth and Provability

Let us look back at the ideas of this chapter and comment on the difference between truth and provability.

An elementary statement like

$$\mathbf{A} = \text{"George is tall."}$$

has a truth value assigned to it. It is either true or false. From the point of view of mathematics, there is nothing to prove about this statement. Likewise for the statement

$$\mathbf{B} = \text{"Barbara is wise."}$$

On the other hand, the statement

$$\mathbf{A} \vee \mathbf{B}.$$

is subject to mathematical analysis. Namely, it is true if at least one of $\mathbf{A}$ or $\mathbf{B}$ is true. Otherwise it is false.

Any statement which is true regardless of the truth value of its individual components is called a *tautology*. An example of a tautology is

$$\mathbf{B} \Rightarrow \left(\mathbf{A} \vee \sim \mathbf{A} \right).$$

This statement is true all the time—regardless of the truth values of $\mathbf{A}$ and $\mathbf{B}$. Set up a truth table to satisfy yourself that this is the case.

Another example of a tautology is

$$\left(\mathbf{A} \Rightarrow \mathbf{B} \right) \Leftrightarrow \left(\sim \mathbf{A} \vee \mathbf{B} \right).$$

Again, you may verify that this is a tautology by setting up a truth table.

So we have two ways to think about whether a certain statement is valid all the time: (i) to substitute in all possible truth values, and (ii) to prove the statement from elementary principles. We have seen two examples of (i). Now let us think about method (ii).

In order to provide an example of a provable statement, we must isolate in advance what are the syllogisms that we assume in advance to be true, and what rules of logic are allowed. In a formal treatment of logic, such as [SUP] or [STO], we would begin on page 1 of the book with these syllogisms and rules of logic and then proceed rigidly, step by step. At each stage we would have to check which rule or syllogism is being applied. The present book is *not* a formal treatment of logic. It is in fact a more intuitive approach. For the remainder of the section, however, we lapse into the formal mode so that we may learn more carefully to distinguish truth from provability.

First, which rules of logic do we allow? There is only one: *modus ponendo ponens* is the only rule of logic (this is the rule that $\mathbf{A} \Rightarrow \mathbf{B}$ together with $\mathbf{A}$ entails $\mathbf{B}$). Now the other assumptions are these: for the present discussion we take $\sim$ and $\vee$ as our only primitive connectives. Then

N1 $\mathbf{A} \Rightarrow \mathbf{B}$ is an abbreviation for $\sim \mathbf{A} \vee \mathbf{B}$.

N2 $\mathbf{A} \wedge \mathbf{B}$ is an abbreviation for $\sim (\sim \mathbf{A} \vee \sim \mathbf{B})$.

Axiom 1 $(\mathbf{C} \vee \mathbf{C}) \Rightarrow \mathbf{C}$

Axiom 2 $\mathbf{C} \Rightarrow (\mathbf{C} \vee \mathbf{B})$

Axiom 3 $(\mathbf{C} \vee \mathbf{B}) \Rightarrow (\mathbf{B} \vee \mathbf{C})$

Axiom 4 $(\mathbf{B} \Rightarrow \mathbf{A}) \Rightarrow ([\mathbf{C} \vee \mathbf{B}] \Rightarrow [\mathbf{C} \vee \mathbf{A}])$

Notice that **Axioms 1–4** are all "intuitively obvious." Any good axiom should have this feature, because we do not verify or prove axioms. The axioms are our starting place; nothing comes before the axioms. We just accept them. For example, let us think about **Axiom 2**: if we assume that $\mathbf{C}$ is true, then it is certainly the case that $\mathbf{C} \vee \mathbf{B}$ is true. In this way we satisfy our intuition that **Axiom 2** is a reasonable axiom. You may check the other axioms for yourself using similar reasoning.

In some more formal treatments, additional rules of logic are enunciated. The Axiom of Substitution (Axiom Schema of Replacement) is also an important rule of logical reasoning. We shall say more about it later.

In a formal treatment of proof theory (see [BUS, p. 5 ff.]), we sometimes specify—in addition to *modus ponens*—a system of logical axioms that allows the inference of "self-evident" tautologies from no hypotheses. One such system is this (see [BUS]):

(i) $p \Rightarrow (q \Rightarrow p)$

(ii) $(p \Rightarrow q) \Rightarrow [(p \Rightarrow \sim q) \Rightarrow \sim p]$

(iii) $(p \Rightarrow q) \Rightarrow [(p \Rightarrow (q \Rightarrow r)) \Rightarrow (p \Rightarrow r)]$

(iv) $(\sim\sim p) \Rightarrow p$

(v) $p \Rightarrow (p \vee q)$

(vi) $(p \wedge q) \Rightarrow p$

(vii) $q \Rightarrow (p \vee q)$

(viii) $(p \wedge q) \Rightarrow q$

(ix) $(p \Rightarrow r) \Rightarrow [(q \Rightarrow r) \Rightarrow ((p \vee q) \Rightarrow r)]$

(x) $p \Rightarrow [q \Rightarrow (p \wedge q)]$

together with two axiom schemes for the quantifiers:

(xi) $A(t) \Rightarrow \exists x, A(x)$

(xii) $\forall x, A(x) \Rightarrow A(t)$

and two quantifier rules of inference:

(xiii) $[C \Rightarrow A(x)] \Rightarrow [C \Rightarrow \forall x, A(x)]$

(xiv) $[A(x) \Rightarrow C] \rightarrow [\exists x, A(x) \Rightarrow C]$

Let us refer to the system of the first ten of these logical axioms as $\mathcal{F}$, in honor of Gottlob Frege (1848–1925). It is a remarkable fact that $\mathcal{F}$ is complete in the sense that any tautological statement of the propositional calculus can be proved using $\mathcal{F}$.

For the purposes of the present book, *modus ponendo ponens* will be the primary rule of reasoning. The reader can safely worry about no others. Any assertion that we assert to be *provable* must be derivable, using the logical rule *modus ponendo ponens* (which we shall abbreviate MPP), from our notational conventions and these axioms. As an illustration, let us prove the statement $\sim (\mathbf{B} \wedge \sim \mathbf{B})$. [Note that you can easily check this with a truth table; so it *is* a tautology. But now we want to *prove* it from (i) our *definitions*, (ii) our *axioms*, and (iii) our *rules of logic*.]

Now **N2** above shows that the statement that we wish to prove is just $\sim \mathbf{B} \vee \mathbf{B}$. [We have used here the logical equivalence of $\sim\sim \mathbf{B}$ and $\mathbf{B}$. The details of this equivalence are left to you.] It is more natural to prove *this* statement since our axioms are formulated in terms of the connective $\vee$. Here is our proof:

(1) $(\mathbf{B} \vee \mathbf{B}) \Rightarrow \mathbf{B}$ by **Axiom 1**

(2) $[(\mathbf{B} \vee \mathbf{B}) \Rightarrow \mathbf{B}] \Rightarrow$
           $([\sim \mathbf{B} \vee [\mathbf{B} \vee \mathbf{B}]] \Rightarrow [\sim \mathbf{B} \vee \mathbf{B}])$ by **Axiom 4**

(3) $([\sim \mathbf{B} \vee [\mathbf{B} \vee \mathbf{B}]] \Rightarrow [\sim \mathbf{B} \vee \mathbf{B}])$ by MPP applied to (1), (2)

(4) $(\mathbf{B} \Rightarrow (\mathbf{B} \vee \mathbf{B})) \Rightarrow (\mathbf{B} \Rightarrow \mathbf{B})$ applying **N1** to (3)

(5) $\mathbf{B} \Rightarrow (\mathbf{B} \vee \mathbf{B})$ by **Axiom 2**

(6) $\mathbf{B} \Rightarrow \mathbf{B}$ by MPP applied to (4), (5)

(7) $\sim \mathbf{B} \vee \mathbf{B}$ applying **N1** to (6)

That completes the proof.

Implicit in this last discussion is the question of why we can restrict attention to just the connectives $\sim$ and $\vee$. In fact all the other connectives can be expressed in terms of just these two. As an instance, $\mathbf{A} \wedge \mathbf{B}$ is logically equivalent to $\sim (\sim \mathbf{A} \vee \sim \mathbf{B})$. Likewise, $\mathbf{A} \Rightarrow \mathbf{B}$ is logically equivalent with $\sim \mathbf{A}) \vee \mathbf{B}$. These statements can be checked with truth tables. It can also be shown that $\sim$ and $\wedge$ can be used to generate all the other connectives. Some combinations are not possible: $\vee$ and $\wedge$ *cannot* be used to form a statement that is equivalent with $\sim$. Again, you can use truth tables to confirm this assertion.

It is natural to ask whether every tautology is provable (that every provable statement is a tautology is an elementary corollary of our logical structure, or see [STO, p. 152]). That this is so is Frege's theorem. This statement is summarized by saying that elementary sentential logic is *complete*.

In fact Gödel proved in 1930 that the so-called first order predicate calculus is complete. The first order predicate calculus is essentially the logic that we have described in the present chapter: it includes elementary connectives, the quantifiers $\forall$ and $\exists$, and statements P with one or more (but finitely many) variables $x_1, \ldots, x_k$. Thus, according to Gödel, any provable statement in this logic is true and, more profoundly, any true statement is provable. Gödel went on to construct a model for any consistent system of axioms. Interestingly, his proof requires the Axiom of Choice (Appendix 4).

Gödel's more spectacular contribution to modern thought is that, in any logic that complex enough to contain arithmetic, there are sensible statements that cannot be proved either true or false. For example, Peano's arithmetic contains statements that cannot be proved either true or false. A rigorous discussion of this celebrated "incompleteness theorem" is beyond the scope of the present book. Suffice it to say that Gödel's proof consists of making an (infinite) list of all provable statements, enumerating with a system of "Gödel numbers," and then constructing a new statement that differs from each of these. Since the constructed statement could not be on the list, it also cannot be provable. For further discussion of Gödel's ideas, see [DAV], [NAN], and [SMU].

Theoretical computer scientists have shown considerable interest in the incompleteness theorem. For a computer language—even an expert system— can be thought of as a logical theory. Gödel's theorem says, in effect, that there will be statements formulable in any sufficiently complex language that cannot be established through a sequence of logical steps from first principles. For more on this matter, see [KAR], [SHO], and [STO].

Appendix 2: Principles of Set Theory

A2.1 Undefinable Terms

Even the most elementary considerations in logic may lead to conundrums. Suppose that we wish to define the notion of "line." We might say that it is the shortest path between two points. This is not completely satisfactory because we have not yet defined "path" or "point." And when we say "the shortest path" do we mean that there is just one unique shortest path? And why does it exist? Every new definition is, perforce, formulated in terms of other ideas. And those ideas in terms of other ones. Where does the regression cease?

The accepted method for dealing with this problem is to begin with certain terms (as few as possible) that are agreed to be "undefinable." These terms should be so simple that there can be little argument as to their meaning. But it is agreed in advance that these undefinable terms simply cannot be defined in terms of ideas that have been previously defined. Our undefined terms are our starting place.

In modern mathematics it is customary to use "set" and "element of" as undefinables. A *set* is declared to be a collection of objects. (Please do not ask what an "object" is or what a "collection" is; when we say that the term "set" is an undefinable, then we mean just that.) If S is a set then we say that x is an element of S, and we write $x \in S$ or $S \ni x$ precisely when x is one of the objects that compose the set S. For example, we write $5 \in \mathbb{N}$ to indicate that the number 5 is an element of the set of natural numbers. We write $-7 \notin \mathbb{N}$ to specify that -7 is *not* an element of the set of natural numbers.

Definition A2.1.1 We say that two sets S and T are equal precisely when they have the same elements. We write $S = T$.

As an example of equality of sets, if $S = \{x \in \mathbb{N} : x^2 > 3\}$ and $T = \{x \in \mathbb{N} : x \geq 2\}$ then $S = T$.

Incidentally, the method of specifying a set with the notation $\{x : P(x)\}$, where P denotes a property, is the most common method in mathematics of defining a set. This is sometimes called "setbuilder notation."

We shall endeavor, in what follows, to formulate all of our set-theoretic notions in a rigorous and logical fashion from the undefinables "set" and "element of." If at any point we arrive at an untenable position, or a logical contradiction, or a fallacy, then we know that the fault lies with either our method of reasoning or with our undefinables or with our axioms. One of the advantages of the way that we do mathematics is that if there is ever trouble, then we know where the trouble must lie.

However, it should be stressed that basic mathematics is *known* to be—indeed has been *proved to be*—logically consistent. [Strictly speaking, all notions of consistency in mathematics are relative to a higher order system; you learn about these ideas in a course on formal mathematical logic. We shall not give a rigorous treatment of consistency in the present book.] The strict way in which we organize the subject is an important step in establishing this consistency. We shall say more about consistency, and also about independence, in Chapter 5.

A2.2 Elements of Set Theory

Beginning in this section, we will be doing mathematics in the way that it is usually done. That is, we shall define terms and we shall state and prove properties that they satisfy. In earlier chapters we were careful, but we were less mathematical. Sometimes we even had to say, "This is the way we do it; don't worry." Many of the topics in Chapters 1 and 2 are really only best understood from the advanced perspectives of mathematical logic. Now, and for the rest of this book, it is time to show how mathematics is done in practice.

We have already said what theorems, propositions, lemmas, and proofs are. Another formal ingredient of mathematical exposition is the "definition." A definition usually introduces a new piece of terminology or a new idea and *explains what it means in terms of ideas and terminology that have already been presented.* As you read this chapter, pause frequently to check that we are following this paradigm.

Definition A2.2.1 Let S and T be sets. We say that S is a *subset* of T, and we write $S \subset T$ or $T \supset S$, if

$$x \in S \Rightarrow x \in T.$$

We do not prove our definitions. There is *nothing to prove.* A definition introduces you to a new idea, or piece of terminology, or piece of notation.

EXAMPLE A2.1 Let $S = \{x \in \mathbb{N} : x > 3\}$ and $T = \{x \in \mathbb{N} : x^2 > 4\}$. Determine whether $S \subset T$ or $T \subset S$.
Solution: The key to success and clarity in handling subset questions is to *use the definition.* To see whether $S \subset T$ we must check whether $x \in S$ implies $x \in T$. Now if $x \in S$ then $x > 3$ hence $x^2 > 9$ so certainly $x^2 > 4$. Our syllogism is proved, and we conclude that $S \subset T$.

The reverse inclusion is false. For example, the number 3 is an element of T but is certainly not an element of S. We write $T \not\subset S$.

EXAMPLE A2.2 Let $\mathbb{Z}$ denote the system of integers. Let $S = \{-2, 3\}$. Let $T = \{x \in \mathbb{Z} : x^3 - x^2 - 6x = 0\}$. Determine whether $S \subset T$ or $T \subset S$.
Solution: To see whether $S \subset T$ we must check whether $x \in S$ implies $x \in T$. Let $x \in S$. Then either $x = -2$ or $x = 3$. If $x = -2$ then $x^3 - x^2 - 6x = (-2)^3 - (-2)^2 - 6(-2) = 0$. Also, if $x = 3$ then $x^3 - x^2 - 6x = (3)^3 - (3)^2 - 6(3) = 0$. This verifies the syllogism $x \in S$ implies $x \in T$. Therefore $S \subset T$.

The reverse inclusion fails, for $0 \in T$ but $0 \notin S$.

EXAMPLE A2.3 Let $S = \{x \in \mathbb{N} : x \geq 4\}$ and $T = \{x \in \mathbb{N} : x < 9\}$. Is it true that either $S \subset T$ or $T \subset S$?
Solution: Both inclusions are false. For $10 \in S$ but $10 \notin T$ and $2 \in T$ but $2 \notin S$.

Proposition A2.2.1 *Let S and T be sets. Then $S = T$ if and only if both $S \subset T$ and $T \subset S$.*

Proof: If $S = T$ then, by definition, S and T have precisely the same elements. In particular, this means that $x \in S$ implies $x \in T$ and also $x \in T$ implies $x \in S$. That is, $S \subset T$ and $T \subset S$.

Now suppose that both $S \subset T$ and $T \subset S$. Seeking a contradiction, suppose that $S \neq T$. Then either there is some element of S that is not an element of T or there is some element of T that is not an element of S. The first eventuality contradicts $S \subset T$ and the second eventuality contradicts $T \subset S$. We conclude that $S = T$. $\qquad\qquad\square$

Definition A2.2.2 We let $\emptyset$ denote the set that contains no elements. That is, $\forall x, x \notin \emptyset$. We call $\emptyset$ the *empty set*.

EXAMPLE A2.4   If S is any set, then $\emptyset \subset S$. To see this, notice that the statement "If $x \in \emptyset$, then $x \in S$" *must* be true because the hypothesis $x \in \emptyset$ is false. [Check the truth table for "if–then" statements.] This verifies that $\emptyset \subset S$.

EXAMPLE A2.5   Let $S = \{x \in \mathbb{N} : x + 2 \geq 19 \text{ and } x < 3\}$. Then S is a sensible set. There are no internal contradictions in its definition. But $S = \emptyset$. There are no elements in S.

Definition A2.2.3 Let S and T be sets. We say that x is an element of $S \cap T$ if both $x \in S$ *and* $x \in T$. We say that x is an element of $S \cup T$ if either $x \in S$ *or* $x \in T$.

We call $S \cap T$ the *intersection* of the sets S and T. We call $S \cup T$ the *union* of the sets S and T.

EXAMPLE A2.6   Let $S = \{x \in \mathbb{N} : 2 < x < 9\}$ and $T = \{x \in \mathbb{N} : 5 \leq x < 14\}$. Then $S \cap T = \{x \in \mathbb{N} : 5 \leq x < 9\}$, for these are the points common to both sets. And $S \cup T = \{x \in \mathbb{N} : 2 < x < 14\}$, for these are the points that are either in S or in T or in both.

Remark A2.2.2 Observe that the use of "or" in the definition of set union justifies our decision to use the "inclusive 'or' " rather than the "exclusive 'or' " in mathematics.

EXAMPLE A2.7 Let $S = \{x \in \mathbb{N} : 1 \leq x \leq 5\}$ and $T = \{x \in \mathbb{N} : 8 < x \leq 12\}$. Then $S \cap T = \emptyset$, for the sets S and T have no elements in common. On the other hand, $S \cup T = \{x \in \mathbb{N} : 1 \leq x \leq 5 \text{ or } 8 < x \leq 12\}$.

Definition A2.2.4 Let S and T be sets. We say that $x \in S \setminus T$ if both $x \in S$ and $x \notin T$. We call $S \setminus T$ the *set-theoretic difference* of S and T.

EXAMPLE A2.8 Let $S = \{x \in \mathbb{N} : 2 < x < 7\}$ and $T = \{x \in \mathbb{N} : 5 \leq x < 10\}$. Then $S \setminus T = \{x \in \mathbb{N} : 2 < x < 5\}$ and $T \setminus S = \{x \in \mathbb{N} : 7 \leq x < 10\}$.

Definition A2.2.5 Suppose that we are studying subsets of a fixed set X. If $S \subset X$ then we use the symbol ^cS to denote $X \setminus S$. In this context we sometimes refer to X as the *universal set*. We call ^cS the *complement* of S (in X).

EXAMPLE A2.9 Let $\mathbb{N}$ be the universal set. Let $S = \{x \in \mathbb{N} : 3 < x \leq 20\}$. Then

$$^cS = \{x \in \mathbb{N} : 1 \leq x \leq 3\} \cup \{x \in \mathbb{N} : 20 < x\}.$$

Proposition A2.2.3 *Let X be the universal set and $S \subset X$, $T \subset X$. Then*

(a) $^c(S \cup T) = {}^cS \cap {}^cT$;
(b) $^c(S \cap T) = {}^cS \cup {}^cT$.

Proof: We shall present this proof in detail since it is a good exercise in understanding both our definitions and our method of proof.

We begin with the proof of **(a)**. It is often best to treat the proof of the equality of two sets as two separate proofs of containment. [This is why Proposition A2.2.1 is important.] That is what we now do.

Let $x \in {}^c(S \cup T)$. Then, by definition, $x \notin (S \cup T)$. Thus x is neither an element of S nor an element of T. So both $x \in {}^cS$ and $x \in {}^cT$. Hence $x \in {}^cS \cap {}^cT$. We conclude that $^c(S \cup T) \subset {}^cS \cap {}^cT$. Conversely, if $x \in {}^cS \cap {}^cT$

then $x \notin S$ and $x \notin T$. Therefore $x \notin (S \cup T)$. As a result, $x \in {}^{c}(S \cup T)$. Thus ${}^{c}S \cap {}^{c}T \subset {}^{c}(S \cup T)$. Summarizing, we have ${}^{c}(S \cup T) = {}^{c}S \cap {}^{c}T$.

The proof of part **(b)** is similar, but we include it for practice. Let $x \in {}^{c}(S \cap T)$. Then, by definition, $x \notin (S \cap T)$. Thus x is not both an element of S and an element of T. So either $x \in {}^{c}S$ or $x \in {}^{c}T$. Hence $x \in {}^{c}S \cup {}^{c}T$. We conclude that ${}^{c}(S \cap T) \subset {}^{c}S \cup {}^{c}T$. Conversely, if $x \in {}^{c}S \cup {}^{c}T$ then either $x \notin S$ or $x \notin T$. Therefore $x \notin (S \cap T)$. As a result, $x \in {}^{c}(S \cap T)$. Thus ${}^{c}S \cup {}^{c}T \subset {}^{c}(S \cap T)$. Summarizing, we have ${}^{c}(S \cap T) = {}^{c}S \cup {}^{c}T$.   □

The two formulas in the last proposition are often referred to as "de Morgan's laws."

A2.3   Venn Diagrams

We sometimes use a *Venn diagram* to aid our understanding of set-theoretic relationships. In a Venn diagram, a set is represented as a region in the plane (for convenience, we use rectangles). The intersection $A \cap B$ of two sets A and B is the region common to the two domains (we have shaded that region with dots in Figure 1):

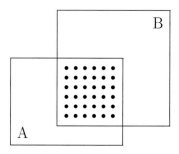

Figure 1

Now let A, B, and C be three sets. The Venn diagram in Figure 2 makes it easy to see that $A \cap (B \cup C) = (A \cap B) \cup (A \cap C)$.

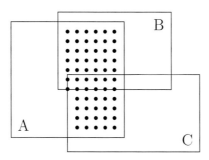

Figure 2

The Venn diagram in Figure 3 illustrates the fact that

$$A \setminus (B \cup C) = (A \setminus B) \cap (A \setminus C)$$

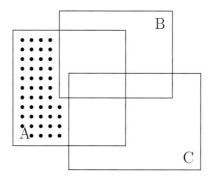

Figure 3

A Venn diagram is not a proper substitute for a rigorous mathematical proof. However it can go a long way toward guiding our intuition.

A2.4 Further Ideas in Elementary Set Theory

Now we learn some new ways to combine sets.

Definition A2.4.1 Let S and T be sets. We define $S \times T$ to be the set of all ordered pairs (s, t) such that $s \in S$ and $t \in T$. The set $S \times T$ is called the *set-theoretic product* of S and T.

It is worth pausing a moment to consider this last definition. Strictly speaking, it is not entirely satisfactory because we have not defined "ordered pair." Any attempt to do so using phrases like "first element" and "second element" seems to lead to more questions than answers. A rigorous and elegant way to define the ordered pair (a, b) is that it is equal to the set $\{a, \{a, b\}\}$. This definition exhibits directly that the ordered pair contains both the elements a and b and that the element a is distinguished.

EXAMPLE A2.10   Let $S = \{1, 2, 3\}$ and $T = \{a, b\}$. Then

$$S \times T = \{(1, a), (1, b), (2, a), (2, b), (3, a), (3, b)\}.$$

It is no coincidence that, in the last example, the set S has 3 elements, the set T has 2 elements, and the set $S \times T$ has $3 \times 2 = 6$ elements. In fact one can prove that if S has k elements and T has ℓ elements, then $S \times T$ has $k \cdot \ell$ elements.

Notice that $S \times T$ is a different set from $T \times S$. With S and T as in the last example,

$$T \times S = \{(a, 1), (b, 1), (a, 2), (b, 2), (a, 3), (b, 3)\}.$$

The phrase "ordered pair" means that the pair $(a, 1)$, for example, is distinct from the pair $(1, a)$.

If S is a set, then the *power set* of S is the set of all subsets of S. We denote the power set by $\mathcal{P}(S)$.

EXAMPLE A2.11   Let $S = \{1, 2, 3\}$. Then

$$\mathcal{P}(S) = \{\{1\}, \{2\}, \{3\}, \{1, 2\}, \{2, 3\}, \{1, 3\}, \{1, 2, 3\}, \emptyset\}.$$

If the concept of power set is new to you, then you might have been surprised to see $\{1, 2, 3\}$ and $\emptyset$ as elements of the power set. But they are both subsets of S, and they must be listed.

Proposition A2.4.1 Let $S = \{s_1, \ldots, s_k\}$ be a set. Then $\mathcal{P}(S)$ has 2^k elements.

Proof: We prove the assertion by induction on k.

P(1) is true. In this case, $S = \{s_1\}$ and $\mathcal{P}(S) = \{\{s_1\}, \emptyset\}$. Notice that S has $k = 1$ element and $\mathcal{P}(S)$ has $2^k = 2$ elements.

P(k) $\Rightarrow$ P(k + 1). Assume that any set of the form $S = \{s_1, \ldots, s_k\}$ has power set with 2^k elements. Now let $T = \{t_1, \ldots, t_k, t_{k+1}\}$. Consider the subset $T' = \{t_1, \ldots, t_k\}$ of T. Then $\mathcal{P}(T)$ certainly contains $\mathcal{P}(T')$ (that is, every subset of T' is also a subset of T). But it also contains each of the sets that is obtained by adjoining the element t_{k+1} to each subset of T'. Thus the total number of subsets of T is

$$2^k + 2^k = 2^{k+1}.$$

Notice that we have indeed counted all subsets of T, since any subset either contains t_{k+1} or it does not.

Thus, assuming the validity of our assertion for k, we have proved its validity for $k + 1$. That completes our induction and the proof of the proposition. $\qquad\square$

We have seen that the operation of set-theoretic product corresponds to the arithmetic product of natural numbers. And now we have seen that the operation of taking the power set corresponds to exponentiation.

A2.5 Indexing and Extended Set Operations

Frequently we wish to manipulate infinitely many sets. Perhaps we will take their intersection or union. We require suitable notation to perform these operations.

If $S_1, S_2, \ldots$ are sets then we define

$$\bigcup_{j=1}^{\infty} S_j \equiv \{x : \exists j \text{ such that } x \in S_j\}.$$

Similarly, we define

$$\bigcap_{j=1}^{\infty} S_j \equiv \big\{x : \forall j, x \in S_j\big\}.$$

Notice that we employ the common mathematical notation $\equiv$ to mean "is defined to be." Other texts use the notation $\stackrel{\text{def}}{=}$ or $=:$ or $\doteq$.

EXAMPLE A2.12   Let $\mathbb{Q}$ be the rational numbers and let $S_j = \{x \in \mathbb{Q} : 0 < x < 1 + 1/j\}$, $j = 1, 2, \ldots$. Let us describe $\cup_{j=1}^{\infty} S_j$ and $\cap_{j=1}^{\infty} S_j$.

Notice that $S_1 \supset S_j$ for every j, hence $\cup_{j=1}^{\infty} S_j = S_1 = \{x \in \mathbb{Q} : 0 < x < 2\}$.

Next, notice that if $x \in \mathbb{Q}$ and $x > 1$ then if we select $j > 1/(x-1)$ then $x \notin S_j$. It follows that $x \notin \cap_{j=1}^{\infty} S_j$. On the other hand, $\{x \in \mathbb{Q} : 0 < x \le 1\} \subset S_j$ for every j. It follows that $\cap_{j=1}^{\infty} S_j = \{x \in \mathbb{Q} : 0 < x \le 1\}$.

EXAMPLE A2.13   It is entirely possible for *nested*, non-empty sets to have empty intersection. Let $S_j = \{x \in \mathbb{Q} : 0 < x < 1/j\}$. Certainly each S_j is non-empty, for it contains the point $1/(2j)$. Next, $S_1 \supset S_2 \supset \cdots$. Finally, for any positive integer k,

$$\bigcap_{j=1}^{k} S_j = S_k \ne \emptyset.$$

However

$$\bigcap_{j=1}^{\infty} S_j = \emptyset.$$

To verify this last assertion, notice that if $x > 0$ and $j > 1/x$ then $x \notin S_j$ hence $x \notin \cap_{j=1}^{\infty} S_j$. However if $x \le 0$ then x is not an element of any S_j. As a result, no x lies in the intersection. The intersection is empty.

In the examples given thus far, the "index set" has been the natural numbers. That is, we let the index j range over $\{1, 2, \ldots\}$. It is frequently useful to use a larger index set, such as the real numbers or some unspecified index set. Usually we specify an index set with the letter A and we denote a specific index by $\alpha \in A$.

EXAMPLE A2.14 For each real number α we let $S_\alpha = \{x \in \mathbb{R} : \alpha \leq x < \alpha + 1\}$. Thus each S_α is an "interval" of real numbers, and we may speak of

$$\bigcap_{\alpha \in A} S_\alpha \equiv \{x : \forall \alpha, x \in S_\alpha\}$$

and

$$\bigcup_{\alpha \in A} S_\alpha \equiv \{x : \exists \alpha, x \in S_\alpha\}.$$

For the sets S_α that we have specified,

$$\bigcap_{\alpha \in A} S_\alpha = \emptyset.$$

This is so because if $x \in \mathbb{R}$ then $x \notin S_{x+1}$ hence certainly $x \notin \cap_\alpha S_\alpha$.

On the other hand,

$$\bigcup_{\alpha \in A} S_\alpha = \mathbb{R}$$

since every real x lies in $S_{x-1/2}$.

Proposition A2.5.1 *Fix a universal set X. Let A be an index set and, for each $\alpha \in A$, let S_α be a subset of X. Then*

(a) $\displaystyle {}^c\!\left(\bigcap_{\alpha \in A} S_\alpha\right) = \bigcup_{\alpha \in A} {}^c S_\alpha;$

(b) $\displaystyle {}^c\!\left(\bigcup_{\alpha \in A} S_\alpha\right) = \bigcap_{\alpha \in A} {}^c S_\alpha.$

Proof: The proof is similar to that of Proposition A2.2.3 of this chapter. We leave the details to the interested reader. $\square$

Further properties of intersection and union over arbitrary index sets are explored in the exercises. These are some of the most important exercises in the chapter.

A2.6   Countable and Uncountable Sets

One of the most profound ideas of modern mathematics is Georg Cantor's theory of the infinite (Georg Cantor, 1845–1918). Cantor's insight was that infinite sets can be compared by size, just as finite sets can. For instance, we think of the number 2 as *less* than the number 3; so a set with two elements is "smaller" than a set with three elements. We would like to have a similar notion of comparison for infinite sets. In this section we will present Cantor's ideas; we will also give precise definitions of the terms "finite" and "infinite."

Definition A2.6.1   Let A and B be sets. We say that A and B have the *same cardinality* if there is a function f from A to B which is both one-to-one and onto (that is, f is a bijection from A to B). We write $\mathrm{card}(A) = \mathrm{card}(B)$.

EXAMPLE A2.6.2   Let $A = \{1, 2, 3, 4, 5\}$, $B = \{\alpha, \beta, \gamma, \delta, \varepsilon\}$, and $C = \{a, b, c, d, e, f\}$. Then A and B have the same cardinality because the function

$$f = \{(1, \alpha), (2, \beta), (3, \gamma), (4, \delta), (5, \varepsilon)\}$$

is a bijection of A to B. This function is not the *only* bijection of A to B (can you find another?), but we are only required to produce one.

On the other hand, A and C do not have the same cardinality; neither do B and C.

Notice that if $\mathrm{card}(A) = \mathrm{card}(B)$ via a function f_1 and $\mathrm{card}(B) = \mathrm{card}(C)$ via a function f_2 then $\mathrm{card}(A) = \mathrm{card}(C)$ via the function $f_2 \circ f_1$.

Definition A2.6.3   Let A and B be sets. If there is a one-to-one function from A to B but no bijection between A and B, then we will write

$$\mathrm{card}(A) < \mathrm{card}(B).$$

This notation is read "A has smaller cardinality than B."

We use the notation

$$\mathrm{card}(A) \leq \mathrm{card}(B)$$

to mean that either $\mathrm{card}(A) < \mathrm{card}(B)$ or $\mathrm{card}(A) = \mathrm{card}(B)$.

Notice that $\text{card}(A) \leq \text{card}(B)$ and $\text{card}(B) \leq \text{card}(C)$ imply that $\text{card}(A) \leq \text{card}(C)$. Moreover, if $A \subset B$ then the inclusion map $i(a) = a$ is a one-to-one function of A into B; therefore $\text{card}(A) \leq \text{card}(B)$.

The next theorem gives a useful method for comparing the cardinality of two sets.

Theorem A2.6.1 (Schroeder-Bernstein) *Let A, B, be sets. If there is a one-to-one function $f : A \to B$ and a one-to-one function $g : B \to A$, then A and B have the same cardinality.*

Remark A2.6.2 This remarkable theorem says that if we can find an injection from A to B and an injection from B to A, then in fact there exists a single map that is a *bijection* of A to B. Observe that the two different injections may be completely unrelated. Often it is much easier to construct two separate injections than it is to construct a single bijection.

Proof of the Theorem: It is convenient to assume that A and B are disjoint; we may arrange this if necessary by replacing A by $\{(a, 0) : a \in A\}$ and B by $\{(b, 1) : b \in B\}$. Let D be the image of f and let C be the image of g. Let us define a *chain* to be a sequence of elements of either A or B—that is, a function $\phi : \mathbb{N} \to (A \cup B)$—such that

- $\phi(1) \in B \setminus D$;

- If for some j we have $\phi(j) \in B$, then $\phi(j + 1) = g(\phi(j))$;

- If for some j we have $\phi(j) \in A$, then $\phi(j + 1) = f(\phi(j))$.

We see that a chain is a sequence of elements of $A \cup B$ such that the first element is in $B \setminus D$, the second in A, the third in B, and so on. Obviously each element of $B \setminus D$ occurs as the first element of at least one chain.

Define $\mathcal{S} = \{a \in A : a \text{ is some term of some chain}\}$. It is helpful to note that

$$\mathcal{S} = \{x \in A : x \text{ can be written in the form}$$
$$x = g(f(g(\cdots f(g(y))\ldots))) \text{ for some } y \in B \setminus D\}. \qquad (*)$$

Observe that $\mathcal{S} \subset C$.

We set

$$k(x) = \begin{cases} f(x) & \text{if } x \in A \setminus \mathcal{S} \\ g^{-1}(x) & \text{if } x \in \mathcal{S} \end{cases}$$

Note that the second half of this definition makes sense because $\mathcal{S} \subset C$ and because g is one-to-one. Then $k : A \to B$. We shall show that in fact k is a bijection.

First notice that f and g^{-1} are one-to-one. This is not quite enough to show that k is one-to-one, but we now reason as follows: If $f(x_1) = g^{-1}(x_2)$ for some $x_1 \in A \setminus \mathcal{S}$ and some $x_2 \in \mathcal{S}$ then $x_2 = g(f(x_1))$. But, by $(*)$, the fact that $x_2 \in \mathcal{S}$ now implies that $x_1 \in \mathcal{S}$. That is a contradiction. Hence k is one-to-one.

It remains to show that k is onto. Fix $b \in B$. We seek an $x \in A$ such that $k(x) = b$.

Case A: If $g(b) \in \mathcal{S}$ then $k(g(b)) \equiv g^{-1}(g(b)) = b$ hence the x that we seek is $g(b)$.

Case B: If $g(b) \notin \mathcal{S}$ then we claim that there is an $x \in A$ such that $f(x) = b$. Assume this claim for the moment.

Now the x that we just found must lie in $A \setminus \mathcal{S}$. For if not then x would be in some chain. Then $f(x)$ and $g(f(x)) = g(b)$ would also lie in that chain. Hence $g(b) \in \mathcal{S}$, and that is a contradiction. But $x \in A \setminus \mathcal{S}$ tells us that $k(x) = f(x) = b$. That completes the proof that k is onto. Hence k is a bijection.

To prove the claim that we made in Case B, notice that if there is no $x \in A$ with $f(x) = b$ then $b \in B \setminus D$. Thus some chain would begin at b. So $g(b)$ would be a term of that chain. Hence $g(b) \in \mathcal{S}$ and that is a contradiction.

The proof of the Schroeder-Bernstein theorem is complete. □

Now it is time to look at some specific examples.

EXAMPLE A2.6.4　Let $\mathcal{E}$ be the set of all even integers and $\mathcal{O}$ the set of all odd integers. Then

$$\text{card}(\mathcal{E}) = \text{card}(\mathcal{O}).$$

Indeed, the function

$$f(j) = j + 1$$

is a bijection from $\mathcal{E}$ to $\mathcal{O}$.

EXAMPLE A2.6.5   Let $\mathcal{E}$ be the set of even integers. Then

$$\text{card}(\mathcal{E}) = \text{card}(\mathbb{Z}).$$

The function

$$g(j) = j/2$$

is a bijection from $\mathcal{E}$ to $\mathbb{Z}$.

This last example is a bit surprising, for it shows that a set (namely, $\mathbb{Z}$, the integers) can be put in one-to-one correspondence with a proper subset (namely $\mathcal{E}$, the even integers) of itself.

EXAMPLE A2.6.6 We have

$$\text{card}(\mathbb{Z}) = \text{card}(\mathbb{N}).$$

We define the function f from $\mathbb{Z}$ to $\mathbb{N}$ as follows:

- $f(j) = -(2j + 1)$ if j is negative

- $f(j) = 2j + 2$ if j is positive or zero

The values that f takes on the negative integers are $1, 3, 5, \ldots$, on the positive integers are $4, 6, 8, \ldots$, and $f(0) = 2$. Thus f is one-to-one and onto.

Definition A2.6.7   If a set A has the same cardinality as $\mathbb{N}$, then we say that A is *countable*.

By putting together the preceding examples, we see that the set of even integers, the set of odd integers, and the set of all integers are countable sets.

EXAMPLE A2.6.8   The set of all ordered pairs of positive integers

$$S = \mathbb{N} \times \mathbb{N} = \{(j, k) : j, k \in \mathbb{N}\}$$

is countable.

To see this we will use the Schroeder-Bernstein theorem. The function

$$f(j) = (j, 1)$$

is a one-to-one function from $\mathbb{N}$ to S. Also, the function

$$g(j, k) = j \cdot 10^{j+k} + k$$

is a function from S to $\mathbb{N}$. Let n be the number of digits in the number k. Notice that $g(j, k)$ is obtained by writing the digits of j, followed by $j + k - n$ zeros, then followed by the digits of k. For instance,

$$g(23, 714) = 23\underbrace{000\ldots000}_{734}714,$$

where there are $23 + 714 - 3 = 734$ zeros between the 3 and the 7. It is clear that g is one-to-one. By the Schroeder-Bernstein theorem, S and $\mathbb{N}$ have the same cardinality; hence S is countable.

Since there is a bijection f of the set of *all* integers $\mathbb{Z}$ with the set $\mathbb{N}$, it follows from the last example that the set $\mathbb{Z} \times \mathbb{Z}$ of all pairs of integers (positive *and* negative) is countable. Indeed the map $(f \times f)(x, y) = (f(x), f(y))$ is a bijection of $\mathbb{Z} \times \mathbb{Z}$ to $\mathbb{N} \times \mathbb{N}$. Let h be the bijection, provided by Example 4.41, from $\mathbb{N} \times \mathbb{N}$ to $\mathbb{N}$. Then $h \circ (f \times f)$ is a bijection of $\mathbb{Z} \times \mathbb{Z}$ to $\mathbb{N}$.

Notice that the word "countable" is a good descriptive word: if S is a countable set, then we can think of S as having a first element (the one corresponding to $1 \in \mathbb{N}$), a second element (the one corresponding to $2 \in \mathbb{N}$), and so forth. Thus we write $S = \{s_1, s_2, \ldots\}$.

Definition A2.6.9 A set S is called *finite* if it is either empty or else there is a bijection of S with a set of the form $I_n \equiv \{1, 2, \ldots, n\}$ for some positive integer n. If S is not empty and if no such bijection exists, then the set is called *infinite*.

Remark A2.6.3 The empty set is a finite set, but not one of any particular interest. Nevertheless we must account for it, so we include it explicitly in the definition of "finite set."

In some treatments a different approach is taken to the concepts of "finite" and "infinite" sets. In fact one defines an infinite set to be one which can be put in one-to-one correspondence with a proper subset of itself. For instance, Example 4.39 shows that the set $\mathbb{Z}$ of all integers can be put in one-to-one correspondence with the set $\mathcal{E}$ of all even integers (and of course $\mathcal{E}$ is a proper subset of $\mathbb{Z}$). By contrast, a finite set *cannot* be put in one-to-one correspondence with a proper subset of itself. This last assertion amounts to verifying that I_k cannot be put in one-to-one correspondence with I_n when $k < n$. This claim can be verified by contradiction: if there were a one-to-one correspondence $f : I_k \to I_n$ then, by rearranging, it could be assumed that $f(i) < f(j)$ when $i < j$. But then it could be assumed that the sequence $\{f(1), f(2), \ldots, f(k)\}$ does not skip any integers—it is just a natural sequence of positive integers beginning with 1. But then the image of f is just the set $\{1, 2, \ldots, k\}$. And that is a *proper* subset of $\{1, 2, \ldots, n\}$. Thus f cannot be onto.

An important property of the natural numbers $\mathbb{N}$ is that any subset $S \subset \mathbb{N}$ has a least element. See the discussion in Section 4.2. This is known as the Well Ordering Principle, and is studied in a course on logic (see also Section 6.1). In the present chapter we take the properties of the natural numbers as given (see Section 6.1 for more on the natural numbers). We use some of these properties in the next proposition.

Proposition A2.6.4 *If S is a countable set and R is a subset of S, then either R is finite or R is countable.*

Proof: Assume that R is not empty.

Write $S = \{s_1, s_2, \ldots\}$. Let j_1 be the least positive integer such that $s_{j_1} \in R$. Let j_2 be the least integer following j_1 such that $s_{j_2} \in R$. Continue in this fashion. If the process terminates at the nth step, then R is finite and has n elements.

If the process does not terminate, then we obtain an enumeration of the elements of R :

$$1 \longleftrightarrow s_{j_1}$$
$$2 \longleftrightarrow s_{j_2}$$
$$\ldots$$

etc.

All elements of R are enumerated in this fashion since $j_\ell \geq \ell$. Therefore R is countable. □

A set is called *countable* if it is countably infinite, that is if it can be put in one-to-one correspondence with the natural numbers $\mathbb{N}$. A set is called *denumerable* if it is either finite or countable. In actual practice, mathematicians use the word "countable" to describe sets which are either finite or countable. In other words, they use the word "countable" interchangeably with the word "denumerable." We shall also indulge in this slight imprecision in this book when no confusion can arise as a result.

The set $\mathbb{Q}$ of all rational numbers consists of all expressions

$$\frac{a}{b},$$

where a and b are integers and $b \neq 0$. Thus $\mathbb{Q}$ can be identified with the set of all pairs (a, b) of integers. After discarding duplicates, such as $\frac{2}{4} = \frac{1}{2}$, and using Proposition 4.5 and the fact that $\mathbb{Z} \times \mathbb{Z}$ is countable, we find that the set $\mathbb{Q}$ is countable. We shall deal with the rational number system in a much more precise manner in Chapter 6.

THEOREM A2.6.5   *Let S_1, S_2 be countable sets. Set $\mathcal{S} = S_1 \cup S_2$. Then $\mathcal{S}$ is countable.*

Proof: Let us write

$$S_1 = \{s_1^1, s_2^1, \ldots\}$$
$$S_2 = \{s_1^2, s_2^2, \ldots\}.$$

If $S_1 \cap S_2 = \emptyset$ then the function

$$s_j^k \mapsto (j, k)$$

is a bijection of $\mathcal{S}$ with a subset of $\{(j, k) : j, k \in \mathbb{N}\}$. We proved earlier (Example 4.41) that the set of ordered pairs of elements of $\mathbb{N}$ is countable. By Proposition 4.5, $\mathcal{S}$ is countable as well.

If there exist elements which are common to S_1, S_2, then discard any duplicates. The same argument (use Proposition 4.5) shows that $\mathcal{S}$ is countable. □

Proposition A2.6.6 *If S and T are each countable sets, then so is*

$$S \times T = \{(s,t) : s \in S, t \in T\}.$$

Proof: Since S is countable there is a bijection f from S to $\mathbb{N}$. Likewise there is a bijection g from T to $\mathbb{N}$. Therefore the function

$$(f \times g)(s,t) = (f(s), g(t))$$

is a bijection of $S \times T$ with $\mathbb{N} \times \mathbb{N}$, the set of ordered pairs of positive integers. But we saw in Example 4.41 that the latter is a countable set. Hence so is $S \times T$. □

Corollary A2.6.7 *If $S_1, S_2, \ldots, S_k$ are each countable sets, then so is the set*

$$S_1 \times S_2 \times \cdots \times S_k = \{(s_1, \ldots, s_k) : s_1 \in S_1, \ldots, s_k \in S_k\}$$

consisting of all ordered k-tuples $(s_1, s_2, \ldots, s_k)$ with $s_j \in S_j$.

Proof: We may think of $S_1 \times S_2 \times S_3$ as $(S_1 \times S_2) \times S_3$. Since $S_1 \times S_2$ is countable (by the Proposition) and S_3 is countable, then so is $(S_1 \times S_2) \times S_3 = S_1 \times S_2 \times S_3$ countable. Continuing in this fashion (i.e., inductively), we can see that any finite product of countable sets is also a countable set. □

Corollary A2.6.8 *The countable union of countable sets is countable.*

Proof: Let $A_1, A_2, \ldots$ each be countable sets. If the elements of A_j are enumerated as $\{a_k^j\}_{k=1}^\infty$ and if the sets A_j are pairwise disjoint, then the correspondence

$$a_k^j \longleftrightarrow (j, k)$$

is one-to-one between the union of the sets A_j and the countable set $\mathbb{N} \times \mathbb{N}$. This proves the result when the sets A_j have no common element. If some of the A_j have elements in common, then we discard duplicates in the union and use Proposition 4.5. □

Proposition A2.6.9 *The collection $\mathcal{P}$ of all polynomials $p(x)$ with integer coefficients is countable.*

Proof: Let $\mathcal{P}_k$ be the set of polynomials of degree k with integer coefficients. A polynomial p of degree k has the form

$$p(x) = p_0 + p_1 x + p_2 x^2 + \cdots + p_k x^k.$$

The identification

$$p(x) \longleftrightarrow (p_0, p_1, \ldots, p_k)$$

identifies the elements of $\mathcal{P}_k$ with the $(k+1)$-tuples of integers. By Corollary 4.8, it follows that $\mathcal{P}_k$ is countable. But then Corollary 4.9 implies that

$$\mathcal{P} = \bigcup_{j=0}^{\infty} \mathcal{P}_j$$

is countable. □

Definition A2.6.10 Let x be a real number. We say that x is *algebraic* if there is a polynomial p with integer coefficients such that $p(x) = 0$.

EXAMPLE A2.6.11   The number $\sqrt{2}$ is algebraic because it satisfies the polynomial equation $x^2 - 2 = 0$. The number $\sqrt{3} + \sqrt{2}$ is also algebraic. This assertion is less obvious, but in fact the number satisfies the polynomial equation $x^4 - x^2 + 1 = 0$. The numbers π and e are *not* algebraic, but this assertion is extremely difficult to prove. We say that π and e are transcendental. In the next proposition we give an elegant method for showing that most real numbers are transcendental without actually saying what any of them are.

Georg Cantor's remarkable discovery is that *not all infinite sets are countable*. We next give an example of this phenomenon.

In what follows, a *sequence* on a set S is a function from $\mathbb{N}$ to S. We usually write such a sequence as $s(1), s(2), s(3), \ldots$ or as $s_1, s_2, s_3, \ldots$.

EXAMPLE A2.6.12   There exists an infinite set which is not countable (we call such a set *uncountable*). Our example will be the set S of all sequences on the set $\{0, 1\}$. In other words, S is the set of all infinite sequences of 0's and 1's.

To see that S is uncountable, assume the contrary—that is, we assume that S is countable. Then there is a first sequence

$$\mathcal{S}^1 = \{s_j^1\}_{j=1}^\infty,$$

a second sequence

$$\mathcal{S}^2 = \{s_j^2\}_{j=1}^\infty,$$

and so forth. This will be a complete enumeration of all the members of S. But now consider the sequence $\mathcal{T} = \{t_j\}_{j=1}^\infty$, which we construct as follows:

- If $s_1^1 = 0$, then set $t_1 = 1$; if $s_1^1 = 1$, then set $t_1 = 0$;

- If $s_2^2 = 0$, then set $t_2 = 1$; if $s_2^2 = 1$, then set $t_2 = 0$;

- If $s_3^3 = 0$, then set $t_3 = 1$; if $s_3^3 = 1$, then set $t_3 = 0$;

$$\cdots$$

- If $s_j^j = 0$, then set $t_j = 1$; if $s_j^j = 1$, then make $t_j = 0$;

$$\text{etc.}$$

Now the sequence $\mathcal{T}$ differs from the first sequence $\mathcal{S}^1$ in the first element: $t_1 \neq s_1^1$.

The sequence $\mathcal{T}$ differs from the second sequence $\mathcal{S}^2$ in the second element: $t_2 \neq s_2^2$.

And so on: the sequence $\mathcal{T}$ differs from the jth sequence $\mathcal{S}^j$ in the jth element: $t_j \neq s_j^j$. So the sequence $\mathcal{T}$ is not in the set S. But $\mathcal{T}$ is *supposed* to be in the set S because it is a sequence of 0's and 1's and all of these are supposed to have been enumerated.

This contradicts our assumption, so S must be uncountable.

Example A2.6.13   Consider the set of all decimal representations of numbers strictly between 0 and 1—both terminating and non-terminating. Here a terminating decimal is one of the form

$$0.43926$$

while a non-terminating decimal is one of the form

$$0.14159265\ldots.$$

In the case of the non-terminating decimal, no repetition is implied; the decimal simply continues without cease.

Now the set of all those decimals containing only the digits 0 and 1 can be identified in a natural way with the set of sequences containing only 0 and 1 (just put commas between the digits). And we just saw that the set of such sequences is uncountable.

Since the set of all decimal numbers is an even bigger set, it must be uncountable also. [Put a different way, if the set of all decimal numbers *were* countable, then any of its infinite subsets would be countable—that is the content of Proposition 4.5. Thus the collection of decimal numbers containing only the digits 0 and 1 would be countable, and that is a contradiction.]

As you may know, the set of all decimals identifies with the set of all real numbers. [Many real numbers have two decimal representations—one terminating and one not. Think for a moment about which numbers these are, and why this observation does not invalidate the present discussion.] We find then that the set $\mathbb{R}$ of all real numbers is uncountable. (Contrast this with the situation for the rationals.) In Chapter 6 we will learn more about how the real number system is constructed using just elementary set theory.

Definition A2.6.14   Let x be a real number. We say that x is *algebraic* if there is a polynomial p with integer coefficients such that $p(x) = 0$.

EXAMPLE A2.6.15   The number $\sqrt{2}$ is algebraic because it satisfies the polynomial equation $x^2 - 2 = 0$. The number $\sqrt{3} + \sqrt{2}$ is also algebraic. This assertion is less obvious, but in fact the number satisfies the polynomial equation $x^4 - x^2 + 1 = 0$. The numbers π and e are *not* algebraic, but this assertion is extremely difficult to prove. We say that π and e are transcendental. In the next proposition we give an elegant method for showing that most real numbers are transcendental without actually saying what any of them are.

Proposition A2.6.10   *The set of all algebraic real numbers is countable. The set of all transcendental numbers is uncountable.*

Proof: Let $\mathcal{P}$ be the collection of all polynomials with integer coefficients. We have already noted in Proposition A2.6.9 that $\mathcal{P}$ is a countable set. If $p \in \mathcal{P}$ then let s_p denote the set of real roots of p. Of course s_p is finite, and the number of elements in s_p does not exceed the degree of p. Then the set A of algebraic real numbers may be written as

$$A = \cup_{p \in \mathcal{P}} s_p \,.$$

This is the countable union of finite sets so of course it is countable.

Now that we know that the set of algebraic numbers is countable, we can notice that the set T of transcendental numbers must be uncountable. For $\mathbb{R} = A \cup T$. If T were countable then, since A is countable, it would follow that $\mathbb{R}$ is countable. But that is not so. □

Our last result in this section is a counterpoint to Definition 4.42 and the discussion leading up to it.

Proposition A2.6.11 *Let S be any infinite set. Then S has a subset T that is countable.*

Proof: Let $t_1 \in S$ be any element. Now let $t_2 \in S$ be any element that is distinct from t_1. Continue this procedure. It will not terminate, because that would imply that S is finite. And it will produce a countable set T that is a subset of S. □

To repeat the main point of this section, the natural numbers have a cardinality that we call *countable* and the real numbers have a cardinality that we call *uncountable*. These cardinalities are distinct. In fact the real numbers form a larger set because there is an injective mapping of the natural numbers into the reals but not the other way around. We refer to the cardinality of the natural numbers as "countable" and to that of the real numbers as "the cardinality of the continuum."

It is natural to ask whether there is a set with cardinality strictly between countable and the continuum. G. Cantor posed this question one hundred years ago, and his failed attempts to resolve the question tormented his final years. We discussed the resolution of this "continuum hypothesis" in Chapter 5.

It is an important result of set theory (due to Cantor) that, given any set S, the set of all subsets of S (called the *power set* of S) has strictly greater cardinality than the set S itself. As a simple example, let $S = \{a, b, c\}$. Then the set of all subsets of S is

$$\{\emptyset, \{a\}, \{b\}, \{c\}, \{a, b\}, \{a, c\}, \{b, c\}, \{a, b, c\}\}.$$

The set of all subsets has eight elements while the original set has three.

We stress that this result is true not just for finite sets, but also for infinite sets: if S is an infinite set, then the set of all its subsets (the power set) has greater cardinality than S itself. Thus there are infinite sets of arbitrarily large cardinality. In other words, there is no "greatest" cardinal. This fact is so important that we now formulate it as a theorem.

Theorem A2.6.12 (Cantor) *Let S be any set. Then the power set $\mathcal{P}(S)$, consisting of all subsets of S, has cardinality greater than the cardinality of S. In other words,*

$$\mathrm{card}(S) < \mathrm{card}(\mathcal{P}(S)).$$

Proof: First observe that the function

$$
\begin{aligned}
f : S &\longrightarrow \mathcal{P}(X) \\
s &\longmapsto \{s\}
\end{aligned}
$$

is one-to-one. Thus we see that $\mathrm{card}(S) \leq \mathrm{card}(\mathcal{P}(S))$. We need to show that there is no function from S *onto* $\mathcal{P}(S)$. Let $g : S \to \mathcal{P}(S)$. We will produce an element of $\mathcal{P}(S)$ that cannot be in the image of this mapping.

Define $T = \{s \in S : s \notin g(s)\}$. Assume, seeking a contradiction, that $T = g(z)$ for some $z \in S$. By definition of T, the element $z \in T$ if and only if $z \notin g(z)$; thus $z \in T$ if and only if $z \notin T$. That is a contradiction. We see that g cannot map S onto $\mathcal{P}(S)$, therefore $\mathrm{card}(S) < \mathrm{card}(\mathcal{P}(S))$. $\square$

In some of the examples in this section we constructed a bijection between a given set (such as $\mathbb{Z}$) and a proper subset of that set (such as E, the even integers). It follows from the definitions that this is possible only when the sets involved are infinite. In fact any infinite set can be placed in a set-theoretic isomorphism with a proper subset of itself. We explore this assertion in the exercises.

Put in other words, we have come upon an intrinsic characterization of infinite sets. We state it (without proof) as a proposition:

Proposition A2.6.13 *Let S be a set. The set S is infinite if and only if it can be put in one-to-one correspondence with a proper subset of itself.*

Appendix 3: The Real Numbers

A3.1   The Real Number System

Now that we are accustomed to the notion of equivalence classes, the construction of the integers and of the rational numbers seems fairly natural. In fact, equivalence classes provide a precise language for declaring certain objects to be equal (or for identifying certain objects). We can now use the integers and the rationals as we always have done, with the added confidence that they are not simply a useful notation but that they have been *constructed*.

We turn next to the real numbers. We know from calculus that for many other purposes the rational numbers are inadequate. It is important to work in a number system which is closed with respect to all the operations we shall perform. While the rationals are closed under the usual arithmetic operations, they are not closed under the operation of taking *limits*. For instance, the sequence of rational numbers $3, 3.1, 3.14, 3.141, \ldots$ consists of terms that seem to be getting closer and closer together, *seem* to tend to some limit, and yet there is no rational number which will serve as a limit (of course it turns out that the limit is π—an "irrational" number).

We will now deal with the real number system, a system which contains all limits of sequences of rational numbers (as well as all limits of sequences of real numbers!). In fact our plan will be as follows: in this section we shall discuss all the requisite properties of the reals. The actual construction of the reals is rather complicated, and we shall put in the next section.

Definition A3.1.1   Let A be an ordered set and X a subset of A. The set X is called *bounded above* if there is an element $b \in A$ such that $x \le b$ for all

$x \in X$. We call the element b an *upper bound* for the set X.

EXAMPLE A3.1   Let $A = \mathbb{Q}$ with the usual ordering. The set $X = \{x \in \mathbb{Q} : 2 < x < 4\}$ is bounded above. For example, the number 15 is an upper bound for X. So are the numbers 12 and 4. It is interesting to observe that no element of this particular X can be an upper bound for X. The number 4 is a good candidate, but 4 is not an element of X. In fact if $b \in X$, then $(b+4)/2 \in X$ and $b < (b+4)/2$, so b could not be an upper bound for X.

It turns out that the most convenient way to formulate the notion that the real numbers have "no gaps" (i.e., that all sequences which seem to be converging actually have something to converge to) is in terms of upper bounds.

Definition A3.1.2   Let A be an ordered set and X a subset of A. An element $b \in A$ is called a *least upper bound* (or *supremum*) for X if b is an upper bound for X and there is no upper bound b^* for X which is less than b.

By its very definition, if a least upper bound exists, then it is unique.

EXAMPLE A3.2   In the last example, we considered the set X of rational numbers strictly between 2 and 4. We observed there that 4 is the least upper bound for X. Note that this least upper bound is not an element of the set X.

The set $Y = \{y \in \mathbb{Z} : -9 \leq y \leq 7\}$ has least upper bound 7. In this case, the least upper bound *is* an element of the set Y.

Notice that we may define a lower bound for a subset of an ordered set in a fashion similar to that for an upper bound: $l \in A$ is a lower bound for $X \subset A$ if $x \geq l$ for all $x \in X$. A *greatest lower bound* (or *infimum*) for X is then defined to be a lower bound ℓ such that there is no lower bound ℓ^* with $\ell^* > \ell$.

EXAMPLE A3.3   The set X in the last two examples has lower bounds -20, $0, 1, 2$, for instance. The greatest lower bound is 2, which is *not* an element of the set.

The set Y in the last example has lower bounds $-53, -22, -10, -9$, to name just a few. The number -9 is the greatest lower bound. It *is* an element of Y.

EXAMPLE A3.4 Let $S = ZZ \subset \mathbb{R}$. Then S does not have an upper bound.

The purpose that the real numbers will serve for us is as follows: they will contain the rationals, they will still be an ordered field, and *every subset which has an upper bound will have a least upper bound.* We formulate this property as a theorem.

THEOREM A3.1.1 *There exists an ordered field $\mathbb{R}$ which (i) contains $\mathbb{Q}$ and (ii) has the property that any non-empty subset of $\mathbb{R}$ which has an upper bound has a least upper bound.*

The last property described in this theorem is called the Least Upper Bound Property of the real numbers. As mentioned previously, this theorem will be proved in the next section. Now we begin to realize why it is so important to *construct* the number systems that we will use. We are endowing $\mathbb{R}$ with a great many properties. Why do we have any right to suppose that there exists a number system with all these properties? We must produce one!

Let us begin to explore the richness of the real numbers. The next theorem states a property which is certainly not shared by the rationals. It is fundamental in its importance.

THEOREM A3.1.2 *Let x be a real number such that $x > 0$. Then there is a positive real number y such that $y^2 = y \cdot y = x$.*

Proof: We will use throughout this proof the fact that if $0 < a < b$ then $a^2 < b^2$. The interested reader may prove this for himself.

Let
$$S = \{s \in \mathbb{R} : s > 0 \ \text{ and } \ s^2 < x\}.$$

Then S is not empty since $x/2 \in S$ if $x < 2$ and $1 \in S$ otherwise. Also S is bounded above since $x + 1$ is an upper bound for S. By Theorem A3.1.1, the set S has a least upper bound. Call it y. Obviously $0 < \min\{x/2, 1\} \leq y$

hence y is positive. We claim that $y^2 = x$. To see this, we eliminate the other two possibilities.

If $y^2 < x$ then set $\varepsilon = (x - y^2)/[4(x + 1)]$. Then $\varepsilon > 0$ and

$$
\begin{aligned}
(y + \varepsilon)^2 &= y^2 + 2 \cdot y \cdot \varepsilon + \varepsilon^2 \\
&= y^2 + 2 \cdot y \cdot \frac{x - y^2}{4(x + 1)} + \frac{x - y^2}{4(x + 1)} \cdot \frac{x - y^2}{4(x + 1)} \\
&< y^2 + 2 \cdot \frac{y}{x + 1} \cdot \frac{x - y^2}{4} + \frac{x - y^2}{4} \cdot \frac{x}{4x} \\
&< y^2 + \frac{x - y^2}{2} + \frac{x - y^2}{16} \\
&< y^2 + (x - y^2) \\
&= x.
\end{aligned}
$$

Thus $y + \varepsilon \in S$, and y cannot be an upper bound for S. This contradiction tells us that $y^2 \not< x$.

Similarly, if it were the case that $y^2 > x$ then we set $\varepsilon = (y^2 - x)/[4(x + 1)]$. A calculation like the one we just did then shows that $(y - \varepsilon)^2 \geq x$. Hence $y - \varepsilon$ is also an upper bound for S, and y is therefore not the *least* upper bound. This contradiction shows that $y^2 \not> x$.

The only remaining possibility is that $y^2 = x$. $\square$

A similar proof shows that if n is a positive integer and x a positive real number, then there is a positive real number y such that $y^n = x$.

We next use the Least Upper Bound Property of the Real Numbers to establish two important qualitative properties of the Real Numbers:

THEOREM A3.1.3   *The set $\mathbb{R}$ of real numbers satisfies the Archimedean Property:*

> *Let a and b be positive real numbers. Then there is a natural number n such that $na > b$.*

THEOREM A3.1.4   *The set $\mathbb{Q}$ of rational numbers satisfies the following Density Property:*

Let $c < d$ be real numbers. Then there is a rational number q with $c < q < d$.

Proof of Theorem A3.1.3: Suppose the Archimedean Property to be false. Then $S = \{na : n \in \mathbb{N}\}$ has b as an upper bound. Therefore S has a finite supremum β. Since $a > 0, \beta - a < \beta$. So $\beta - a$ is not an upper bound for S, and there must be a natural number n^* such that $n^* \cdot a > \beta - a$. But then $(n^* + 1)a > \beta$, and β cannot be the supremum for S. This contradiction proves the theorem. $\qquad\square$

Proof of Theorem A3.1.4: Let $\lambda = d - c > 0$. By the Archimedean Property, choose a positive integer N such that $N \cdot \lambda > 1$. Again the Archimedean Property gives a natural number P such that $P > N \cdot c$ and another Q such that $Q > |-N \cdot c|$. Then $Q > -N \cdot c$ and we see that Nc falls between the integers $-Q$ and P; therefore there must be an integer M between $-Q$ and P (inclusive) such that

$$M - 1 \le Nc < M.$$

Thus $c < M/N$. Also

$$M \le Nc + 1 \quad \text{hence} \quad \frac{M}{N} \le c + \frac{1}{N} < c + \lambda = d.$$

So M/N is a rational number lying strictly between c and d. $\qquad\square$

One of the most profound and useful properties of the real numbers, and one that is equivalent to the Least Upper Bound Property, is the Intermediate Value Property:

THEOREM A3.1.5 *Let f be a continuous, real-valued function with domain the interval $[a, b]$. If $f(a) = \alpha$, $f(b) = \beta$, and if $\alpha < \gamma < \beta$ then there is a value $t_0 \in (a, b)$ such that $f(t_0) = \gamma$.*

Proof: Let

$$S = \{x \in [a, b] : f(x) < \gamma\}.$$

Then $S \neq \emptyset$ since $a \in S$. Moreover S is bounded above by b. So $t_0 = \sup S$ exists as a finite real number. We claim that $f(t_0) = \gamma$.

Clearly $f(t_0) \leq \gamma$ since t_0 is the limit of numbers at which f takes values less than γ (we use the continuity of f here). Suppose, seeking a contradiction, that $f(t_0) < \gamma$. Let $\epsilon = \gamma - f(t_0)$. By the continuity of f, we may select $\delta > 0$ such that $|t - t_0| < \delta$ implies that $|f(t) - f(t_0)| < \epsilon/2$. But then, for $t \in (t_0 - \delta, t_0 + \delta)$, $f(t) < f(t_0) + \epsilon/2 < \gamma$. It follows that $(t_0 - \delta, t_0 + \delta) \subset S$, so t_0 cannot be the supremum of S. That is a contradiction. Therefore $f(t_0) = \gamma$. $\qquad\qquad\square$

As an application, we prove the following special case of a theorem of Brouwer:

THEOREM A3.1.6 *Let $f : [0, 1] \to [0, 1]$ be a continuous function. Then f has a fixed point, in the sense that there is a point $c \in [0, 1]$ such that $f(c) = c$.*

Proof: Seeking a contradiction, we suppose not. Then, in particular, $f(0) > 0$ and $f(1) < 1$. Now set $g(x) = x - f(x)$. We see that $g(0) = 0 - f(0) < 0$ and $g(1) = 1 - f(1) > 0$. By the Intermediate Value Property, there must therefore be a point c between 0 and 1 such that $g(c) = 0$. But this says that $f(c) = c$, as required. $\qquad\qquad\square$

The set of all decimal representations of numbers is uncountable. It follows that the set of all real numbers is uncountable. In fact the same proof shows that the set of all real numbers in the interval $(0, 1)$, or in any non-empty open interval (c, d), is uncountable.

The set $\mathbb{R}$ of real numbers is uncountable, yet the set $\mathbb{Q}$ of rational numbers is countable. It follows that the set $\mathbb{R} \setminus \mathbb{Q}$ of *irrational* numbers is uncountable. In particular, it is non-empty. Thus we may see with very little effort that there exist a great many real numbers which cannot be expressed as a quotient of integers. However, given any particular real number (such as π or e or $\sqrt{2}^{\sqrt{2}}$), it can be quite difficult to see whether it is irrational.

We conclude by recalling the "absolute value" notation:

Definition A3.1.3   Let x be a real number. We define

$$|x| = \begin{cases} x & \text{if } x > 0 \\ 0 & \text{if } x = 0 \\ -x & \text{if } x < 0 \end{cases}$$

The absolute value of a real number x measures the distance of x to 0. It is left as an exercise for you to verify the important *triangle inequality*:

$$|x + y| \le |x| + |y|.$$

A3.2   Construction of the Real Numbers

There are several techniques for constructing the real number system $\mathbb{R}$ from the rational number system $\mathbb{Q}$. We use the method of Dedekind (Julius W. R. Dedekind, 1831-1916) cuts because it uses a minimum of new ideas and is fairly brief.

Keep in mind that, throughout this subsection, our universe is the system of rational numbers $\mathbb{Q}$. We are *constructing* the new number system $\mathbb{R}$.

Definition A3.2.1   A *cut* is a subset $\mathcal{C}$ of $\mathbb{Q}$ with the following properties:

1. $\mathcal{C} \ne \emptyset$

2. If $s \in \mathcal{C}$ and $t < s$, then $t \in \mathcal{C}$

3. If $s \in \mathcal{C}$, then there is a $u \in \mathcal{C}$ such that $u > s$

4. There is a rational number x such that $c < x$ for all $c \in \mathcal{C}$

You should think of a cut $\mathcal{C}$ as the set of all rational numbers to the left of some point in the real line (that is, it is an open half-line of rational numbers). For example, the set $\{x \in \mathbb{Q} : x^2 < 2\} \cup \{x \in \mathbb{Q} : x < 0\}$ is a cut. Roughly speaking, it is the set of rational numbers to the left of $\sqrt{2}$. [Take care to note that $\sqrt{2}$ does not exist as a rational number; so we are using a circuitous method to specify this set.] Since we have not constructed the real line yet, we cannot define this cut in that simple way; we have to make the construction more indirect. But if you consider the four properties of a cut, they describe a set that looks like a "rational left half-line."

Notice that if $\mathcal{C}$ is a cut and $s \notin \mathcal{C}$, then any rational $t > s$ is also not in $\mathcal{C}$. Also, if $r \in \mathcal{C}$ and $s \notin \mathcal{C}$ then it must be that $r < s$.

Definition A3.2.2 If $\mathcal{C}$ and $\mathcal{D}$ are cuts, then we say that $\mathcal{C} < \mathcal{D}$ provided that $\mathcal{C}$ is a subset of $\mathcal{D}$ but $\mathcal{C} \neq \mathcal{D}$.

Check for yourself that "$<$" is a strict, simple ordering on the set of all cuts. We note that $\mathcal{C} = \mathcal{D}$ if and only if $\mathcal{C} \subset \mathcal{D}$ and $\mathcal{D} \subset \mathcal{C}$.

Now we introduce operations of addition and multiplication which will turn the set of all cuts into a field.

Definition A3.2.3 If $\mathcal{C}$ and $\mathcal{D}$ are cuts, then we define

$$\mathcal{C} + \mathcal{D} = \{c + d : c \in \mathcal{C}, d \in \mathcal{D}\}.$$

We define the cut $\hat{0}$ to be the set of all negative rationals.

The cut $\hat{0}$ will play the role of the additive identity. We are now required to check that field axioms A1 - A5 hold.

For **A1**, we need to see that $\mathcal{C} + \mathcal{D}$ is a cut. Obviously $\mathcal{C} + \mathcal{D}$ is not empty. If s is an element of $\mathcal{C} + \mathcal{D}$ and t is a rational number less than s, write $s = c + d$, where $c \in \mathcal{C}$ and $d \in \mathcal{D}$. Then $t - c < s - c = d \in \mathcal{D}$ so $t - c \in \mathcal{D}$; and $c \in \mathcal{C}$. Hence $t = c + (t - c) \in \mathcal{C} + \mathcal{D}$. A similar argument shows that there is an $r > s$ such that $r \in \mathcal{C} + \mathcal{D}$. Finally, if x is a rational upper bound for $\mathcal{C}$ and y is a rational upper bound for $\mathcal{D}$, then $x + y$ is a rational upper bound for $\mathcal{C} + \mathcal{D}$. We conclude that $\mathcal{C} + \mathcal{D}$ is a cut.

Since addition of rational numbers is commutative, it follows immediately that addition of cuts is commutative. Associativity follows in a similar fashion. That takes care of **A2** and **A3**.

Now we show that if $\mathcal{C}$ is a cut then $\mathcal{C} + \hat{0} = \mathcal{C}$. For if $c \in \mathcal{C}$ and $z \in \hat{0}$ then $c + z < c + 0 = c$ hence $\mathcal{C} + \hat{0} \subset \mathcal{C}$. Also, if $c^* \in \mathcal{C}$ then choose a $d^* \in \mathcal{C}$ such that $c^* < d^*$. Then $c^* - d^* < 0$ so $c^* - d^* \in \hat{0}$. And $c^* = d^* + (c^* - d^*)$. Hence $\mathcal{C} \subset \mathcal{C} + \hat{0}$. We conclude that $\mathcal{C} + \hat{0} = \mathcal{C}$. This is **A4**.

Finally, for Axiom **A5**, we let $\mathcal{C}$ be a cut and set $-\mathcal{C}$ to be equal to $\{d \in \mathbb{Q} : \exists d^* > d$ such that $c + d^* < 0$ for all $c \in \mathcal{C}\}$. If x is a rational upper bound for $\mathcal{C}$, then $-x \in -\mathcal{C}$ so $-\mathcal{C}$ is not empty. It is also routine to check that $-\mathcal{C}$ is a cut. By its very definition, $\mathcal{C} + (-\mathcal{C}) \subset \hat{0}$.

Further, if $z \in \hat{0}$ then there is a $z^* \in \hat{0}$ such that $z < z^*$. Choose an element $c \in \mathcal{C}$ such that $c + (z^* - z) \notin \mathcal{C}$ (why is this possible?). Let $c^* \in \mathcal{C}$ be such that $c < c^*$. Set $c^{**} = z - c^*$. Then $d^* = z - c > c^{**}$. We claim that $\widetilde{c} + d^* < 0$ for all $\widetilde{c} \in \mathcal{C}$. Suppose for the moment that this claim has been proved. Then this shows that $c^{**} \in -\mathcal{C}$. Then $z = c^* + c^{**} \in \mathcal{C} + (-\mathcal{C})$ so that $\hat{0} \subset \mathcal{C} + (-\mathcal{C})$. We then conclude that $\mathcal{C} + (-\mathcal{C}) = \hat{0}$, and Axiom **A5** is established.

It remains to prove the claim. So let d^* be defined as above and select $\widetilde{c} \in \mathcal{C}$. Then

$$d^* + \widetilde{c} = z + (-c + \widetilde{c}) < z + (z^* - z) = z^* < 0.$$

Here we have used the choice of c. This establishes the claim and completes the proof of **A5**.

Having verified the axioms for addition, we turn now to multiplication.

Definition A3.2.4 If $\mathcal{C}$ and $\mathcal{D}$ are cuts, then we define the product $\mathcal{C} \cdot \mathcal{D}$ as follows:

- If $\mathcal{C}, \mathcal{D} > \hat{0}$, then $\mathcal{C} \cdot \mathcal{D} = \{q \in \mathbb{Q} : q < c \cdot d \text{ for some } c \in \mathcal{C}, d \in \mathcal{D} \text{ with } c > 0, d > 0 \}$

- If $\mathcal{C} > \hat{0}, \mathcal{D} < \hat{0}$, then $\mathcal{C} \cdot \mathcal{D} = -(\mathcal{C} \cdot (-\mathcal{D}))$

- If $\mathcal{C} < \hat{0}, \mathcal{D} > \hat{0}$, then $\mathcal{C} \cdot \mathcal{D} = -((-\mathcal{C}) \cdot \mathcal{D})$

- If $\mathcal{C}, \mathcal{D} < \hat{0}$, then $\mathcal{C} \cdot \mathcal{D} = (-\mathcal{C}) \cdot (-\mathcal{D})$

- If either $\mathcal{C} = \hat{0}$ or $\mathcal{D} = \hat{0}$, then $\mathcal{C} \cdot \mathcal{D} = \hat{0}$.

Notice that, for convenience, we have defined multiplication of negative numbers just as we did in high school. The reason is that the definition that we use for the product of two positive numbers cannot work when one of the two factors is negative (exercise).

We have said what the additive identity is in this realization of the real numbers. Of course the multiplicative identity is the cut corresponding to 1, or

$$\hat{1} \equiv \{t \in \mathbb{Q} : t < 1\}.$$

We leave it to the reader to verify that if $\mathcal{C}$ is any cut then $\widehat{1} \cdot \mathcal{C} = \mathcal{C} \cdot \widehat{1} = \mathcal{C}$.

It is now routine to verify that the set of all cuts, with this definition of multiplication, satisfies field axioms **M1 - M5**. The proofs follow those for **A1 - A5** rather closely.

For the distributive property, one first checks the case when all the cuts are positive, reducing it to the distributive property for the rationals. Then one handles negative cuts on a case-by-case basis.

The two properties of an ordered field are also easily checked for the set of all cuts.

We now know that the collection of all cuts forms an ordered field. Denote this field by the symbol $\mathbb{R}$ and call it the real number system. We next verify the crucial property of $\mathbb{R}$ that sets it apart from $\mathbb{Q}$:

THEOREM A3.2.1 *The ordered field $\mathbb{R}$ satisfies the least upper bound property.*

Proof: Let S be a subset of $\mathbb{R}$ which is bounded above. That is, there is a cut α such that $s < \alpha$ for all $s \in S$. Define

$$\mathcal{S}^* = \bigcup_{\mathcal{C} \in S} \mathcal{C}.$$

Then $\mathcal{S}^*$ is clearly non-empty, and it is therefore a cut since it is a union of cuts. It is also clearly an upper bound for S since it contains each element of S. It remains to check that $\mathcal{S}^*$ is the least upper bound for S.

In fact if $\mathcal{T} < \mathcal{S}^*$ then $\mathcal{T} \subset \mathcal{S}^*$ and there is a rational number q in $\mathcal{S}^* \setminus \mathcal{T}$. But, by the definition of $\mathcal{S}^*$, it must be that $q \in \mathcal{C}$ for some $\mathcal{C} \in S$. So $\mathcal{C} > \mathcal{T}$, and $\mathcal{T}$ cannot be an upper bound for S. Therefore $\mathcal{S}^*$ is the least upper bound for S, as desired. $\square$

We have shown that $\mathbb{R}$ is an ordered field which satisfies the least upper bound property. It remains to show that $\mathbb{R}$ contains (a copy of) $\mathbb{Q}$ in a natural way. In fact, if $q \in \mathbb{Q}$ we associate to it the element $\varphi(q) = \mathcal{C}_q \equiv \{x \in \mathbb{Q} : x < q\}$. Then $\mathcal{C}_q$ is obviously a cut. It is also routine to check that

$$\varphi(q + q^*) = \varphi(q) + \varphi(q^*) \quad \text{and} \quad \varphi(q \cdot q^*) = \varphi(q) \cdot \varphi(q^*).$$

Therefore we see that ϕ is a ring homomorphism (see [LAN]) and hence represents $\mathbb{Q}$ as a "subfield" of $\mathbb{R}$.

Appendix 4: The Axiom of Choice and Its Implications

The Axiom of Choice, first enunciated by Zermelo, is one of the most subtle of the axioms of set theory, and it has profound and mysterious implications. The books [JEC], [RR1], and [RR2] treat the Axiom of Choice in great detail.

A4.1   Well Ordering

It is not difficult to see that the well ordering of a set is closely related to the Axiom of Choice. For if S is a set that is well ordered by a relation $\mathcal{R}$, then we may let $f : \mathcal{P}(S) \to S$ be the function that assigns to each $A \in \mathcal{P}(S)$ the (perforce unique) minimal element of A. The converse statement is true as well; but its proof involves ordinal arithmetic and transfinite induction and cannot be treated at this time.

Here is a way to specify a well ordering of any countable set, such as the rationals. The method seems artificial; but it *has* to be. The natural ordering on these sets will not do the job.

Let S be a countable set. Let $\phi : S \to \mathbb{N}$ be a set-theoretic isomorphism. If $x, y \in S$ are distinct elements, then we say that $(x, y) \in \mathcal{R}$ if $\phi(x) < \phi(y)$. Check for yourself that this creates a strict, simple order on S that well orders S.

As previously noted, it is impossible explicitly to give a well ordering of the real numbers; however, it is a theorem that *any* set can be well ordered (see [SUP]). The proof uses the Axiom of Choice.

A4.2   The Continuum Hypothesis

The Continuum Hypothesis is the assertion that there are no cardinalities strictly between the cardinality of the integers and the cardinality of the continuum (the cardinality of the reals). Call this statement C. In 1938 Gödel showed that C could be added to the axioms of set theory and no contradiction would ensue. That is to say, there is a "model" for set theory in which the usual axioms of set theory are true and so is C. In 1963 Paul Cohen showed that instead $\sim C$ could be added to the axioms of set theory and no contradiction would ensue. That is, there is a model for set theory in which the usual axioms of set theory are true but C is false (i.e., $\sim C$ is true).

In logical terms, we say that the Continuum Hypothesis is *independent* from the other axioms of set theory, in particular it is independent from the Axiom of Choice. The assertion C can never be proved as a theorem from the other axioms, nor can $\sim C$ be proved. Georg Cantor's inability to resolve the truth or falsity of C was a strong contributing factor to his debilitating mental illness at the end of his life. Sadly, mathematical logic was not sufficiently developed in Cantor's time for him to have been able to understand the ultimate resolution of the problem.

There are a number of very standard and useful tools in mathematics that are consequences of the Axiom of Choice. We now enunciate two of them.

A4.3   Zorn's Lemma

In modern mathematics, especially in algebra, Zorn's lemma plays a central role. It is used to prove the existence of maximal ideals, of bases for vector spaces, and of other "maximal sets."

We need two pieces of terminology in order to formulate Zorn's lemma. First, if $(S, \leq)$ is a partially ordered set then a *chain* in S is a subset $C \subseteq S$ that is linearly ordered (i.e., any two elements are comparable). An element u is an upper bound of the chain C if $c \leq u$ for every $c \in C$.

A typical enunciation of Zorn's lemma is this:

> Let $(S, \leq)$ be a non-empty, partially ordered set with the property that every chain in S has an upper bound. Then S has a maximal element, i.e., an element x such that $s \leq x$ for every $s \in S$.

Zorn's lemma is equivalent to the Axiom of Choice.

Zorn's lemma is commonly applied in algebra to prove the existence of various objects. For example, it can be used to establish the existence of a basis for any vector space:

Proposition A4.3.1 *Let V be a vector space over the field F. Then V has a basis.*

Proof: Let $\mathcal{S}$ be the collection of all linearly independent sets in V. Partially order $\mathcal{S}$ as follows: if $A, B \in \mathcal{S}$ then $A \leq B$ if $A \subset B$. Now each chain clearly has a maximal element (or upper bound) since we may simply take the union of all the elements of the chain to be the upper bound. We then apply Zorn's lemma to conclude that the entire collection $\mathcal{S}$ has a maximal element. Call that maximal element Φ. We claim that Φ is a basis for V.

If the claim is not true, then there is an element x in V that is not in the span of Φ. But then x may be added to Φ, thereby contradicting the maximality of Φ. We conclude that Φ is the basis we seek. $\square$

Likewise, Zorn's lemma is used to establish the existence of maximal ideals in a ring, of algebraic closures, and of many other basic algebraic constructs. You will learn about these ideas in your course on abstract algebra.

A4.4 The Hausdorff Maximality Principle

The Hausdorff maximality principle is a variant of Zorn's lemma, also commonly used in algebraic applications.

> If $\mathcal{R}$ is a transitive relation on a set S, then there exists a maximal subset of S which is linearly ordered by $\mathcal{R}$.

Hausdorff's principle is equivalent to the Axiom of Choice.

A4.5 The Banach-Tarski Paradox

Just for fun, we conclude this section with a description of one of the most dramatic paradoxes in mathematics. This paradox stems from the Axiom of

Choice, and is called the *Banach/Tarski paradox*. A version of this paradox is as follows:

> **Banach/Tarski Paradox [HJE], [JEC]:** It is possible to partition the solid ball in $\mathbb{R}^3$, of diameter one inch, into seven (disjoint) pieces in such a way that these seven pieces may be reassembled into a life-sized replica of the Statue of Liberty.

We refer the reader to the Bibliography (in particular [JEC]) for a detailed discussion of this paradox. We should note that the paradox fails in dimension two; it holds in dimension three in part because of the complexity of the three-dimensional rotation group (as contrasted with the rather simple two-dimensional rotation group).

Of course the seven pieces into which we break the unit ball in the Banach/Tarski paradox are extremely pathological. The subject of measure theory was invented, in part, to rule out sets such as these. Measure theory is another subject, like axiomatic set theory, in which there are very specific rules limiting the ways in which sets may be created.

Appendix 5: Ideas from Algebra

Our treatment of algebraic topology (homotopy theory and homology theory) uses certain fundamental ideas from algebra, and we very briefly review them here. Nice references for the reader requiring further detail are [HER] and [HUN].

A5.1 Groups

A *group* is a set G equipped with a binary operation (which we usually denote by $\cdot$) satisfying these properties:

(a) If $g, h \in G$, then $g \cdot h \in G$.

(b) If $g, h, k \in G$, then
$$(g \cdot h) \cdot k = g \cdot (h \cdot k).$$

(c) There is an element $e \in G$ such that, for all $g \in G$,
$$e \cdot g = g \qquad \text{and} \qquad g \cdot e = g.$$

(d) If $g \in G,$[1] then there is an element $g^{-1} \in G$ such that
$$g \cdot g^{-1} = g^{-1} \cdot g = e.$$

It is a remarkable fact that these simple axioms give rise to a rich and powerful theory that occurs in subjects ranging from topology to quantum mechanics to folkdancing.

If a group has the property that $g \cdot h = h \cdot g$ for all $g, h \in G$ then we call the group *abelian* (in honor of Niels Henrik Abel (1802–1829)). Otherwise the group is called *non-abelian*.

If G, H are groups then a mapping

$$\varphi : G \to H$$

is called a *homomorphism* if

$$\varphi(g \cdot h) = \varphi(g) \cdot \varphi(h) .$$

It is easily shown that, for such a φ, $\varphi(e) = e$ and $\varphi(g^{-1}) = [\varphi(g)]^{-1}$ for any $g \in G$. If φ is one-to-one and onto then it is called an *isomorphism*.

Self-maps of a given group G are also a matter of some interest. An *inner automorphism* of G, induced by a fixed element $a \in G$, is a mapping of the form

$$g \mapsto a \cdot g \cdot a^{-1} .$$

A cyclic group G is one that has an element h with the property that every element of G has the form

$$h^k = \underbrace{h \cdot h \cdot \cdots \cdot h}_{k \text{ times}} .$$

Certainly the integers $\mathbb{Z}$ under the binary operation of addition form a group. The positive real numbers under multiplication form a group. And the invertible 2×2 matrices (under matrix multiplication) form a group. The first two of these are abelian, and the third is not.

A5.2   Rings

A *ring* is a set R equipped with two binary operations, which we usually denote by $\cdot$ and $+$. These are assumed to satisfy the following axioms:

(a) The set R is an abelian group with respect to addition.

(b) Multiplication is associative, and has a unit element.

(c) For all $x, y, z \in R$, we have

$$(x + y) \cdot z = x \cdot z + y \cdot z \qquad \text{and} \qquad z \cdot (x + y) = z \cdot x + z \cdot y .$$

The unit element for addition is denoted by 0 and the unit element for multiplication is denoted by 1.

A *left ideal* in a ring R is a subset $I \subseteq R$ which is closed under addition and such that if $r \in R$ and $i \in I$ then $r \cdot i \in I$. For a right ideal, we require instead that $i \cdot r \in I$ for every $i \in I$ and $r \in R$. A *maximal ideal* is one that is contained in no larger ideal of R.

A ring R is said to be *commutative* if the multiplication operation is commutative. The integers $\mathbb{Z}$, with the standard operations of multiplication and addition, form a ring. The 2×2 matrices with real entries, using the standard matrix operations of multiplication and addition, form a ring. The collection of all real-coefficient polynomials of the single variable x, with the usual operations of addition and multiplication, form a ring.

A5.3   Fields

A *field* is a ring k equipped with two binary operations, which we usually denote by $\cdot$ and $+$. We mandate that $1 \neq 0$ and that the collection of nonzero elements of k form a group under multiplication.

A field has no ideals except $\{0\}$ and the entire field k. If R is a ring and I a maximal ideal then R/I is a field.

The rational numbers $\mathbb{Q}$, the real numbers $\mathbb{R}$, and the complex numbers $\mathbb{C}$ all are fields.

A5.4   Modules

A *left module* M over a ring R is an abelian group together with an operation of scalar multiplication by elements of R. We require that

$$(a + b)x = ax + bx \qquad \text{and} \qquad a(x + y) = ax + ay$$

for $a, b \in R$ and $x, y \in M$. A right module is defined similarly. In this book we consider modules which are both left and right, and we simply call those *modules*.

The collection of 2×2 matrices with integer coefficients, over the ring $\mathbb{Z}$, is a module. The collection of polynomials with integer coefficients, over the ring $\mathbb{Z}$, is a module.

A5.5   Vector Spaces

A vector space V over a field k is a module over k.

Certainly $\mathbb{R}^N$ over the field $\mathbb{R}$ is a vector space. The collection of polynomials with real coefficients, over the field $\mathbb{R}$, is a vector space. The collection of 2×2 matrices with rational entries, over the field $\mathbb{Q}$, is a vector space.

Solutions of Selected Exercises

Chapter 1

1. We may check directly that the intersection of any pair of the sets in $\mathcal{U}$ is still an element of $\mathcal{U}$. Likewise, the union of any pair of sets in $\mathcal{U}$ is still in $\mathcal{U}$. Finally, note that the space X and the empty set are elements of $\mathcal{U}$. It follows that $\mathcal{U}$ is a topology.

2. Unions and intersections of sets of irrational numbers are, in turn, sets of irrational numbers. Everything else is obvious.

4. Let $U_j = \{j\}$ for $j = 1, 2, \ldots$. Each of these sets, according to the definition, is open. However, $\cup_j U_j = \{1, 2, 3, \ldots\}$, which is not a finite set and hence not open. Thus the open sets are not closed under union.

5. The possible topologies are

$$\begin{aligned} \mathcal{U} &= \{\emptyset, X\} \\ \mathcal{U} &= \{\emptyset, \{a\}, X\} \\ \mathcal{U} &= \{\emptyset, \{b\}, X\} \\ \mathcal{U} &= \{\emptyset, \{a\}, \{b\}, X\} \end{aligned}$$

6. Let the topological space be $X = \{p_1, p_2, \ldots, p_n\}$. Fix an index j. Then, for each $\ell \neq j$, there is a neighborhood U_ℓ of p_j that does not contain p_ℓ and there is a neighborhood V_ℓ of p_ℓ that does not contain p_j. Then the set $\mathcal{U} = U_1 \cap U_2 \cap \cdots U_{j-1} \cap U_{j+1} \cap \cdots U_n$ is open and disjoint from $p_1, p_2, \ldots p_{j-1}, p_{j+1}, \ldots p_n$. Thus $\mathcal{U}$ must coincide with $\{p_j\}$ and we see that the singleton $\{p_j\}$ is open. Hence the space is discrete.

8. If f is any continuous function on $[0,1]$ and U is an open neighborhood of f, then U contains a set of the form $V = \{g \text{ continuous} : \max_{x \in [0,1]} |f(x) - g(x)| < \epsilon\}$ for some $\epsilon > 0$. The Weierstrass approximation theorem guarantees that there is a polynomial in V. Hence the polynomials are dense in the continuous functions.

If $x \in \mathbb{R}$ and U is a neighborhood of x, then U contains a set of the form $V = (x - \epsilon, x + \epsilon)$ for some $\epsilon > 0$. Choose a positive integer N so large that $N > 10/\epsilon$. Then one of the points j/N for $j \in \mathbb{Z}$ must lie in V. Thus the rationals are dense in the reals.

Let $x = 1/2$ be a real number. Then $U = (1/2 - 0.1, 1/2 + 0.1)$ is a neighborhood of x that does not contain any integer. Hence the integers are not dense in the reals.

10. Let E be open in X and e a point of E. By definition of open set, there is a neighborhood U of e so that $U \subseteq E$. However, this says that e is in the interior $\overset{\circ}{E}$ of E. Hence $E \subseteq \overset{\circ}{E}$. For the converse, let $f \in \overset{\circ}{E}$. It follows from the definition of interior then that $f \in E$. Hence $\overset{\circ}{E} \subseteq E$. We have shown that, when E is open, then $E = \overset{\circ}{E}$.

For the converse, suppose that $E \subseteq X$ and $E = \overset{\circ}{E}$. If $e \in E$, then $e \in \overset{\circ}{E}$, so there is a neighborhood U of e that lies entirely in E. We conclude that E is open.

12. Let X be $\mathbb{R}^2$. Let $K = \{(x,y) : y = 1/x \text{ and } x > 0\}$ and $E = \{(x,y) : y = -1/x \text{ and } x < 0\}$. Then K and E are closed but the distance of the two sets is 0.

13. This does not form a topology. Let $U_1 = \mathbb{R} \setminus [0,1]$ and $U_2 = \mathbb{R} \setminus [3,4]$. Then $U_1 \cap U_2 = \mathbb{R} \setminus ([0,1] \cup [3,4])$, which is not open according to the definition. So the open sets are not closed under finite intersection.

15. Let X be the real numbers. Declare a set to be open if it is an interval of the form (a, ∞) or the entire real line or the empty set. These sets clearly form a topology on X. Then the closed sets are intervals of the form $(-\infty, b]$. However, the set $E = [0,1] \subseteq X$ is clearly compact since any open covering $\{U_\alpha\}_{\alpha \in A}$ will contain one set U_α that has an element that is less than 0, and that single set will cover E. However, E is not closed, and of course X is not Hausdorff.

17. No, $f^{-1}(K)$ need not be compact. For example, equip the interval $X = (-1, 1)$ with the usual Euclidean topology and define $f : X \to X$ by $f(x) = 1 - x^2$. Then the inverse image under f of $[0, 1/2]$ is $(-1, -1/\sqrt{2}] \cup [1/\sqrt{2}, 1)$, and that set is not compact.

19. For $j = 1, 2, \ldots$, let A_j be the sequence with a 1 in the jth position and 0s in all other slots. Then the A_j are plainly elements of X and linearly independent, and there are infinitely many of them. The space is therefore infinite dimensional.

21. The complement of S is clearly open so S is closed. Thus the closure of S is S. If $s \in S$ and U is any neighborhood of s, then U will contain points not in S (i.e., points with non-integral coordinates). Thus S has no interior. For the same reason, the boundary of S is S itself.

23. Declare a set $U \subseteq \mathbb{R}^2$ to be open if it is open in the usual Euclidean sense and, in addition, U contains the real axis. Clearly the collection of all such sets is closed under finite intersection and arbitrary union. Together with the total space and the empty set, these open sets form a topology. Now if P is any point of $\mathbb{R}^2$ and W is a neighborhood of P, then, by definition, W will contain the real line. Thus we conclude that the real line is dense in $\mathbb{R}^2$.

25. Clearly E is a complete metric space when equipped with the usual Euclidean metric. If E were countable, then write $E = \{p_j\}_{j=1}^{\infty}$. Then each singleton $\{p_j\}$ is nowhere dense (because it is a limit point). Thus we have written a complete metric space as the countable union of nowhere dense sets. That is a contradiction.

29. Let U be a connected open set in $\mathbb{R}^N$. Let $P \in U$. Define

$$S = \{X \in U : P \text{ and } X \text{ can be connected by a path}\}.$$

Clearly S is open for, if $X \in S$, then a point in a small ball about X can certainly be connected to X, and that gives a path to the new point. Similar reasoning shows that S is closed and certainly S is nonempty, for $P \in S$. It follows from connectivity that $S = U$. That proves the result.

31. The product of compact spaces is compact. The product of connected spaces is connected. Finally, the product of Hausdorff spaces is Hausdorff. That establishes the result.

Chapter 2

1. Finite intersections of the closed intervals just give closed intervals. Arbitrary unions of closed intervals give closed intervals, open intervals, and half-open intervals. Thus the resulting topology consists of *all* intervals, together with the entire line and the empty set.

3. Let $X = [0,1]$ in the real line and $Y = [0,1]$ in the real line. Then $X \times Y$ is the closed square box in the plane, and $\partial(X \times Y)$ is the square. On the other hand, $\partial X = \partial Y = \{0,1\}$ so $\partial X \times \partial Y = \{(0,1),(0,1),(1,0),(1,1)\}$. This is a different set. In fact

$$\partial(X \times Y) = (X \times \partial Y) \cup (\partial X \times Y),$$

as this example illustrates.

6. This is the topology of pointwise convergence at the point 0. Thus two functions are close if their values at 0 are close.

8. Certainly X is a metric space with metric

$$d(f,g) = \max_{x \in [0,1]} |f(x) - g(x)|.$$

If p is a point of X, then the sets

$$U_j = \{q \in X : d(q,p) < 1/j\}$$

form a countable neighborhood base for p.

Certainly the polynomials with rational coefficients are a countable, dense subset of X. The collection of all sets

$$S_{p,\epsilon} \equiv \{q \in X : d(p,q) < 1/j\},$$

for p such a polynomial, forms a countable basis for the topology on X.

10. Certainly $C \cap (1/2, 3/4)$ is a relatively open set in C. If p and q are points of C, then there is a removed interval from the Cantor construction that lies between them. So no point of C is connected to an other. The connected components are singletons. Hence the space is totally disconnected.

12. Let $X = (x, y)$ and $X' = (x', y')$ be elements of X. Then $X \sim X'$ if and only if $x - x'$ is an integer and $y - y'$ is an integer. Thus the set

$$\mathcal{A} = \{(x, y) : 0 \le x < 1, 0 \le y < 1\}$$

consitutes the representatives for the equivalence classes. Clearly the left edge is glued to the right edge (since they are identified in the quotient) and, likewise, the bottom edge is glued to the top edge. The result is a torus. The quotient topology is the usual manifold topology on the torus: it is locally Euclidean.

14. A subset U of $\mathbb{R}$ pulls back to an open set in X if and only if it is an open interval. Hence the quotient topology is the usual Euclidean topology.

15. Let $f(x, y, z) = (x^2 + y^2 + z^2, 1)$. If $K \subseteq \mathbb{R}^2$ is compact, then $K' = \{x \in \mathbb{R} : (x, 1) \in K\}$ is also compact. Hence $L = \{(x, y, z) \in \mathbb{R}^3 : x^2 + y^2 + z^2 \in K'\}$ is closed and bounded, and thus compact. Since $f^{-1}(K) = L$, we are done.

18. Define

$$\eta(t) = \begin{cases} -t^2 + 1 & \text{if} \quad |t| \le 1 \\ 0 & \text{if} \quad |t| > 1. \end{cases}$$

Now set

$$\varphi_1(x, y) = \eta(|(x, y) - (1, 0)|),$$
$$\varphi_2(x, y) = \eta(|(x, y) - (0, 1)|),$$
$$\varphi_1(x, y) = \eta(|(x, y) - (-1, 0)|),$$
$$\varphi_1(x, y) = \eta(|(x, y) - (0, -1)|).$$

Then define

$$T(x, y) = \varphi_1(x, y) + \varphi_2(x, y) + \varphi_3(x, y) + \varphi_4(x, y)$$

and let, for $j = 1, 2, 3, 4$,

$$\psi_j(x, y) = \frac{\varphi_j(x, y)}{T(x, y)}.$$

It is immediate that each ψ_j is a positive function on U_j and $\sum_j \psi_j = 1$ on the union of the discs. Thus we have a partition of unity.

22. Define
$$f(x) = \begin{cases} -x & \text{if} \quad x \leq 0 \\ x & \text{if} \quad x > 0. \end{cases}$$

Then, if $K \subseteq \mathcal{L}$ is compact we see that $f^{-1}(L) = K \cup -K$, where $-K = \{-x : x \in K\}$. Thus $f^{-1}(K)$ is compact.

23. Of course the unit interval $[0, 1]$ is compact, and its inverse image must be the entire real line, which is noncompact.

24. Consider the topology of pointwise convergence on the continuous functions $C([0, 1])$ on $[0, 1]$. Hence a sub-basis for the topology is sets of the form (for f a function, x a point, and ϵ a positive number)

$$\mathcal{S}_{f,x,\epsilon} \equiv \{g \in C([0, 1]) : |f(x) - g(x)| < \epsilon\}.$$

It is easy to see that, since there are uncountably many points in $[0, 1]$, it is impossible to find a countable neighborhood basis for any element of the space.

26. An element of the long line is indexed by the pair (s, x). Define a map from the long line to $\mathbb{R}$ by $(s, x) \mapsto s + x$. It is easy to check that this function is onto. The function is obviously continuous.

28. The quotient will be a torus, for a reason similar to that in Exercise 12. The quotient topology on this torus will be the usual manifold topology—locally Euclidean.

29. As usual, let P be a point *not on* the long line. Let the open sets be open subsets of the long line as described in the text together with sets of the form ${}^c K \cup \{P\}$ for K compact in the long line. This topology, in effect, ties together the two "ends" of the long line to create a long circle.

Chapter 3

1. Let $\gamma : [0, 1] \rightarrow S^2$ be a closed loop. Then, by perturbing γ slightly, it may be arranged that the image of γ lies in a patch $U \subseteq S^2$ that is homeomorphic to the unit disc in the plane. Thus the loop may surely be shrunk to a point. Thus all loops are trivial, and the fundamental group of the sphere is just the singleton $\{e\}$.

3. Consider the mapping $(x, y) \rightarrow (-y, x)$. This is just rotation counter-clockwise through an angle of $\pi/2$. Then there is no fixed point. If we endeavor to solve
$$(x, y) = (-y, x),$$
then we find that $x = y = 0$. Of course the origin is not a point of the annulus.

5. The fundamental group of S^1 is $\mathbb{Z}$ while the fundamental group of the sphere S^2 is the trivial group $\{e\}$ (see the solution to Exercise 1 above). Thus the two spaces cannot be homotopy equivalent.

9. The open annulus $A = \{(x, y) \in \mathbb{R}^2 : 1 < x^2 + y^2 < 4\}$ and the circle $S^1 = \{(x, y) \in \mathbb{R}^2 : x^2 + y^2 = 1\}$ both have fundamental group $\mathbb{Z}$. However, they cannot be homeomorphic since S^1 is compact while A is not.

10. If $^c U$ has more than two connected components, then one of the components is unbounded and the others are all bounded. There will be at least two bounded components—call them E_1 and E_2. Then a loop that encircles E_1 once counterclockwise will generate a subgroup of the fundamental group that is isomorphic to $\mathbb{Z}$, and a loop that encircles E_2 once countrclockwise will generate a *separate* subgroup of the fundamental group that is isomorphic to $\mathbb{Z}$. Thus the fundamental group of U cannot be just $\mathbb{Z}$. We conclude that the complement of U can have just two connected components—one of them bounded and one of them not. Now a classical theorem in complex function theory (see [AHL] or [KRA2]) shows that there is a conformal mapping from U to an annulus.

11. The alternating group on three letters is non-abelian.

12. Just as an instance, suppose that U has complement with three connected components—two bounded and one unbounded. Then U has the same homotopy as a Figure 8.

13. By the Riemann mapping theorem, the domain is conformally equivalent to the open unit disc.

15. Let φ be the mapping that assigns to each point on the sphere (which is the surface of the earth) the ordered pair consisting of the wind speed and the temperature. Then this is a continuous mapping of S^2 into $\mathbb{R}^2$. By Borsuk-Ulam, there is a pair of diametrically opposite points at which φ agrees. That gives the result.

17. Certainly the strip $S = \{(x, y) : \ln 2 < x < \ln 4\}$ can be thought of as the universal covering space, with the covering mapping defined by

$$S \ni (x, y) \rightarrow (e^x \cos y, e^x \sin y) \ .$$

19. Certainly the map $\pi : X \rightarrow X$, which is the identity, is a covering map and satisfies the definition of universal covering space.

20. If X is the unit circle and Y the interval, then the universal covering space of X is the line (as discussed in the text) and the universal covering space for $X \times Y$ is the line cross the interval (see our solution to Exercise 17 above). If now X is arbitrary with universal covering $\pi : \widehat{X} \rightarrow X$ and Y is simply connected, then $\widehat{X} \times Y$ will be a universal covering space for $X \times Y$ with the obvious map $\pi \times \mathrm{id}$.

24. The unit disc D in the plane and a singleton $\{(0,0)$ in the plane are not homeomorphic. One set is uncountable and the other is finite.

28. Let $\gamma : [0, 1] \rightarrow X \times Y$. Let $\pi_1 : X \times Y \rightarrow X$ be projection on the first factor and $\pi_2 : X \times Y \rightarrow Y$ be projection on the second factor. Then $\pi_1 \circ \gamma$ is homotopic to a point by a homotopy Γ_1 and $\pi_2 \circ \gamma$ is homotopic to a point by a homotopy Γ_2. We may define a homotopy of γ to a point in $X \times Y$ by

$$\Gamma(s, t) = (\Gamma_1(s, t), \Gamma_2(s, t)) \ .$$

Thus $X \times Y$ is simply connected.

29. Let $\gamma : [0, 1] \to \mathcal{U}$ be a closed loop. Denote the image of γ by $\widehat{\gamma}$. Then the open sets $\{U_j\}$ form an open cover of $\widehat{\gamma}$, and $\widehat{\gamma}$ is a compact set. Hence there is a finite subcover $U_{j_1}, U_{j_2}, \ldots, U_{j_k}$. The one of these with the greatest index will then cover the image of the loop all by itself (since the sets U_j are nested). Then γ can be deformed to a point inside that single, simply connected open set. Thus it can certainly be deformed to a point inside $\mathcal{U}$. Thus $\mathcal{U}$ is simply connected.

30. Let S_1 be the sphere in $\mathbb{R}^3$ with center $(0, 0, 0)$ and radius 1. Let S_2 be the sphere in $\mathbb{R}^3$ with center $(1, 0, 0)$ and radius 1. Each of these spaces is simply connected. However, their intersection is a circle and that is not simply connected.

31. Let $E = \{re^{i\theta} \in \mathbb{C} : -\pi/6 \leq \theta \leq 7\pi/6\}$ and $F = \{re^{i\theta} \in \mathbb{C} : \pi \leq \theta \leq 2\pi\}$. Then each of E and F is simply connected. However, $E \cup F$ is the unit circle, and that is not simply connected.

Chapter 4

1. Let P be a point of this quotient—so P is some (x, y, z) in the sphere paired with its antipodal opposite $(-x, -y, -z)$. Now certainly there is a small neighborhood U of (x, y, z) in the sphere that is homeomorphic to the unit disc in the plane via some mapping φ. Let π be the mapping that sends each spherical point to its antipodal opposite. Then $\{[((x, y, z), \pi(x, y, z))] : (x, y, z) \in U\}$ is a neighborhood of P in the quotient space. A coordinate map is just φ acting on (x, y, z).

3. Let $(m, n) \in M \times N$. Then m has a coordinate neighborhood $U \subseteq M$ with coordinate map π and n has a coordinate neighborhood $V \subseteq N$ with coordinate map μ. It follows that $\pi \times \mu$ is a coordinate map for the neighborhood $U \times V$ of (m, n).

6. Parametrize each line in the plane by the ordered pair (y, θ) consisting of its y-intercept together with the angle it makes with the x-axis (where θ ranges from $-\pi$ to π). [It is convenient here to take the one-point compactification of the y-axis so that a vertical line will have a y-intercept—at the point at infinity.] If ℓ is a line in the plane and $p_\ell = (y_\ell, \theta_\ell)$ is its parametrization, then a neighborhood of p_ℓ in the Euclidean plane induces a neighborhood of ℓ in the collection of all

lines. Of course the identification mapping (as just described) is the coordinate map, hence the collection of all lines in the plane forms a manifold.

7. We may identify each plane in 3-space with its unit normal vector, together with its intercept with the z axis. [Again—see the preceding Exercise solution—it is convenient to take the one-point compactification of the z-axis.] Of course the set of unit normal vectors lives in the unit sphere, and the sphere cross a line (or a compactified line) is certainly a manifold. Thus it follows that the set of planes in space is a manifold.

8. Let $\mathbf{d}$ be the closed disc in the torus T. Then $T \backslash \mathbf{d}$ is open. If $P \in T \backslash \mathbf{d}$ and U is a coordinate chart at P with coordinate map φ then we may let $(T \backslash \mathbf{d}) \cap U$ be a coordinate chart for $T \backslash \mathbf{d}$ and φ restricted to that set be the coordinate map. Hence $T \backslash \mathbf{d}$ is a manifold.

9. We think of an element of this space as an ordered pair $(P, \mathbf{v})$, where $\mathbf{v}$ is a unit direction vector for the tangent line. Of course $\mathbf{v}$ depends on P, and it does so in a smooth (continuously differentiable) fashion. Now P is parametrized by the interval and $\mathbf{v}$ is parametrized by the circle. The interval cross the circle is certainly a manifold. Hence so is the "tangent bundle" that we are analyzing here.

11. Without loss of generality consider the point $P = (0, 0, 1)$ in the sphere. A coordinate patch is

$$U = \{(x, y, z) : x^2 + y^2 + z^2 = 1, x^2 + y^2 < 1/2, z > 0\}$$

and a coordinate map is $\varphi : U \to D(0, 1/\sqrt{2})$ given by $(x, y, z) \mapsto (x, y)$.

If $Q = (a, b, c)$ is a nearby point on the sphere, then we may take a coordinate patch at Q to be

$$V = \{(x, y, z) : x^2 + y^2 + z^2 = 1, (x - a)^2 + (y - b)^2 + (z - c)^2 < 1/2\}$$

and a coordinate map is $\psi : V \to (x, y)$.

Then the mapping $\varphi \circ \psi^{-1} : \varphi(U) \cap \psi(V) \to \varphi(U) \cap \psi(V)$ is simply the identity. Thus it is certainly C^k for every k. We conclude then that the sphere is a C^k manifold for each k.

13. Let $P = (x, f(x))$ be a point on the graph. Take

$$U = \{(t, f(t) : |t - x| < 1\}$$

to be a coordinate chart and let $\varphi : U \to (-1, 1)$ be the function $\varphi(t) = t$. Then certainly φ is a coordinate map, and it is easy to see that the transition maps are all the identity and thus certainly C^k.

14. Without loss of generality consider the point $P = (0, 0, -1)$ on the sphere. Let U be the intersection of a small ball centered at P with the sphere. Then U is of course, under the stereographic projection, the image of a disc in the plane. The same will be true for a nearby point P' with an open set V centered at it in the sphere. The transition map will be the identity hence, of course, holomorphic. Thus the Riemann sphere is a complex manifold.

15. It is convenient to consider the great circle described by

$$S = \{(0, y, z) : y^2 + z^2 = 1\}$$

in the sphere. Consider the point $P = (0, 0, 1)$ on this circle, and the coordinate patch

$$U = \{(x, y, z) : x^2 + y^2 < 1/2, z > 0\}\,.$$

Let $\varphi : U \to D(0, 1/\sqrt{2})$ be given by $(x, y, z) \mapsto (x, y)$. Then the image of $S \cap U$ is $D(0, 1/\sqrt{2}) \cap \{x = 0\}$. This is a line segment. Thus we have exhibited the great circle as a regularly embedded submanifold.

17. Let

$$\mathcal{G} = \{(x, f(x)) : x \in \mathbb{R}\}$$

be the graph of f. If the point $P = (x, f(x))$ lies on $\mathcal{G}$, then let $D(P, 1)$ be the open disc of radius 1 centered at P and let us consider the mapping

$$\varphi : D(P, 1) \to \varphi(D(P, 1))$$

defined by

$$\varphi((x, y)) = (x, y - f(x))\,.$$

We see that the image of $D(P, 1) \cap \mathcal{G}$ is a line segment, and the mapping φ is manifestly C^k. Hence we have exhibited the graph as a regularly embedded C^k submanifold.

19. Pick a point $P \in M$. We may assume that $\partial F/\partial y(P) \neq 0$. Then the implicit function theorem tells us that there is a locally defined C^1 function $\varphi(x)$ so that $F(x, \varphi(x)) = 0$. Then the set M is described locally as the graph of φ, and we know from the solution of Exercise 13 above that M is a 1-dimensional manifold.

21. Define $\varphi(x, y) = (2x, \sqrt{2}y)$. Then φ clearly maps the circle to the ellipse in a one-to-one and onto fashion. The point $(0, 1)$ on the circle is mapped to the point $(0, \sqrt{2})$ on the ellipse. If we take $U = \{(x, y) : x^2 + y^2 = 1, |x| < 1/2\}$ and $V = \{x^2 + 2y^2 = 4, 1/\sqrt{2}\}$, then φ takes U to V. The coordinate chart on U is $\varphi : (x, y) \mapsto (x, 0)$ and the coordinate chart on V is $\psi : (x, y) \mapsto (x, 0)$. Then one may check directly that $\psi \circ f \circ \varphi^{-1}$ is a C^k mapping for any k.

23. Let X be the x-axis and Y the y-axis. Each of these is plainly a 1-manifold. Yet $X \cup Y$ is *not* a 1-manifold because there is no coordinate chart centered at the origin. Likewise, $X \cap Y$ is *not* a 1-manifold. In fact it is the singleton $\{(0,0)\}$ and that is a 0-manifold.

25. Of course π is a local homeomorphism. If $x \in X$ and $\hat{x}$ an inverse image of x under π, then let U be a neighborhood of x, which is homeomorphic, by way of π, to an open set $\hat{U} \subseteq \hat{X}$ that contains $\hat{x}$. If W is a coordinate chart at x with coordinate map $\varphi : W \to W' \subseteq \mathbb{R}^k$, then set $V = U \cap W$. Also take $\hat{V}$ to be that subset of $\hat{U}$ which projects under π to V. Then $\hat{V}$ is a coordinate chart for $\hat{x}$ and the coordinate map is $\varphi \circ \pi$. Hence $\hat{X}$ is a k-manifold.

27. Here are two ways to think about the matter: **(i)** Imagine cutting each figure with a pair of scissors. The figure on the left obviously then becomes (homeomorphic to) a segment, and the figure on the right, after it is untangled, also becomes (homeomorphic to) a segment. These are homeomorphic. Now if we re-join the cuts, turning the left segment back into a circle and the right segment back into a trefoil, and if we identify the cut points on the two figures, then the homeomorphism extends from the segments to a homeomorphism of the circle to the trefoil. **(ii)** Imagine traveling a circuit counterclockwise around the trefoil. This circuit will end at the point at which it began. One can easily parametrize such a circuit with domain the circle. And that gives the needed homeomorphism.

29. The polynomial $p(x, y) = x \cdot y$ has the zero set $\mathcal{Z} = \{(x, y) : x = 0\} \cup \{(x, y) : y = 0\}$. As explained in the solution to Exercise 23, the set $\mathcal{Z}$ is not a 1-manifold. It is easy to see, however, that ∇p will be nonvanishing on the zero set of a polynomial $p(x, y)$ except on a discrete subset. Then the implicit function theorem (see the argument in the solution of Exercise 19) tells us that the points of the zero set where the gradient does not vanish form a manifold.

Chapter 5

1. A net $\{f_j\}$ converges to a limit f if any given derivative $(d/dx)^\ell f_j$ converges uniformly on any given compact set K to the corresponding derivative $(d/dx)^\ell f$.

3. We may verify that this is a Hausdorff space by using the sets $\mathcal{U}_{f,k,K,\epsilon}$ for any fixed k. As an instance, with $k = 1$, we note that if f, g in X are distinct and do not differ by a constant, then there must be some point $x \in \mathbb{R}$ at which $f'(x) \neq g'(x)$. But then continuity of the derivative tells us that there is a compact interval K centered at x and a positive number ϵ such that $|f'(t) - g'(t)| > \epsilon$ for $t \in K$. But then $\mathcal{U}_{f,1,K,\epsilon/2}$ and $\mathcal{U}_{g,1,K,\epsilon/2}$ are sub-basic open sets, containing f and g respectively, which are disjoint. Thus X is Hausdorff.

4. Let A be a subset of the topological space $(X, \mathcal{U})$. We say that the point a is in the boundary of A if there is a net in A converging to a and also there is a net in $^c A$ converging to a.

7. With this topology, any net is convergent to any point.

11. Fix $f, h \in X$ such that $f \neq h$. Then there is an $x \in \mathbb{R}$ such that $f(x) \neq h(x)$. It follows by continuity that there is a compact K containing x and an $\epsilon > 0$ such that $|f(t) - h(t)| > \epsilon$ for $t \in K$. Then $V_{f,K,\epsilon/2}$ is a neighborhood of f that excludes h, and $V_{h,K,\epsilon/2}$ is a neighborhood of h that excludes f. Thus the space is T_1. In fact these two neighborhoods are disjoint, so the space is T_2.

12. Define

$$f_j(x) = \begin{cases} 0 & \text{if} & x \le -j-1 \\ e^{j^2}(j+1-|x|) & \text{if} & -j-1 < x \le -j \\ e^{x^2} & \text{if} & -j < x < j \\ e^{j^2}(j+1-x) & \text{if} & j \le x \le j+1 \\ 0 & \text{if} & j+1 < x. \end{cases}$$

Certainly this f_j is a bounded, continuous function. Also, if K is a compact set and if j is large enough (depending on K), then $f_j(x) = e^{x^2}$ for $x \in K$. It follows then that f_j converges as a net to e^{x^2}.

14. Define

$$g_j(x) = \begin{cases} |x| - j & \text{if} & x \le -j \\ 0 & \text{if} & -j < x < j \\ x - j & \text{if} & j \le x. \end{cases}$$

Then certainly g is unbounded and continuous. If K is an compact set and if j is large enough (depending on K), then $f_j(x) = 0$ on K. It follows then that f_j converges as a net to the identically 0 function.

15. Let K be any compact interval. Then, for j sufficiently large, f_j takes all values between -1 and 1 inclusive on the set K. The same assertion holds on any subinterval of K. It follows that there is no function (continuous or otherwise) to which $\{f_j\}$ converges as a net.

17. In this topology it is very hard for a net to be convergent. Indeed, since singletons are open and a convergent net must be eventually in every neighborhood of the limit point, a net can converge to a point P only if the net is eventually constantly equal to P.

Chapter 6

1. Let $\epsilon > 0$. We may assume without loss of generality that $f(0) = 0$. By the Weierstrass theorem there is a polynomial q such that $|f'(x) - q(x)| < \epsilon/2$ for $x \in [0, 1]$. Now set

$$p(x) = \int_0^x q(t)\, dt.$$

Then, for any $x \in [0,1]$,

$$
\begin{aligned}
|p(x) - f(x)| &= \left| \int_0^x q(t)\, dt - \int_0^x f'(t)\, dt \right| \\
&\leq \int_0^x |q(t) - f'(t)|\, dt \\
&\leq \epsilon/2 < \epsilon .
\end{aligned}
$$

2. Let f be in $C^k[0,1]$. We may assume without loss of generality (just by subtracting a polynomial from f) that $f(0) = f'(0) = \cdots f^{(k-1)}(0) = 0$. Let $\epsilon > 0$. Now, using the Weierstrass theorem, select a polynomial q so that $|f^{(k)}(x) - q(x)| < \epsilon$ for all $x \in [0,1]$. Then antidifferentiate $k-1$ to obtain the desired result (refer to the solution to Exercise 1).

3. Define $\mathcal{F}(K, U)$ to be the continuous functions g such that $g(K) \subseteq U$. Suppose that f_j converge uniformly on compact sets to a limit function f. Let U be an open set in the range of f. Since the sequence converges uniformly on each interval we may be sure that f is continuous. Hence $f^{-1}(U)$ is open. Hence there is a compact $K \subseteq f^{-1}(U)$. We conclude that $f \in \mathcal{F}(K, U)$.

Now there is an $\epsilon > 0$ such that the Euclidean distance of $f(K)$ to ^cU is greater than 2ϵ. If j is sufficiently large then $|f_j(x) - f(x)| < \epsilon$ for all $x \in K$. We conclude then that $f_j(K) \subseteq U$ for all large j, hence $f_j \in \mathcal{F}(K, U)$ for j large.

This reasoning also runs in the reverse direction.

5. Fix a continuous function f on $\mathbb{R}$. Fix a compact interval $J \subseteq \mathbb{R}$. Let $\epsilon > 0$ and choose an $N > 0$ such that, if $\delta < 1/N$ and $|x - t| < \delta$, then $|f(x) - f(t)| < \epsilon$. If L is the length of J, then divide I into $\lfloor NL \rfloor + 1$ subintervals of equal length. Then certainly each of these subintervals I has length less than $1/N$. Define a new function p_I on I that is constantly equal to the mean of f on I. Then it is immediate that $|p_I(x) - f(x)| < \epsilon$ on I for each I. Also p_I is piecewise constant by construction.

7. If instead the degrees of the p_j were bounded by M, then we write

$$
p_j(x) = a_0^j + a_1^j x + a_2^j x^2 + \cdots + a_M^j x^M .
$$

An easy compactness argument shows that if the p_j converge uniformly on $[0, 1]$, then the a_0^j are bounded, the a_1^j are bounded, and so forth. We may extract a subsequence $\{p_{j_k}\}$ of the p_j so that the coefficients $a_0^{j_k}$ converge, the coefficients $a_1^{j_k}$ converge, etc. Thus the limit function will be a polynomial. Clearly the given f is *not* a polynomial, so the degrees of the p_j cannot be bounded.

9. Let $f_j(x) = \sin jx$. Then $\max |f_j'(x)| = j$, so (by the mean value theorem), the f_j cannot be equicontinuous. On the other hand, if $N \leq 100$ and $|a_j| \leq 1$ for all j then

$$\left| \frac{d}{dx} \sum_{j=-100}^{100} a_j \sin jx \right| \leq \sum_{j=-100}^{100} j \leq 10100 \,.$$

Therefore, by the mean value theorem,

$$|f_j(x) - f_j(t)| \leq 10100 \cdot |x - t|$$

for any x and t. Thus the family is equicontinuous.

11. The statement tells us, for instance, that a family of functions $\{f_\alpha\}_{\alpha \in A}$ of functions on a compact interval $I \subseteq \mathbb{R}$ which satisfies

   (i) $|f_\alpha(x)| \leq M$ for all α and all x;

   (ii) $|f_\alpha(x) - f_\alpha(t)| \leq C \cdot |x - t|$ for all $x, t \in I$

has a convergent subsequence. This is a statement of sequential compactness for this family of functions which is uniformly bounded with uniformly bounded Lipschitz norm.

13. Let
$$g(x) = e^{-x^2} \cdot \sin x \,.$$

Notice that g is a C^∞ function which is bounded by 1 and has all derivatives bounded. Now if p is a polynomial, then the functions

$$f_j(x) = p(x) + \frac{1}{j} \cdot g(x)$$

are *not* polynomials (because g is not a polynomial) and converge uniformly to p.

15. A 1-dimensional space (such as that generated by x^3) is nowhere dense in a complete normed linear space (such as the continuous functions on the unit interval). Thus the space of polynomials less the monomial x^3 will still be dense. A similar remark applies to a finite-dimensional space, so the polynomials less finitely many monomials will still be dense.

17. Let $p(x)$ be a polynomial of degree k. Then we know that

$$\frac{d^{k+1}}{dx^{k+1}}p(x) = 0 \le e^x = \frac{d^{k+1}}{dx^{k+1}}e^x.$$

Now integrate $k + 1$ times to obtain the desired estimate.

19. Fix an f as described in the exercise and let $\epsilon > 0$. Then, by the Weierstrass approximation theorem, there is a polynomial p such that $|f(x) - p(x)| < \epsilon/3$ for all $x \in [0, 1]$. Let $\alpha = f(0) - p(0)$ and $\beta = f(1) - p(1)$. Both these numbers are smaller than $\epsilon/3$. Let $\varphi(x) = (1 - x)\alpha + x\beta$ and set $\widetilde{p}(x) = p(x) + \varphi(x)$. Then it is easy to see that $|f(x) - \widetilde{p}(x)| < \epsilon$ for all x and that $\widetilde{p}(0) = f(0)$, $\widetilde{p}(1) = f(1)$.

21. Let $\epsilon > 0$. Fix x and y. We may choose j so large that $|f_j(x) - f(x)| < \epsilon/2$ and $|f_j(y) - f(y)| < \epsilon/2$. Then

$$
\begin{aligned}
|f(x) - f(y)| &\le |f_j(x) - f_j(y)| + |f_j(x) - f(x)| + |f_j(y) - f(y)| \\
&< |x - y| + \epsilon/2 + \epsilon/2 = |x - y| + \epsilon.
\end{aligned}
$$

Since ϵ was arbitrary, we conclude that

$$|f(x) - f(y)| \le |x - y|.$$

23. Of course f is bounded on the closed unit interval I and so is g. We may conclude then that the f_j and the g_j are uniformly bounded on I by some constant M. Then

$$|f_j \cdot g_j - f \cdot g| \le |f_j - f| \cdot |g_j| + |g_j - g| \cdot |f|.$$

Since $|f| \le M$ and $|g_j| \le M$, we conclude that $f_j \cdot g_j$ converges uniformly to $f \cdot g$.

On an unbounded set, such as the real line, things are different. Let $f_j(x) = x^2 + 1/j$ and $g_j(x) = x^3 + 1/j$. Then $f_j \to x^2$ uniformly on

$\mathbb{R}$ and $g_j \to x^3$ uniformly on $\mathbb{R}$. However, $f_j(x) \cdot g_j(x) = x^5 + x^2/j + x^3/j + 1/j^2$, and this sequence does not converge uniformly to any limit (though it does converge pointwise to x^5).

25. Let $\epsilon > 0$. For each x in the closed unit interval $I \subseteq \mathbb{R}$, choose an $N_x > 0$ such that $j > N$ implies that $|f_j(x) - f(x)| < \epsilon/2$. Fix such a j, and choose by continuity an open interval J_x centered at x such that $|f_j(x) - f(x)| < \epsilon$ for $x \in J_x$. The intervals J_x form an open covering of I, so (by compactness) there is a finite subcover. Thus we have

$$J_{x_1}, J_{x_2}, \ldots, J_{x_k}$$

which cover I, and we may then choose a single j so that $|f_j(x) - f(x)| < \epsilon$ for all $x \in I$. But then, by monotonicity of the convergence of the sequence of functions, the same inequality is true for all f_ℓ with $\ell > j$. That proves the result.

27. Let $f_j(x)$ be defined by

$$f_j(x) = \begin{cases} 0 & \text{if} & 0 \leq x \leq 1/2 - 1/j \\ j(x - (1/2 - 1/j)) & \text{if} & 1/2 - 1/j < x < 1/2 + 1/j \\ 2 & \text{if} & 1/2 + 1/j \leq x \leq 1. \end{cases}$$

Then f is continuous on the interval I and it is easy to see that the sequence f_j converges in the metric d to the function

$$f(x) = \begin{cases} 0 & \text{if} \quad 0 \leq x \leq 1/2 \\ 2 & \text{if} \quad 1/2 < x \leq 1. \end{cases}$$

29. Let f be a continuous function on I. Let $\epsilon > 0$. By uniform continuity, choose a $\delta > 0$ so that $|x - t| < \delta$ implies $|f(x) - f(t)| < \epsilon/2$. Let $N > 1/\delta$ be an integer and divide I into N intervals of equal length. Define a function g on I as follows. If $J_j = [p_j, q_j]$, $j = 1, \ldots, N$ is one of the subintervals of I, then set

$$g(x) = \frac{q_j - x}{q_j - p_j} f(p_j) + \frac{x - p_j}{q_j - p_j} f(q_j)$$

for $x \in J_j$. Also f and g agree at the endpoints of all the intervals J_j. Then g is piecewise linear and continuous, and it is plain that $|f(x) - g(x)| < \epsilon$ for all $x \in I$.

30. Let $f_j(x) = (1/j)\sin j^2 x$. These functions plainly converge uniformly to the 0 function. However, the derivative $f_j'(x) = j\cos j^2 x$ plainly converges at no point.

Chapter 7

1. Let φ be a C^∞ function on $[0, +\infty) \subseteq \mathbb{R}$ that is bounded by 1, takes the value 1 on the interval $[0, 1/4]$, and takes the value 0 on $[1/2, +\infty)$. Let

$$\psi(x, y) = \varphi(\max(|x|, |y|)).$$

Set

$$\Phi(x, y, t) = \psi(x, y) \cdot \mathrm{id} + (1 - \psi(x, y)) \cdot \left[(1 - t) \cdot 1 + t \cdot \frac{\max(|x|, |y|)}{|(x, y)|}\right](x, y).$$

Then Φ is a continuous mapping and, for each fixed $t \in [0, 1]$, it defines a homeomorphism of the plane.

Notice that, when (x, y) is near the origin, then $\Phi(x, y)$ coincides with id, which is the identity mapping. When (x, y) is greater than distance $3/4$ from the origin and $t = 0$, then Φ is still the identity mapping. When (x, y) is greater than distance $3/4$ from the origin and $t = 1$, then Φ is the mapping

$$(x, y) \longmapsto \frac{\max(|x|, |y|)}{|(x, y)|}(x, y),$$

and this is a homeomorphism of the plane that takes the unit square to the unit circle.

3. The essential point is that two knots are equivalent *not* when the knots themselves are homeomorphic—because all knots are homeomorphic to the circle. Rather it is the knots in space that must be homeomorphic. Hence, in particular, their complements in space must be homeomorphic. That is what the ambient isotopy guarantees.

5. For this problem we ignore the stipulation that the first mapping in an ambient isotopy be the identity.

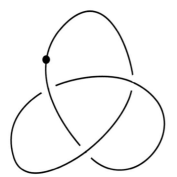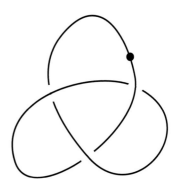

Figure S7.1: Two versions of the trefoil.

Say that two embeddings $f : X \to Y$ and $g : X \to Y$ are related if there is an ambient isotopy from f to g. Clearly the relation is reflexive, for the mapping

$$\Phi(x, t) = f(x)$$

provides the isotopy from f to f. Also, if f is related to g via Φ, then

$$\Lambda(x, t) = \Phi(x, 1 - t)$$

gives an isotopy from g to f. So the relation is symmetric.

For transitivity, suppose there is an ambient isotopy $\Phi(x, t)$ from the embedding f of X into Y to the embedding g of X into Y and an ambient isotopy $\Psi(x, t)$ from the embedding g of X into Y to the embedding h of X into Y. We may then define a new ambient isotopy

$$\Gamma(x, t) = \begin{cases} \Phi(x, 2t) & \text{if} \quad 0 \le t \le 1/2 \\ \Psi(x, 2(t - 1/2)) & \text{if} \quad 1/2 < t \le 1. \end{cases}$$

Then Γ is an isotopy from f to h. Thus the relation is transitive and we have an equivalence relation.

6. Examine Figure S7.1.

Imagine tracing a path in the left-hand knot—beginning at the dot—moving to the left around the curve. At the same time trace a path in the right-hand knot—beginning at the dot—moving to the right around the curve. Then the two knots will match up exactly and the matching of these two paths suggests an equivalence between the two knots. However, the orientations are opposite, and that is not allowed. The knots are not equivalent.

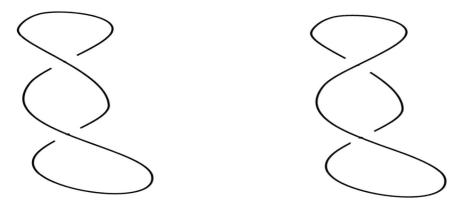

Figure S7.2: Knots with two crossings.

8. The disjoint circles have a different projection from a pair of linked circles. The crossing numbers are different. One cannot be obtained from the other by Reidemeister moves. Hence they cannot be equivalent.

9. There is no ambient isotopy of the two links because some of the intermediate homeomorphisms would involve two circles that intersect—a configuration that cannot be mapped to the two disjoint circles on the right. Thus these are inequivalent.

11. There are only two possibilities for knots with two crossings, and they are exhibited in Figure S7.2. Both of these can easily be unknotted with two twists, so they are trivial. There are no nontrivial knots with just two crossings.

13. Refer to Figures S7.3 and S7.4 for a concordance of the knots that we shall use in this calculation. The right-oriented trefoil is knot B' in our list. Following the calculation for the left-oriented trefoil that we performed in Example 7.3.2, we see that

$$
\begin{aligned}
\langle B'\rangle &= A\langle C'\rangle + A^{-1}\langle D'\rangle \\
&= A\left[A\langle E'\rangle + A^{-1}\langle F'\rangle\right] + A^{-1}\left[A\langle M'\rangle + A^{-1}\langle S'\rangle\right] \\
&= A^2\left[-A^2 - A^{-2}\right]\langle G\rangle + 2\langle G\rangle + A^{-2}\langle J'\rangle .
\end{aligned}
$$

Knot	Name	Bracket Polynomial
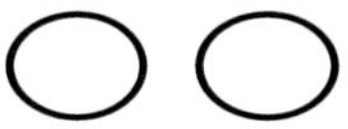	W	$- A^2 - A^{-2}$
	G	$- A^{-3}$
	H	$- A^3$
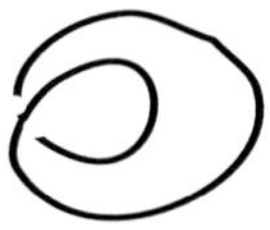	J'	$- A^3$
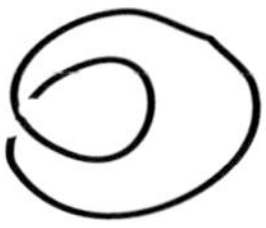	K'	$- A^{-3}$

Figure S7.3: A concordance of knots.

Name	Knot

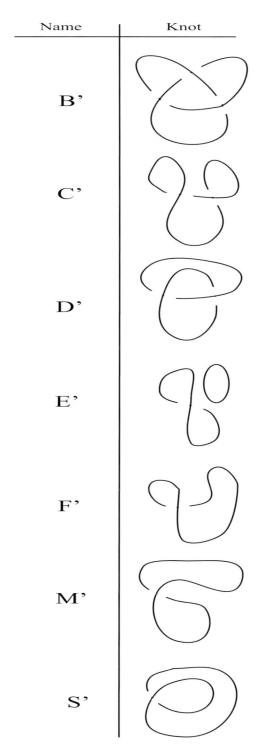

B'

C'

D'

E'

F'

M'

S'

Figure S7.4: Another concordance of knots.

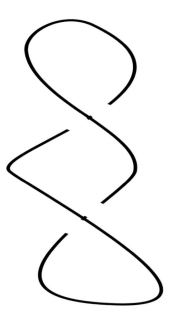

Figure S7.5: The right-oriented trefoil.

The natural thing to do now is to substitute in the bracket polynomials from Figures S7.4 and S7.5. Thus we obtain

$$\begin{aligned}
\langle B' \rangle &= A^2 \left[-A^2 - A^{-2} \right] (-A^{-3}) + 2(-A^{-3}) + (A^{-2})(-A^3) \\
&= -A^{-3}.
\end{aligned}$$

We conclude that the bracket polynomial of the projection B of the right-oriented trefoil is

$$-A^{-3}.$$

That is certainly distinct from the Kauffman polynomial for the left-oriented trefoil in Example 7.3.2.

15. The crossings in a knot with two crossings have opposite orientations so are each assigned one of $+1$ and -1. As a result, the writhe is 0. Furthermore, if we let X' denote the knot in Figure S7.5, then we may calculate that

$$\begin{aligned}
\langle X' \rangle &= A\langle H \rangle + A^{-1}\langle((-A^2 - A^{-2})\langle H \rangle) \\
&= A(-A^3) + A^{-1}((-A^2 - A^{-2})(-A^3)) \\
&= 1.
\end{aligned}$$

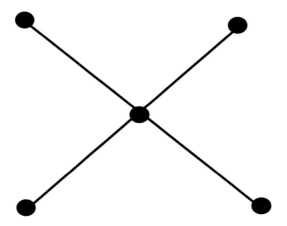

Figure S7.6: A graph on five vertices with no Euler path.

In sum, the Kauffman polynomial for X' is

$$K(X') = (-A^3)^{-w(K')} \cdot \langle X' \rangle = (-A^3)^{-0} \cdot 1 = 1.$$

This is of course also the Kauffman polynomial of the unknot.

20. There is the left-oriented trefoil and the right-oriented trefoil. Any other combination of three crossings simplifies to a knot with fewer crossings.

21. No Reidemeister move will change the number of crossings in the trefoil. Thus the trefoil cannot be equivalent to the unknot.

22. Our analysis of the fundamental group of the circle applies here to see that the fundamental group of the complement in space of the unknot is $\mathbb{Z}$.

Chapter 8

1. The graph in Figure S7.6 has no Euler path. Once a path terminates at one of the four extreme vertices (labeled A, B, C, D) then it must stop.

3. This two-handled torus is shown in Figure S7.7 together with a graph from which we may calculate the Euler number. Note that there are 10 vertices, 16 edges, and 4 faces. So the Euler number is

$$V - E + F = 10 - 16 + 4 = -2.$$

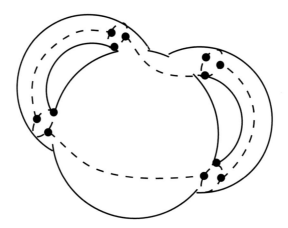

Figure S7.7: The torus with two handles.

5. The complete graph on k vertices has $\binom{k}{2} = \dfrac{k!}{(k-2)!2!}$ edges. It has $\binom{k}{3} = \dfrac{k!}{(k-3)!3!}$ faces.

7. The star has 10 vertices, 15 edges, and 7 faces. Draw a picture so that you can confirm this information. Of course $10 - 15 + 7 = 2$.

9. This would be the Klein bottle. Draw a graph with 2 vertices, 3 edges, and 1 face to confirm this fact.

11. The correct formula is $\chi = 2 - 2g$. This can be proved by induction on the number of handles.

13. Each edge has two endpoints or vertices. Thus the total number of ends of edges is even. In other words, the total of the degrees of all the vertices is even, but clearly the total of the degrees of all the even vertices is even. It follows then that the total of the degrees of all the odd vertices is even. Hence the number of odd vertices must be even.

17. Single out one person from the group of six named Bob. There are either three of the other people whom Bob knows or three whom he does not know. Say there are three that he knows. If two of them know each other, then those two and Bob form a mutually acquainted threesome. If no two of them knows each other, then those three form a mutually unacquainted threesome. The reasoning is similar if there

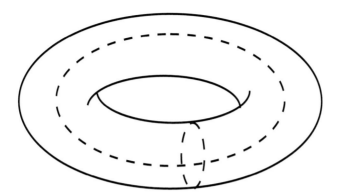

Figure S7.8: The torus.

are three people whom Bob does not know. This question may be construed in the language of graph theory as follows: Let $\mathcal{G}$ be a graph on six vertices. Then either there is a triangle in the graph, or there are three vertices no two of which are connected by an edge.

19. This result can be proved by induction on the number of edges. Obviously a graph on one vertex $\{P\}$ with no edges can be embedded in $\mathbb{R}^3$ with the function

$$P \longmapsto \mathbf{0}.$$

Now suppose that a graph $\mathcal{H}$ with k edges is embedded using a mapping Φ. Let us add an edge e connecting points P and Q of the graph. Then the points $\Phi(P)$ and $\Phi(Q)$ can certainly be joined in space by an arc e'. We extend the embedding by mapping e to e'. The reasoning is similar if the edge has both endpoints at the same vertex, or with any other possible configuration. That completes the induction.

21. The Euler characteristic of the sphere is 2. The Euler characteristic of the torus is 0, as Figure S7.8 shows.

23. Every pair of distinct vertices is either joined by an edge or it is not. No pair is joined twice, and no vertex is "joined" to itself. Thus the number "V choose 2," or $\binom{V}{2}$, is an upper bound on the number of edges.

24. In the complete graph on five vertices, each vertex has degree 4 (that is, four edges meet at each vertex). However, the 5-pointed star has five

Figure S7.9: Two non-isomorphic graphs on four vertices.

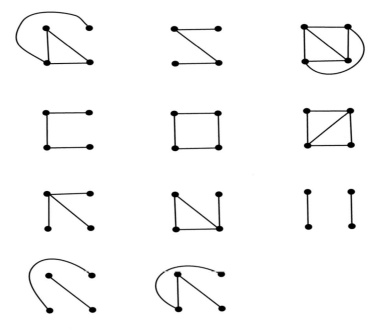

Figure S7.10: Eleven simple graphs on four vertices.

vertices of degree 2. Hence the graphs cannot be isomorphic. Figure S7.9 shows two graphs on 4 vertices that are nonisomorphic for a similar reason.

25. The eleven configurations are shown in Figure S7.10.

26. The solution to Exercise 23 shows that there are, at most, $\binom{V}{2}$ edges. If there are exactly $\binom{V}{2}$ edges, then every vertex must be connected to every other. That gives the complete graph. The argument in the reverse direction is even easier.

27. The 5-pointed star and the complete graph on 5 vertices are both shown in Figures S7.11.

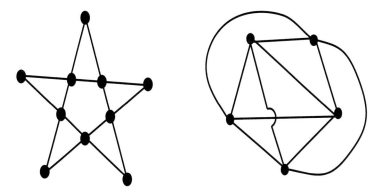

Figure S7.11: The 5-pointed star and the complete graph on 5 vertices.

The 5-pointed star has 5 vertices of degree two. Any permutation of those 5 vertices forces a corresponding permutation of the other 5 vertices (each having degree 4), and of all the connecting edges. This gives rise to an isomorphism; and those are the only isomorphisms.

From the combinatorial point of view, all vertices of the complete graph on 5 vertices are the same. Any permutation of those vertices results in a permutation of all the edges, and that defines an automorphism. Those are all the isomorphisms.

28. Let a graph have k vertices. Any vertex will have 1 or 2 or 3 or ... $(k-1)$ edges emanating from it. Hence there are $k - 1$ possibilities for the number of edges at a vertex. Thus two vertices will have the same number of edges emanating from it. Now if we think of the vertices as people and a connecting edge as "friendship," then two people have the same number of friends.

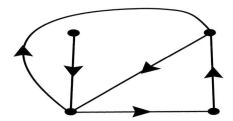

Figure S7.12: An Eulerian graph with 4 vertices and 5 edges.

30. Figure S7.12 exhibits a graph with 4 vertices and 5 edges. The Euler path is indicated.

Chapter 9

1. First we solve the equation

$$\log A = M. \qquad (*)$$

We do so by guessing

$$M = \begin{pmatrix} a & b \\ 0 & c \end{pmatrix},$$

and writing out $(*)$ as

$$A = e^M$$

or

$$\begin{pmatrix} 3 & 1 \\ 0 & 3 \end{pmatrix} = \exp\begin{pmatrix} a & b \\ 0 & c \end{pmatrix}.$$

Expanding the right-hand side in the usual fashion, we find that

$$\begin{aligned} e^a &= 3 \\ e^c &= 3 \\ 3b &= 1. \end{aligned}$$

In conclusion,

$$M = \begin{pmatrix} \ln 3 & 1/3 \\ 0 & \ln 3 \end{pmatrix}.$$

Now our job is to solve the differential equation

$$\frac{d\mathbf{x}}{dt} = M\mathbf{x}(\mathbf{t}),$$

where $\mathbf{x}(t) = (x(t), y(t))$. Of course the solution is

$$\mathbf{x}(t) = e^{Mt}\mathbf{c_0}$$

for some constant vector $\mathbf{c_0} = (c_1, c_2)$. In conclusion,

$$\mathbf{x}(t) = \begin{pmatrix} 3^t & (t/3)\ln t \\ 0 & 3^t \end{pmatrix} \cdot \mathbf{c_0}.$$

This last is the flow that we seek.

2. First we solve the equation

$$\log A = M . \qquad (*)$$

We do so by utilizing the result of Example 9.1.3 to see that if we set

$$M = \begin{pmatrix} 0 & \pi/2 \\ -\pi/2 & 0 \end{pmatrix}$$

then

$$e^M = \begin{pmatrix} 0 & 1 \\ -1 & 0 \end{pmatrix}.$$

Hence

$$\begin{pmatrix} 2 & 0 \\ 0 & 2 \end{pmatrix} \cdot e^M = \begin{pmatrix} 2 & 0 \\ 0 & 2 \end{pmatrix} \cdot \begin{pmatrix} 0 & 1 \\ -1 & 0 \end{pmatrix} = \begin{pmatrix} 0 & 2 \\ -2 & 0 \end{pmatrix}.$$

Therefore

$$\exp \begin{pmatrix} \ln 2 & \pi/2 \\ -\pi/2 & \ln 2 \end{pmatrix} = A .$$

In short,

$$M = \begin{pmatrix} \ln 2 & \pi/2 \\ -\pi/2 & \ln 2 \end{pmatrix}.$$

Now our job is to solve the differential equation

$$\frac{d\mathbf{x}}{dt} = M\mathbf{x}(\mathbf{t}) ,$$

where $\mathbf{x}(t) = (x(t), y(t))$. Of course the solution is

$$\mathbf{x}(t) = e^{Mt}\mathbf{c}_0 = \exp \left[\begin{pmatrix} t\ln 2 & 0 \\ 0 & t\ln 2 \end{pmatrix} + \begin{pmatrix} 0 & t\pi/2 \\ -t\pi/2 & 0 \end{pmatrix} \right] \cdot \mathbf{c}_0$$

for some constant vector $\mathbf{c}_0 = (c_1, c_2)$. In conclusion,

$$\mathbf{x}(t) = \begin{pmatrix} 2^t & \sin(t\pi/2) \\ -\sin(t\pi/2) & 2^t \end{pmatrix} \cdot \mathbf{c}_0 .$$

This last is the flow that we seek.

3. We set

$$
\begin{aligned}
x - y &= x \\
x - y - 4y^2 &= y\,.
\end{aligned}
$$

The only solution is $(0,0)$. The Jacobian matrix of the mapping is

$$
J = \begin{pmatrix} 1 & -1 \\ 1 & -1 - 8y \end{pmatrix}
$$

which, at the point $(0,0)$, is

$$
J = \begin{pmatrix} 1 & -1 \\ 1 & -1 \end{pmatrix}.
$$

The only eigenvalue of this matrix is 0. Thus the fixed point $(0,0)$ is stable.

4. We set

$$
\begin{aligned}
4xy &= x \\
x^2 - y^2 &= y
\end{aligned}
$$

The solutions are $(0,0)$, $(0,-1)$, $(+\sqrt{5}/4, 1/4)$, and $(-\sqrt{5}/4, 1/4)$. The Jacobian matrix of the mapping is

$$
J = \begin{pmatrix} 4y & 4x \\ 2x & -2y \end{pmatrix}.
$$

At $(0,0)$ the matrix is

$$
J(0,0) = \begin{pmatrix} 0 & 0 \\ 0 & 0 \end{pmatrix}
$$

and the only eigenvalue is 0. Thus the fixed point $(0,0)$ is stable.
At $(0,-1)$ the matrix is

$$
J(0,-1) = \begin{pmatrix} -4 & 0 \\ 0 & 2 \end{pmatrix}
$$

and the eigenvalues are -4 and 2. Thus the fixed point $(0, -1)$ is unstable.

At $(+\sqrt{5}/4, 1/4)$ the matrix is

$$J(\sqrt{5}/4, 1/4) = \begin{pmatrix} 1 & \sqrt{5} \\ \sqrt{5}/2 & -1/2 \end{pmatrix}$$

and the eigenvalues are 2 and $-3/2$. Thus the fixed point $(+\sqrt{5}/2, 1/4)$ is unstable.

At $(-\sqrt{5}/4, 1/4)$ the matrix is

$$J(-\sqrt{5}/4, 1/4) = \begin{pmatrix} 1 & -\sqrt{5} \\ -\sqrt{5}/2 & -1/2 \end{pmatrix}$$

and the eigenvalues are 2 and $-3/2$. Thus the fixed point $(-\sqrt{5}/2, 1/4)$ is unstable.

5. Since $\mathbf{p}_{j+1} = f(\mathbf{p}_j)$, we can be sure that the sequence $\{\mathbf{p}_j\}$ is the orbit of the point $\mathbf{p}_0$ under iterates of the mapping f. Of course this orbit may be calculated without limit, and the resulting zigzag gives the staircase representation for this orbit.

6. Figure S7.13 shows the main idea. We see that the fourth iteration of the staircase process brings us to a point on the line $y = x$ that is to the upper right of the initial point, and then the process repeats. The result is a spiral, as shown.

9. The points $0, 1$ are fixed points. Now $f'(0) = 0$ and $f'(1) = 2$. We conclude that 0 is a stable fixed point and 1 is an unstable fixed point.

10. Say that λ is an eigenvalue of A with $|\lambda| > 1$ and that the corresponding eigenvector is $\mathbf{v}$. Of course $(0,0)$ is a fixed point of the mapping since A is linear, and we see that

$$A^j \mathbf{v} = \lambda^j \mathbf{v}$$

which diverges to infinity.

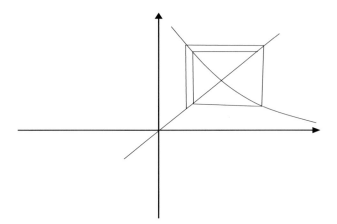

Figure S7.13: The staircase representation on a monotonically decreasing function.

12. (a) We set

$$x - y - 4x^2 = x$$
$$x - y = y.$$

Thus $y = x/2$, so

$$x - \frac{x}{2} - 4x^2 = x,$$

hence $x = 0, -1/8$. This leads to the fixed points $(0,0)$ and $(-1/8, -1/16)$.

Now the Jacobian matrix of the mapping is

$$J = \begin{pmatrix} 1 - 8x & -1 \\ 1 & -1 \end{pmatrix}.$$

At the point $(0,0)$ the Jacobian is

$$J(0,0) = \begin{pmatrix} 1 & -1 \\ 1 & -1 \end{pmatrix}.$$

The only eigenvalue of this matrix is 0. Thus the fixed point $(0,0)$ is stable.

At the point $(-1/8, -1/16)$ the Jacobian matrix is

$$\begin{pmatrix} 2 & -1 \\ 1 & -1 \end{pmatrix}.$$

The eigenvalues of this matrix are $\frac{1 \pm \sqrt{5}}{2}$. One of these eigenvalues is greater than 1, so we conclude that the fixed point $(-1/8, -1/16)$ is unstable.

(b) We set

$$\begin{aligned} x^2 - y^2 &= x \\ 3xy &= y. \end{aligned}$$

Thus $y = 0$ or $x = 1/3$. We thus find the fixed point $(0,0)$. The value $x = 1/3$ leads to no solution for y.

The Jacobian matrix of the mapping is

$$\begin{pmatrix} 2x & -2y \\ 3y & 3x \end{pmatrix}.$$

At $(0,0)$ the Jacobian matrix is the zero matrix, and the only eigenvalue is 0. Thus the fixed point is stable.

18. We think of an arbitrary nonlinear mapping as approximated by its linear part. Let

$$F(x,y) = (x, y + y^2).$$

Then

$$F \circ F(x,y) = F(x, y+y^2) = (x, y+y^2+y^2+2y^3+y^4) = (x, y+2y^2+2y^3+y^4),$$

and

$$\begin{aligned} F \circ F \circ F(x,y) &= F(x, y + 2y^2 + 2y^3 + y^4) \\ &= (x, y + 2y^2 + 2y^3 + y^4 + y^2 + 4y^4 + 4y^6 \\ &\quad + y^8 + 4y^3 + 4y^4 + 2y^5 + 8y^5 + 4y^6 + 4y^7) \\ &= (x, y + 3y^2 + 6y^3 + 9y^4 + 8y^6 + y^8 + 10y^5 + 4y^7), \end{aligned}$$

amd so forth. We see that the orbit of any point with y positive will tend to infinity. Yet the linear approximation is $G(x, y) = (x, y)$, and all orbits of this mapping are constant.

By contrast, consider $F(x, y) = (x, y - y^2)$. Let us calculate the orbit of the point $(1/2, 1/2)$. We see that

$$F(1/2, 1/2) = (1/2, 1/4),$$

$$F(1/2, 1/4) = (1/2, 3/16),$$

$$F(1/2, 3/16) = (1/2, 39/256),$$

and so forth. Clearly the y coordinate is positive and monotone increasing so the orbit converges to a finite point. The linear approximation is (x, y), and this has all orbits constant. So in this case the linear approximation predicts the stability of the orbit.

20. Let $F(x, y) = (2x + y - 1, x^2)$. We seek a fixed point by solving

$$\begin{aligned} x &= 2x + y - 1 \\ y &= x^2. \end{aligned}$$

Substituting the second equation into the first yields

$$x = 2x + x^2 - 1$$

or

$$x^2 + x - 1 = 0.$$

The quadratic formula now tells us that

$$x = \frac{-1 \pm \sqrt{1+4}}{2} = \frac{-1 \pm \sqrt{5}}{2}.$$

Now $x = \frac{-1+\sqrt{5}}{2}$ yields $y = (3 - \sqrt{5})/2$, and $x = \frac{-1-\sqrt{5}}{2}$ yields $y = (3 + \sqrt{5})/2$. Thus there are two fixed points.

21. First take $c = 0.1$ and $z_1 = 0$. Then

$$z_2 = 0.1$$

$$z_3 = 0.11$$

$$z_4 = 0.1121$$

$$z_5 = 0.1126$$

and so forth. It is not hard to see that all the z_j are bounded above by 0.2. So $c = 0.1$ is in the Mandelbrot set.

On the other hand, take $c = 1$ and $z_1 = 0$. Then

$$z_2 = 1$$

$$z_3 = 2$$

$$z_4 = 5$$

$$z_5 = 26$$

and so forth. Hence this orbit diverges to infinity, and we see that 1 is not in the Mandelbrot set.

22. We set

$$
\begin{aligned}
z_1 &= z_1 \\
z_2 &= z_2 + z_1^2 .
\end{aligned}
$$

It follows that $z_1 = 0$ and z_2 is arbitrary. This gives a large family of fixed points.

The Jacobian matrix is

$$J = \begin{pmatrix} 1 & 0 \\ 2z_1 & 1 \end{pmatrix} .$$

At the fixed point this gives the matrix

$$J = \begin{pmatrix} 1 & 0 \\ 0 & 1 \end{pmatrix} .$$

Of course this matrix has eigenvalue 1. We may draw no immediate conclusion about stability.

However, note that, for instance, the point (ϵ, ϵ) near the fixed point $(0,0)$ is mapped to $(\epsilon, \epsilon + \epsilon^2)$. This in turn is mapped to $(\epsilon, \epsilon + 2\epsilon^2)$, and so forth. Hence the orbit of this point tends to infinity and the fixed point is unstable. In fact the behavior will be similar for any point with first coordinate nonzero.

23. The system

$$\frac{dx}{dt} = -y$$
$$\frac{dy}{dt} = x$$

has as a solution the path $x(t) = \cos t$, $y(t) = \sin t$. Certainly that is a closed path: the unit circle.

Bibliography

[**ADF**] C. Adams and R. Franzosa, *Introduction to Topology: Pure and Applied*, Prentice-Hall, Upper Saddle River, NJ, 2008.

[**AHL**] L. Ahlfors, *Complex Analysis*, 3rd ed., McGraw-Hill, New York, 1979.

[**ARM1**] M. A. Armstrong, *Basic Topology*, Springer, New York, 1983.

[**ARM2**] M. A. Armstrong, *The Hauptvermütung according to Lashof and Rothenberg*, Kluwer, Dordrecht and Boston, 1996.

[**BAR**] J. Barwise, ed.*Handbook of Mathematical Logic*, North-Holland, Amsterdam, 1977.

[**BRM**] R. Brooks and J. P. Matelski, The dynamics of 2-generator subgroups of PSL(2, CC). *Riemann Surfaces and Related Topics: Proceedings of the 1978 Stony Brook Conference* (State Univ. New York, Stony Brook, N.Y., 1978), pp. 65–71, *Ann. of Math. Stud.*, 97, Princeton Univ. Press, Princeton, N.J., 1981.

[**BUS**] S. R. Buss, ed., *Handbook of Proof Theory*, Elsevier, Amsterdam, 1998.

[**DAV**] A. Davis, *Gödel's Theorem*, University of Oklahoma preprints, 1964.

[**FUL**] W. Fulton, *Algebraic Topology: A First Course*, Springer-Verlag, New York, 1995.

[**GAG**] T. W. Gamelin and R. E. Greene, *Introduction to Topology*, 2nd ed., Dover, Mineola, New York, 1999.

[**GIH**] S. Givant and P. R. Halmos, *Logic as Algebra*, Mathematical Association of America, Washington, D.C., 1998.

[**GRH**] M. J. Greenberg and J. Harper, *Algebraic Topology: A First Course*, Benjamin/Cummings, Reading, MA, 1981.

[**HAL**] P. R. Halmos, *Algebraic Logic*, Chelsea, New York, 1962.

[**HER**] I. N. Herstein, *Topics in Algebra*, Blaisdell Publishing, New York, 1964.

[**HRJ**] K. Hrbacek and T. Jech, *Introduction to Set Theory*, 3rd ed., Marcel Dekker, New York, 1999.

[**HU**] S. Hu, *Elements of General Topology*, Holden-Day, Inc., San Francisco, CA, 1964.

[**HUN**] T. Hungerford, *Algebra*, Springer-Verlag, New York, 1980.

[**HUW**] W. Hurewicz and H. Wallman, *Dimension Theory*, Princeton University Press, Princeton, NJ, 1941.

[**JEC**] T. Jech, *The Axiom of Choice*, North-Holland, New York, 1973.

[**KAR**] R. Karp, The probabilistic analysis of some combinatorial search problems, *Algorithms and Complexity (Proc. Sympos., Carnegie Mellon Univ., Pittsburgh, Pa. 1976)*, 1-19; Academic Press, New York, 1976.

[**KEL**] J. L. Kelley, *General Topology*, Van Nostrand, Princeton, NJ, 1955.

[**KKM**] E. Khalimsky, R. Kopperman, and P. R. Meyer, Computer graphics and connected topologies on finite ordered sets, *Topology and its Applications* 36(1990), 1–17.

[**KIS**] C. Kiselman, Digital Jordan curve theorems, in Borgefors, G., Nyström, I., and Sanniti di Baja, G., eds., *Discrete Geometry for Computer Imagery*, DGCI 2000 Proceedings, Lecture Notes in Computer Science, Volume 1953, New York: Springer-Verlag, 2000, 46–56.

[**KOKM**] T. Y. Kong, R. Kopperman, and P. R. Meyer, A topological approach to digital topology, *American Mathematical Monthly* 98(1991), 901–917.

[**KRA1**] S. G. Krantz, *Real Analysis and Foundations*, 2nd ed., CRC Press, Boca Raton, FL, 2005.

[**KRA2**] S. G. Krantz, *Cornerstones of Geometric Function Theory: Explorations in Complex Analysis*, Birkhäuser Publishing, Boston, 2006.

[**KRP1**] S. G. Krantz and H. R. Parks, *A Primer of Real Analytic Functions*, 2nd ed., Birkhäuser Publishing, Boston, MA, 2002.

[**KRP2**] S. G. Krantz and H. R. Parks, *The Implicit Function Theorem*, Birk- häuser Publishing, Boston, MA, 2002.

[**LAN**] S. Lang, *Algebra*, 3rd ed., Addison-Wesley, Reading, MA, 1993.

[**MIL**] J. Milnor, *Morse Theory*, Princeton University Press, Princeton, NJ, 1963.

[**MUN**] J. Munkres, *Topology*, 2nd ed., Prentice-Hall, Upper Saddle River, NJ, 2000.

[**NAN**] E. Nagel and J. R. Newman, *Gödel's Proof*, New York University Press, New York, 1958.

[**NMP**] National Mapping Program Technical Instructions, `http://rockyweb.cr.usgs.gov/nmpstds/suppti.html`.

[**ROS**] A. Rosenfeld, Digital topology, *American Mathematical Monthly* 86(1979), 621–630.

[**RR1**] H. Rubin and J. Rubin, *Equivalents of the Axiom of Choice*, 1st ed., North-Holland, Amsterdam, 1963.

[**RR2**] H. Rubin and J. Rubin, *Equivalents of the Axiom of Choice*, 2nd ed., North-Holland, Amsterdam, 1985.

[**RUD**] M. E. Rudin, A new proof that metric spaces are paracompact, *Proc. AMS* 20(1969), 603.

[**RUDW**] W. Rudin, *Functional Analysis*, 2nd ed., McGraw-Hill, New York, 1991.

[**SEY**] P. Seymour, Progress on the 4-color theorem, *Proceedings of the ICM* (Zürich, 1994), 183–195, Birkhäuser, Basel, 1995.

[**SHO**] J. Shoenfield, *Mathematical Logic*, Addison-Wesley, Reading, 1967.

[**SIN**] S. Singh and J. Lynch, *Fermat's Enigma: The Epic Quest to Solve the World's Greatest Mathematical Problem*, Anchor, New York, 1998.

[**SKR**] G. F. Simmons and S. G. Krantz, *Differential Equations: Theory, Technique, and Practice*, McGraw-Hill, New York, 2006.

[**SMU**] R. Smullyan, *Gödel's Incompleteness Theorems*, Oxford University Press, New York, 1992.

[**SPA**] E. Spanier, *Algebraic Topology*, Springer, New York, 1981.

[**STO**] R. R. Stoll, *Sets, Logic, and Axiomatic Theories*, W. H. Freeman and Company, San Francisco, 1961.

[**SUP**] P. Suppes, *Axiomatic Set Theory*, Dover Publications, New York, 1972.

[**WIL**] S. Willard, *General Topology*, Addison-Wesley, Reading, MA, 1970.

Index